Developments in Landscape Management and Urban Planning, 6C

Planning for agroforestry

FURTHER TITLES IN THIS SERIES

1. R.J.A. Goodland and H.S. Irwin
Amazon Jungle Green Hell to Red Desert?

2. B.B. Greenbie
Design for Diversity

3. K. McCoy
Landscape Planning for a New Australian Town

4. D. Bennett and J.F. Thomas (Editors)
On Rational Grounds

5. H. Hengeveld and C. de Vocht (Editors)
Role of Water in Urban Ecology

6A. R. Burnell Held and D.W. Visser
Rural Land Uses and Planning

6B. F.R. Steiner and H.N. van Lier (Editors)
Land Conservation and Development

ISOMUL

ISOMUL = International Studygroup on Multiple Use of Land. This studygroup encourages, guides and publishes studies in the field of the multiple uses of land for rural and urbanized areas. Members are found in several Western European and North-American countries.

Developments in Landscape Management and Urban Planning, 6C

Planning for agroforestry

Edited by

W.W. Budd
Program in Environmental Science and Regional Planning, Washington State University, Pullman, WA 99164-6410, U.S.A.

I. Duchhart
Department of Landscape Planning, Agricultural University Wageningen, 6703 BJ Wageningen, The Netherlands

L.H. Hardesty
Department of Natural Resource Science, Washington State University, Pullman, WA 99164-6410, U.S.A.

F. Steiner
Department of Planning, Arizona State University, Tempe, AZ 85287-2005, U.S.A.

ELSEVIER
Amsterdam — Oxford — New York — Tokyo 1990

ELSEVIER SCIENCE PUBLISHERS B.V.
Sara Burgerhartstraat 25
P.O. Box 211, 1000 AE Amsterdam, The Netherlands

Distributors for the United States and Canada:

ELSEVIER SCIENCE PUBLISHING COMPANY INC.
655, Avenue of the Americas
New York, NY 10010, U.S.A.

Library of Congress Cataloging-in-Publication Data

Planning for agroforestry / edited by W. Budd ... [et al.].
p. cm. -- (Developments in landscape management and urban planning ; 6C)
Papers presented at an international symposium held at Washington State University, Pullman on April 24-27, 1989.
Includes bibliographical references and index.
ISBN 0-444-88634-6
1. Agroforestry--Planning--Congresses. 2. Agroforestry--Developing countries--Planning--Congresses. I. Budd, W. (William), 1925- . II. Series.
S494.5.A45P52 1990
338.1'7499'091724--dc20 90-3670
CIP

ISBN 0-444-88634-6

Although the research described in this book has been funded wholely or in part by the Environmental Protection Agency, contract No. 68-W8-0113 to ICF Incorporated, it does not necessarily reflect the views of the Agency and no official endorsement should be inferred.

This book is printed on acid-free paper.

Printed in The Netherlands

TABLE OF CONTENTS

PREFACE: AGROFORESTRY AS A MULTIPLE LAND-USE STRATEGY

H. N. VAN LIER
International Studygroup On Multiple Use of Land (ISOMUL), Agricultural University Wageningen, Gen Foulkesweg 134, 6703 BJ, Wageningen (The Netherlands)

Agroforestry as a system tries to overcome serious social and environmental problems. A large variety of systems, known collectively as agroforestry, currently exist. In all cases these systems are integrated within the rural land uses and especially within farming methods. Because agroforestry techniques require a relatively long time to implement fully, careful planning is required. Development agencies and research organizations have designed planning methods for implementing these techniques, as illustrated by several extant agroforestry projects. A review of these planning methods in current use and under consideration was needed. F. Steiner, along with W. Budd and others, took the intiative to start such an overview. After compiling a bibliography (Steiner et al., 1988), field visits were made to among others the International Council for Research in Agroforestry (ICRAF) in Kenya, the Centro Agronomico Tropical de Investigacion y Ensenanza (CATIE) in Costa Rica and the Agricultural University Wageningen (AUW) in The Netherlands. At AUW a planning approach for agroforestry is evolving. This approach is emerging through the interaction of the Forestry and Landscape Architecture Departments at AUW and ICRAF. On the basis of the information collected, an international symposium was organized in 1987. This international symposium, "Planning for Agroforestry", was held at Washington State University, Pullman, Washington on April 24-27, 1989. Selected contributions from this symposium are published in this book, under the auspices of ISOMUL, although the final responsibility of each contribution resides with the individual authors.

ISOMUL, the International Studygroup on Multiple Use of Land, was formed by a small group of scientists from various nations who have an interest in the field of rural land-use planning. It seeks to establish an exchange of information and experience by encouraging and guiding studies in land-use planning. Because of a lag in the application of research, there is a need the transfer of knowledge and understanding about planning for the future uses of land in rural areas, and particularly that of rural lands that are the most susceptible to the changing influences of socio-economic development, technologies, metropolitan areas and others. Each country, dealing with its own problems, has specific knowledge that is valuable to others. The exchange of this type of information is a necessary and valuable activity and is the major objective of ISOMUL. It, therefore, encourages studies which focus on specific planning problems such as multiple land-use planning methods, forecasting and prediction methods, evaluation methods and impact analyses.

The first book of ISOMUL, written by R. Burnell Held (Colorado State University, United States of America) and D.W. Visser (Governmental Service for Land and Water Use, The Netherlands) dealt with a comparison of land-use developments and land-use planning policies between the two countries. The second multi-authored book, edited by F.R. Steiner (Washington State University, United States of America) and H.N. van Lier (Agricultural

University Wageningen, The Netherlands) dealt with conservation and development of land resources.

It is hoped that this new book, "Planning for Agroforestry", will help all those who are concerned with the problems of establishing and maintaining sustainable uses of land. Agroforestry can play an important role in achieving this end.

Acknowledgments

A project of this scope owes its success to a large number of individuals and organizations. The papers contained in this volume were part of an international symposium on the planning and implementation of agroforestry held at Washington State University on April 24-27, 1989. The sponsoring organizations provided financial and in-kind support as well as critical encouragement and ideas. All those who participated in the symposium, by presenting papers or through participation in discussions, have added to the development of agroforestry planning, and their contributions should be recognized even though it is not possible to name them all. The contributions in this volume reflect the broad contributions of all those involved in that effort.

Symposium Sponsors

International Science and Education Council
Washington State University, The Agroforestry Consortium
Agricultural University Wageningen
American Forest Council
American Planning Association: Small Town and Rural Planning Division
American Planning Association: Environmental Planning Division
American Planning Association: International Planning Division
Environmental Protection Agency
IUFRO; Extension Forestry Working Group
Tuskegee University
University of Colorado at Denver, School of Architecture and Planning
USDA Extension Service
USDA Forestry Support Program
USDA Forest Service International Forestry
USDA Office of International Cooperation and Development USDA Soil Conservation Service

The authors also wish to extent a special thanks to Ir. Freerk Wiersum for his support and comments during the development of this planning volume. Jan Noel, Genevieve Smith-Thompson, and Jim Henson of the International Development Office, Washington State University, help initiate the Agroforestry Consortium which conceived and provided leadership in the organization of the symposium. The U.S. Environmental Protection Agency provided support for the final production of the book. We appreciate the help of Ken Andrasko in this regard. The authors wish to thank Alan Salazar and U.S. Senator Ted Wirth of Colorado for their assistance in supporting this effort. We wish to thank Gerald Young for his assistance in proofing the final text. The authors also wish to extend a special acknowledgment to Vicki Spencer, Assistant Director of Research Administration, University of Colorado at Denver and Marlene Baldwin, Administrative Assistant, Program in Environmental Science and Regional Planning for their efforts in producing this volume. Without their support this task would not have succeeded.

INTRODUCTION: INTEGRATED PLANNING FOR AGROFORESTRY

I. DUCHHART and R. VAN HAERINGEN
Department of Landscape Planning, Agricultural University Wageningen, 6703BJ Wageningen (The Netherlands)

F. STEINER
Department of Planning, Arizona State University, Tempe, Arizona 85287-2005 (United States of America)

ABSTRACT

General global trends --such as dramatic population growth and world-wide environmental degradation-- require new approaches to land-use planning. Agroforestry can be seen as such a new approach. This book provides an exchange of ideas to contribute appropriate, synthetic approaches to planning for agroforestry in developing countries. Most of the agroforestry planning methods and case studies presented in this volume have incorporated well established methods to agricultural research. Therefore, in this introductory chapter some of these methods --land evaluation, farming systems research, and agro-ecosystem analysis-- are briefly reviewed. In addition integration among scale levels, sectors, and processes seems of an utmost importance for the success of agroforestry projects. In this introductory chapter, it is argued that planning for agroforestry projects on an intermediate or local-scale level may offer the best opportunities for such an integration, and that design might be a good integrative technique. Along these lines, the following chapters in this book are introduced.

THE NEED FOR INTEGRATED PLANNING

The need for integrated planning is induced by several trends in human settlement and their consequences on a global scale. The magnitude and rate of the world population growth is one of the most significant changes. This population growth, coupled with other social and political problems, has resulted in a widespread, international degradation of environmental quality (Argawal, 1986) and a related urgency for sustainable land use. Another trend is that public participation and identification of local knowledge are more and more seen as prerequisites for successful planning and implementation of development projects.

1.1 Need for sustainable land use

The rate of population growth in many developing countries is extremely high. In Kenya, for example, the population was 16 million people in 1979. Now this number has grown to over 22 million. In many countries, such increases in population have caused pronounced changes in land uses. In addition to increasing population density, Wiersum (Chapter 1) attributes changes in land use to the influence of the world economy on local situations. One response to such transformations is land-use intensification through multiple cropping systems. There is, however, a minimum farm size and a maximum intensification level. When this level is reached, farm families divide their land and open up new areas. As a result, often marginal, environmentally-sensitive land is converted to cultivation. Van den Hoek and Bekkering (Chapter 4) describe the growing imbalance between population size and the carrying capacity of agricultural and forest land on Java in Indonesia, where farming on steep slopes has resulted in high erosion rates and degradation of topsoil. Forest resources are degraded and can no longer play their traditional role in the farmer's land-use system. Another example of the impact of population growth is provided by Saad and his colleagues (Chapters 9 and 10). They describe the competition for resources between sedentary farmers, transhumants, and nomads in the western Sudan. All other authors in this volume are concerned with the dramatic impacts of over-cultivation.

The population growth, the intensification of land use, and the concurrent destruction of natural resources has led to a predominantly scientific recognition that the earth's natural systems supporting life have their limits (Brundtland, 1987). This new understanding of nature might result in an age of more sensitive adjustments to environmental conditions. A concern about sustainable development is one indication of this new age. Sustainable development meets the needs of the present without compromising the ability of future generations to meet their own needs (Brundtland Cie, 1987; also see Wiersum, Chapter 1). Therefore, to achieve sustainable land use an integrated approach is necessary. Such an approach implies attention to different processes, sectors, and scales concerned with land use. Processes refer to an understanding of land use, for example from an ecological or socio-economic perspective. A sector is a land-use type, such as agriculture and forestry. The contributors to this book view agroforestry as a promising technique to achieve sustainable land use. In addition, in many developing countries trees and shrubs play a vital role in the life of rural poor, by providing building materials, food, fodder, firewood, natural exudates, and medicine (Noel *et al*., Chapter 9; Baldwin and Bandhu, Chapter 13). Agroforestry, being a combination of tree and food products, therefore can contribute to the fulfillment of many of the basic needs of the local population (Wiersum, Chapter 1; Murray, Chapter 7).

1.2 Decentralization of planning processes

Over the last three decades the focus of development aid changed from mainly financial to technical support and from sectoral, large-scale to integrated small-scale projects. Meanwhile, some governments have decentralized some of their decision-making responsibilities and administrative functions in order to achieve effective and efficient management of their resources (Fowler, 1987). Although management of government resources normally refers to financial and managerial capacity, decentralization is even more important when management of natural resources is involved. The implementation and success of most agroforestry and other development programmes depend largely on the interest and the involvement of the individual farm households. National governments, district committees, and local authorities can give directives or financial support, but the actual managers of these programmes are the farmers. They understand best why they were forced to damage the environment on which they depend, and they are the key to rebuilding (Timberlake, 1985). Decentralization is one of the means to tap this local knowledge.

Individual households, thus, are important and knowledgeable stewards of natural resources, but if everybody works for their personal benefits, the overall result may not be satisfactory. For example, if water conservation measures are not coordinated, the protection measures taken by one farmer can destroy the land of a neighbour. In order to be able to link individual actions to global or national concerns, and thus achieve sustainable land use, an intermediate scale level --such as village, community, or watershed--becomes important.

All contributors to this book have responded to this challenge, and have searched for fresh approaches to planning for agroforestry. This exploration has occurred in different planning fields, including:

(a) National policies for agroforestry research and development (Getahun, Chapter 8; Noel *et al*., Chapter 9; Saka *et al*., Chapter 11).

(b) Planning for the development of agroforestry as a sustainable land-use system (Wiersum, Chapter 1; Raintree, Chapter 2; Beer *et al*. Chapter 3; Van den Hoek and Bekkering, Chapter 4; Buck, Chapter 5; Jansens, Chapter 12).

(c) Project planning and management of agroforestry projects (Murray, Chapter 7; Baldwin and Bandhu, 13).

As a scientific field, agroforestry is relatively young (Raintree, Chapter 2). Agroforestry is an interface between agriculture and forestry. Of the two fields, agricultural research especially influenced the search for planning approaches to agroforestry. In the following sections, some methods frequently used in agriculture, with a special focus on integrated land-use research and development, are reviewed. Land evaluation and farming systems research are well-tested and widely-used methods; agro-ecosystem analysis has been developed more recently. After this review, the chapters which follow are introduced in relation to the global trends and the impacts of these trends on developing countries and to the selected methods.

2 SOME CURRENT METHODS FOR INTEGRATED PLANNING OF LAND USE

An integrated approach to research and development of land use requires that at least two essential concerns be addressed: sustainability of how land is used and fulfillment of the needs of the local population. Several methods have been developed which address these concerns (Van Haeringen, 1988). Although these methods largely concentrate on agriculture, because of their integrated basis they form an excellent character for planning methods for agroforestry. The methods reviewed here are: land evaluation, farming systems research and development, and agro-ecosystem analysis.

2.1 Land evaluation

The objective of land evaluation is to assess the suitability of different tracts of land for selected and specified land-use types (Huizing, 1986). The tracts of land, divided in land units, are assessed with regard to their suitability for selected land-utilization types. In order to accomplish this objective, the land-utilization types are specified in terms of socio-economic, agronomic, and technical attributes. In addition, ecological, management, and conservation requirements are identified. These requirements are expressed in terms of land qualities, which are determined by an analysis of bio-physical properties. By matching the requirements with the qualities, the suitability of the land units is assessed. From here, improved land-use practices can be proposed (Food and Agricultural Organization, 1976). Fig. 1 shows this land evaluation method.

Land evaluation involves the execution and interpretation of basic surveys of climate, soils, vegetation, and other land elements. Bio-physical evaluations are often made by soil survey organizations. Often in these evaluations socio-economic processes are not considered in detail. The main concern becomes the land and not its users. Land evaluations, which integrate bio-physical and socio-economic processes, made by teams of natural resource scientists, agronomists, agro-economists and others, are not common. Land evaluation has long been closely associated with planning for agriculture and forestry.

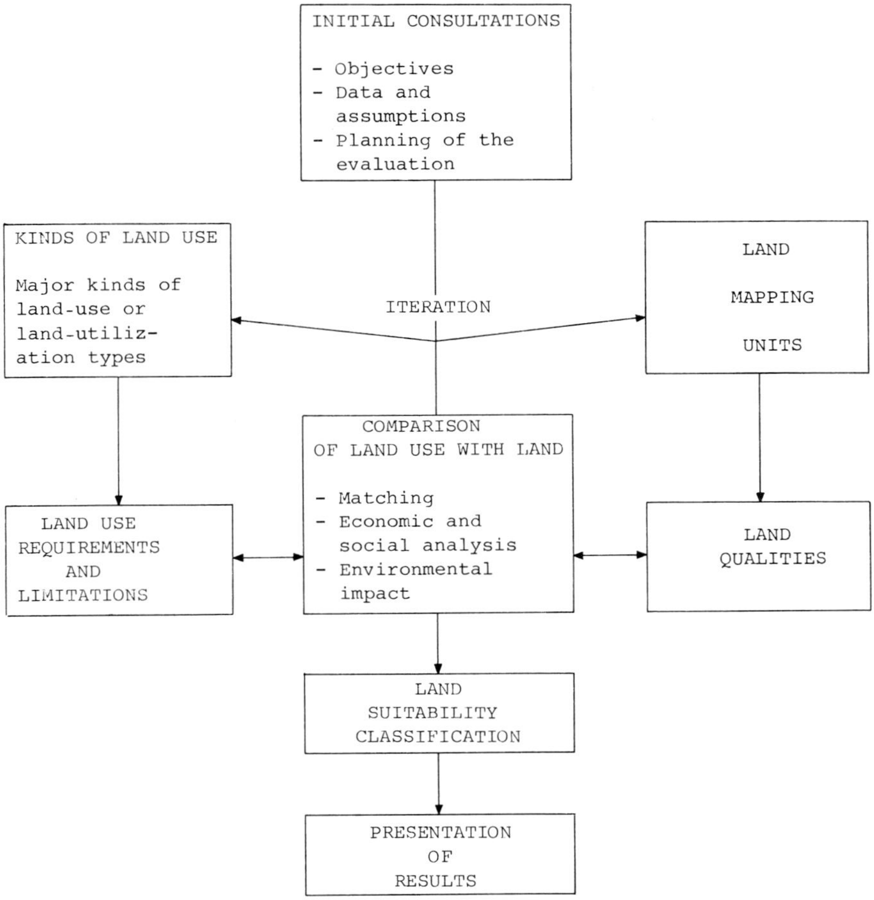

Fig. 1. Schematic representation of activities in land evaluation. (Source: FAO, 1976).

2.2 Farming systems research and development

The primary objective of farming systems research is the improvement of the well-being of individual farming families by increasing the productivity of their farms given the constraints imposed by resources and the environment (Norman and Collinson, 1985). Farming systems research has been a response to the concern about differences between crop yields obtained on research stations and those on farms. It turned out that ecological and socio-economic constraints imposed on the farm system were not sufficiently taken into account by the researchers.

To address this deficiency, different research approaches have been developed that all come under the heading of farming systems research and development. Farming systems research involves two basic approaches that are complementary: an analytical and an experimental approach. In the analytical approach, target groups are identified, who face more or less the same problems and opportunities. A detailed diagnosis of the farm-level constraints, and the farmers' knowledge of climate, soils, plant, and animal species is conducted (Fresco, 1986). This analysis is followed by an experimental planning phase, including technology design, on-farm experimentation, and extension (Fig. 2). The research is carried out by multi-disciplinary groups of researchers in close cooperation with farmers and representatives of the local community. Farming systems research has gained increasing recognition as a cost-effective approach to small-holder agricultural development, particularly by generating technologies suitable for small holders, and thus alleviating some constraints faced by farmers (Grandin, 1986).

2.3 Agro-ecosystem analysis

Agro-ecosystem analysis evolved from the demand for a multi-disciplinary and holistic approach to agricultural research and development (Conway, 1985). Agro-ecosystem analysis has yet to be widely practiced. The approach seeks to generate a set of priorities for research and development.

In this method, the performance of agro-ecosystems at different scale levels, defined socially or physically (including watershed, village, household, or farming system), are analyzed. These agro-ecosystems are described in terms of four properties: sustainability (the ability of an agro-ecosystem to maintain productivity when subject to a major disturbing force), productivity (the output of valued product per unit of resource input), equitability (the evenness of distribution of the productivity of the agro-ecosystem among the human beneficiaries), and stability (the constancy

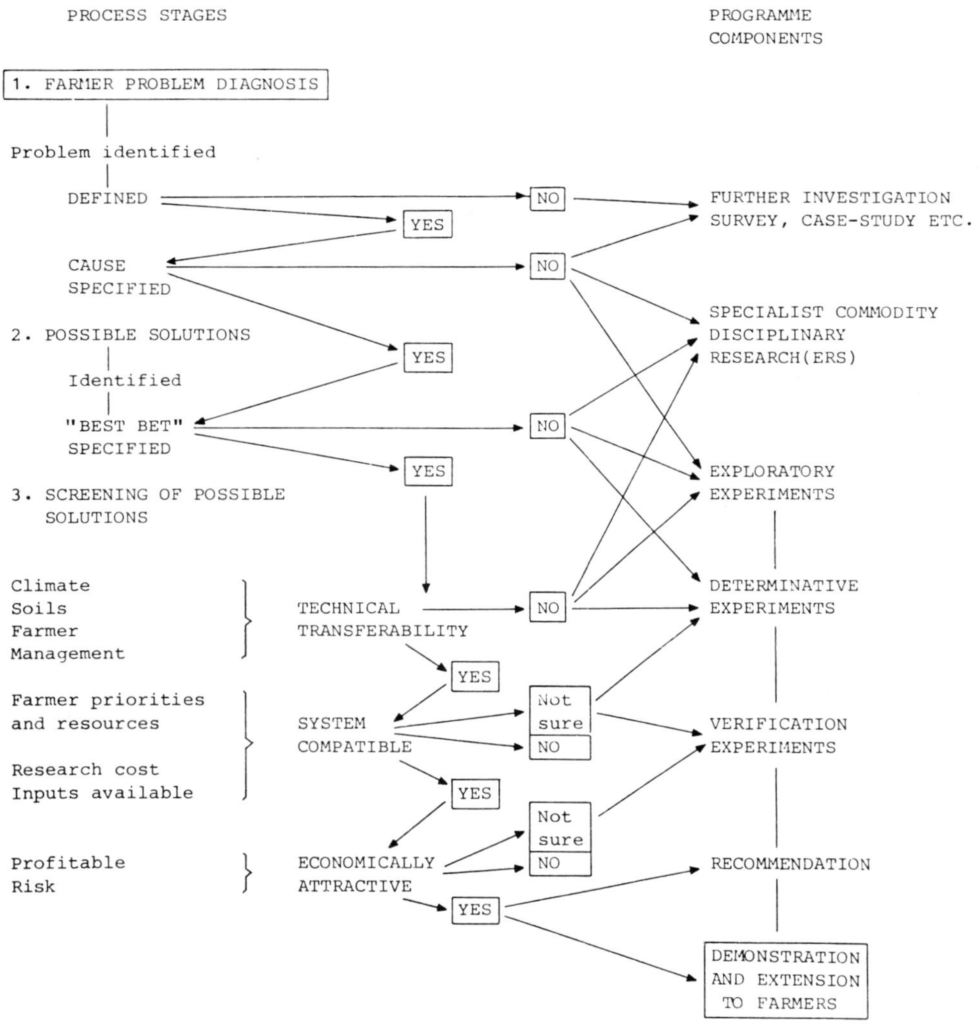

Fig. 2. The process of identifying on farm research components in farming systems research. (Source: Collinson, 1987).

of productivity in the face of small disturbing forces arising from normal fluctuations in the surrounding environment). Through this description, it becomes possible to view agricultural development as a progression of changes. In the analysis four patterns are chosen to reveal the relationships that determine the system properties; space, time, flow, and decision (Fig. 3). They reflect processes in bio-physical aspects and in management of resources (Conway, 1987).

2.4 Conspectus

Agroforestry is seen as a promising technique to achieve sustainability in land use and fulfillment of the needs of the local population. Agroforestry is a land-use type, focused on multipurpose tree planting and agriculture. Agroforestry planning can be conceived as sector planning for rural development (Buck, Chapter 5), and therefore agroforestry planning and research have largely been built on the methods summarized above. In the planning methods for agroforestry introduced in this volume, the selected methods --land evaluation, farming systems research, and agro-ecosystem analysis-- are often used in combination with each other. In this way, the agroforestry practitioners expect to unite the strengths of the methods. The integration among the ecological, socio-economic, and land-use management processes is safeguarded, at least at the farm household level, through a union of agro-ecosystem analysis and farming systems research. A combination of farming systems research (main focus: the farm household), and land evaluation (dominant concern: the regional level), and agro-ecosystem analysis (considers all relevant scale levels), will address a hierarchy of levels of organization. Land evaluation adds a future perspective, whereas farming systems research focuses predominantly on the analysis of the present land use. Efforts are underway to develop an integrated, complementary approach to land evaluation and farming systems research for the purpose of land-use planning. Strengthening their complementarity will allow better anticipation of future changes and perspectives. Agro-ecosystems analysis also seems to have many potentialities to incorporate a future perspective about land use.

The agroforestry planning methods, presented in this volume, have especially been influenced by farming systems research. The International Council for Research in Agroforestry (ICRAF) developed the diagnosis and design (D&D) planning method for agroforestry. In Chapter 2, Raintree presents this D&D method as an agroforestry-specific farming systems research variant. The macro-application of D&D uses the Food and Agriculture

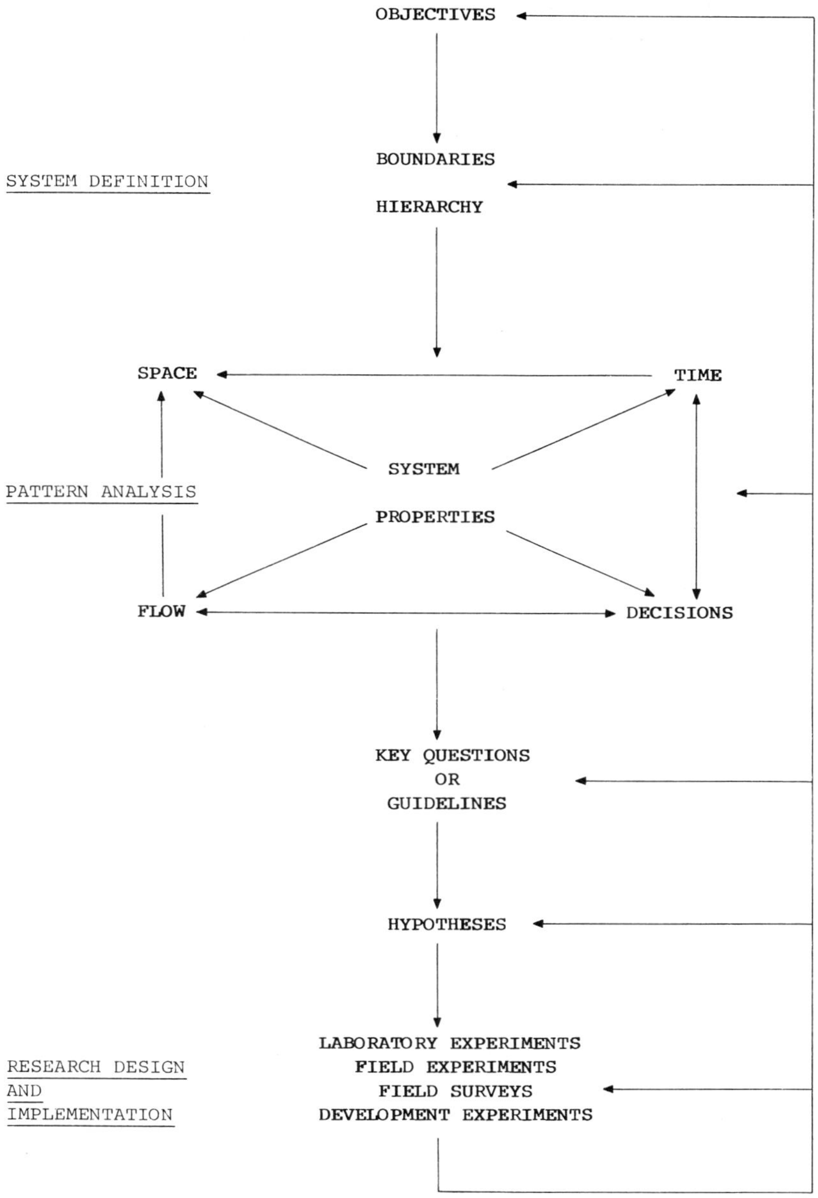

Fig. 3. The steps of the procedure of agro-ecosystem analysis. (Source: Conway, 1985).

Organization (FAO) land-evaluation method (also see Scherr, Chapter 6). CARE International for Kenya adjusted the ICRAF D&D methodology towards planning for agroforestry extension (Buck, Chapter 5). The planning method for agroforestry research in western Sudan, described by Noel et al. and Bunderson et al. (Chapters 9 and 10), adopted a modified farming systems research approach as well. Within the Centro Agronomico Tropical de Investigacion y Ensenanza (CATIE), an on-farm agroforestry research planning method has been developed, which according to Beer et al. (Chapter 3) is based on farming systems research. In the process of monitoring and evaluating the CATIE approach, ICRAF carried out an interdisciplinary D&D exercise, resulting in useful modifications. The approach described by Van den Hoek and Bekkering (Chapter 4) takes agro-ecosystem analysis as a starting point. Originally, the project they discuss was based on land evaluation. It appeared, that the result of this technique was not refined enough to address the large diversity of production systems in the uplands of Java. They expect that agro-ecosystems analysis may overcome this shortcoming.

3 AGROFORESTRY AND PLANNING FOR INTEGRATED LAND USE

Empirical evidence suggests that agroforestry can provide a sound ecological basis for increased crop and animal productivity, more dependable economic returns, and greater diversity in social benefits on a sustained basis (Saka et al. Chapter 11). It is unrealistic, however, to expect that the introduction of agroforestry in itself guarantees all these benefits. Adequate planning for integrated land-use development is necessary.

The concept of sustainable land use is no longer related to the primary land-use units managed by individual farmers, but also to higher hierarchical levels (Wiersum, Chapter 1; Raintree, Chapter 2; Scherr, Chapter 4). An intermediate or local scale level has been acknowledged as a level most suitable to integrate agroforestry technologies in appropriate land-use systems. Related to this acknowledgment, the landscape, a visual and spatial reflection of such land-use systems, has received greater attention as an entity of analysis and intervention. In addition to this attention to local landscapes as a basis for diagnosis and design, there is a growing awareness that a future perspective is necessary. Such a future-oriented vision will be even more important when the goals of agroforestry are interpreted broadly, thus including improvement of nutrition level, watershed management, fire protection, erosion control, regeneration of species diversity, and natural habitats for wildlife, as is described by Murray (Chapter 7). In such a broad approach to agroforestry, the need for an integrative technique becomes even more important. In the D&D methodology of ICRAF, design is used as such a technique; it is however, mainly introduced as a tool to develop agroforestry technologies. In the following sections, the arguments for the importance of such an intermediate scale level, and the potentials of design as an integrative technique on such a scale level, are clarified.

3.1 Intermediate or local-scale level

The intermediate-scale level is often identified as the level that is most appropriate for the integration of individual or local wishes and knowledge with regional and national information. Wiersum (Chapter 1), Raintree (Chapter 2), Scherr (Chapter 4), Buck (Chapter 5), Murray (Chapter 7), as well as Baldwin and Bandhu (Chapter 13) indicate that appropriate agroforestry techniques should begin with the land-use system (including landscape niches and sites), the land user (performances, resilience, adaptability), and individual practices (management practices and capability, input/output) and objectives. Different reasons are given to support this view. Some authors argue that because of the complexity of planning for agroforestry such a decentralized approach is needed. For example, Raintree (Chapter 2) states that agroforestry design is far too complex, varied, and location-specific a process to be addressed by any centralized institution. The only ones capable of developing the full, diversified agroforestry potential of the local landscape are the residents of local communities.

Others indicate that much information about landscape processes and land use is available through the local population. For example, Scherr (Chapter 6) asserts that most information and experience with agroforestry practices is currently concentrated in indigenous farming. She also mentions that for successful implementation of agroforestry projects, technology dissemination networks within a farmer society, are important. In addition, Buck (Chapter 5) adds that in order to enlarge the chance of adaptation of suggested agroforestry interventions, those interventions should be based on what farmers already know.

The intermediate level is often described as a land-use system (Van den Hoek and Bekkering, Chapter 4; Buck, Chapter 5; Scherr, Chapter 6). A land-use system is defined by Scherr (Chapter 6) as a farming system, but more broadly defined to include rangeland, forest, and options for multiple uses (for example for nomads, land-less people, and herbalists). The village level is also seen as an important intermediate (Wiersum, Chapter 1; Scherr, Chapter 6; Noel et al. Chapter 9; Baldwin and Bandhu, Chapter 13). It should be kept in mind, that the intermediate level is not fixed in scale or size, but largely depends on the kind of land-use system, village characteristics, and type of project (Duchhart, 1989). For example, the land use of nomads or transhumants often covers areas of hundreds of square kilometers, while a small agrarian settlement in an oasis might occupy only ten to twenty square kilometers.

As mentioned before, landscapes are the visual reflection of these land-use systems. Among others, Rocheleau (1985; 1987) strongly emphasized the importance of a landscape analysis in the D&D evolution. Also in the CARE International for Kenya methodology, attention is paid to landscape. CARE planners actually expanded the original D&D level of diagnosis to include a landscape level. According to Buck (Chapter 5) such a diagnosis is a motivating and successful method to gain relevant and necessary information. For example, in such a landscape analysis traditional niches for (indigenous) trees in the landscape are located (Buck, Chapter 5) and priority functions for trees or shrubs for land users are identified (Scherr, Chapter 6). These priority functions may be divided in production and protection (or service) functions (Wiersum, Chapter 1). Protection functions can be, for example, water conservation or provision of shade. Although Noel et al. (Chapter 9) do not explicitly mention the different functions or the importance of an intermediate-scale level, their recommendation to research and evaluate water harvesting or run-off supports these statements. All in all, focusing on the intermediate scale level for planning does not exclude the fact that a national governmental involvement is necessary (Wiersum, Chapter 1; Van den Hoek and Bekkering, Chapter 4; Scherr, Chapter 6; Getahun, Chapter 8; Saka et al. Chapter 11).

3.2 Design: An integrative technique

Although, many authors place a high priority on an integrative and pluralistic approach to agroforestry planning, the way to achieve integration is not always made explicit. Some authors look for a combination of planning on different scale levels, which may give a coverage of all relevant scale levels, but it certainly does not guarantee integration between them. For example, Murray (Chapter 7) tries to integrate scales by connecting scale levels and by addressing functional relationships. Within the D&D methodology, the meso- and micro-D&D are nested in the macro-D&D. In addition to these diagnoses at different scale levels, the iterative nature of the D&D might provide sufficient linkages between the scale levels

(Raintree, Chapter 2; Buck, Chapter 5; Scherr, Chapter 6), but still a more specific integrative technique seems necessary.

In D&D, design is used as an integrative technique to develop appropriate agroforestry technologies (Raintree, Chapter 2). Scherr defines design in the D&D method as follows: "Design corresponds to a 'package' or 'intervention' in extensionists' terms. A complete design identifies multipurpose tree and shrub varieties and associated non-woody components (crops/grasses/animals), arranged and managed in specific ways to meet defined land-user objectives in defined types of sites" (Chapter 6). Also, Buck (Chapter 5) mentions that in the CARE International for Kenya approach, design is used for the development of agroforestry techniques. Their approach is patterned on ICRAF D&D technology design, but adapted for extension. Its adaptation involved greater reliance on qualitative data sources, particularly on staff and farmer judgement. Prototype technologies are designed, which are discussed with the farmers, who are stimulated to adjust them to their own insight.

Although design in D&D and in related methods exclusively concerns appropriate technologies, design techniques could be applicable to any situation requiring problem definition, integration of information, and problem solving (Duchhart, 1989, see also Raintree, Chapter 2). For example, Jansens (Chapter 12) uses design as an integrative technique in his development plan for Kakuyuni. This case study came into being through the cooperation between ICRAF and the Agricultural University Wageningen, The Netherlands. As a result, the Kakuyuni case study has D&D characteristics embedded in a landscape planning approach (Fig. 4). The landscape planning approach used by Jansens is being further developed at the Department of Landscape Architecture of the Agricultural University Wageningen (Duchhart, 1989). In this approach, the landscape forms the starting point for analysis and design. The present landscape is visualized in a sketch or map. On the basis of this map and of a landscape analysis, trends can be extrapolated to present an image of the future landscape over 20 or 30 years. Visualization of such a future landscape can help the residents (and often the experts) become aware of processes occurring in their area. For example, the image can show the consequences of a continued devegetation. In addition to the projection of current trends, it is also possible to visually represent a more desirable future image, new ideas, starting points, and/or objectives for improved land use (Kleefmann, 1984). In doing this, it is possible to give an image of a sustainable land-use system by showing an appropriate spatial ordering of different land-use types. Landscape niches for agroforestry interventions can thus be indicated. With the help of such maps, the results of these interventions for different sectors and scale levels can be evaluated. Jansens, for example, developed two different scenarios based on possible social land-use behaviour (Chapter 12). It remains, however, up to the farmers to determine what the real future will bring.

Van den Hoek and Bekkering worked along the same landscape planning approach, but combined it with agro-ecosystem analysis. They, too, made use of a landscape map to develop a land-use plan. According to these authors, linkages between land use and terrain conditions can be localized in such maps. The land use plan will improve the accessibility on a local level of data and planning results from district and provincial levels. Actually Jansens in Kenya and Van den Hoek and Bekkering in Indonesia, used a design approach similar to that of CARE International for Kenya, but applied the prototype notion on a land-use system level, instead of on a plot level.

All designs show "best bets". Rather than final designs, they simply are a tool in the planning process. The process of evaluation, monitoring, and re-design will offer the opportunity to integrate farmers' innovations and improvements of the proposed technologies in order to achieve improved, sustainable land use in the end (Raintree, Chapter 2; Van den Hoek and Bekkering, Chapter 4; Buck, Chapter 5; Scherr, Chapter 6; Jansens, Chapter 12). Moreover, the discussion of these designs will help to strengthen the image of the type of landscape that the participants want their children to inherit, and therefore assist in defining the problem statement and objectives for land-use improvement (Buck, Chapter 5).

4 EPILOGUE

Throughout this book, agroforestry is presented as an important tool to achieve sustainable land use. In the search for suitable approaches to planning for specific situations, several questions need to be answered. The first question to be posed is: Planning for whom? Who is the client? The urban and the rural poor, the middle or large farmer, or the government? The second question to be addressed is: Planning for what? Is planning meant for community development, for a research programme, for national policies, or for agroforestry technologies? The third question to be addressed, is: What is the planning context? What are the specific site conditions? The last question to be posed is: How should people plan? How should the perceptions of farmers and those of researchers be linked? How can the intangible aspects of human life, such as the desire for quality, culture, human stress,

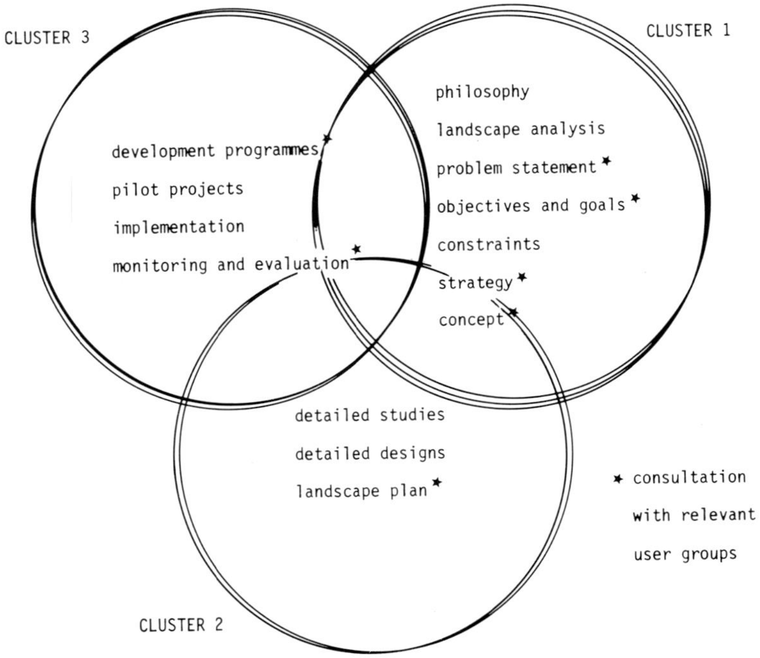

Fig. 4. Schematic representation of the AUW method.

traditions, be integrated with project planning where quantified results are highly valued? How can the different time horizons of the farmer (weeks, maybe months), the ecologist (hundreds of years, maybe for ever), and the politician (4 to 5 years) for project results be combined? How can the short-, mid-, and long-term objectives be matched with each other?

All these questions cannot be addressed with one planning method. Therefore, in this volume different methods can be found, some focused on project planning, others on planning for research. In this introduction the responses to several of these questions have been highlighted. Attention has been paid to the general acknowledgment of the importance of decentralized planning. Nearly all authors recognize the importance of socio-cultural processes. The values and beliefs of the local community may be difficult to quantify, but sensitivity to human systems is a prerequisite to successful project implementation.

Furthermore, an integrated, more holistic philosophy to planning for agroforestry is emerging. Within this context, landscape at an intermediate level comes to the fore as a planning basis. An important notion in this approach is the realization that besides a production function, trees and shrubs also have a protection or service function, such as soil and water conservation or the provision of shade. Such an integrated approach, however, needs to be more than a collection and analysis of data and the development of appropriate technologies. Integrated design, illustrating optimum scenarios for sustainable land use, should be part of the planning process.

Another interesting development is the evolution of design and redesign as an integrative technique within agroforestry planning. Design has always been integral to the D&D method, but exclusively concerned the design of technologies. The use of design as a technique to define problems and to search for solutions for sustainable land-use systems on an intermediate or land-use system level is a promising concept for agroforestry planning. The design of appropriate technologies is an important facet of agroforestry planning, but such an activity needs to be integrated into a broader vision of design, one that involves a future perspective on sustainable arrangements of land uses as a part of a living landscape. In planning for agroforestry project management, explicit attention should be given to this design role.

The chapters that follow illustrate a diversity of planning approaches. Some of the examples focus on research projects, others are extension efforts, and still others are oriented toward sustainable land-use development. Wiersum sets the stage with his discussion of agroforestry for sustainable land use. Raintree, Beer et al., Van den Hoek and Bekkering, and Buck present a variety of agroforestry planning strategies. Specific cases of agroforestry interventions on different scale levels are described by Scherr, Murray, Getahun, Noel et al., Bunderson et al., Saka et al., and Jansens. Budd et al. conclude with a conspectus for agroforestry.

5 REFERENCES

Argawal, B. 1986. Cold Hearths and Barren Slopes, The Woodfuel Crisis in The Third World. Zed Books, London, 209 pp.

Brundtland, Gro Harlem Committee, 1987. Our Common Future, The World Commission on Environment and Development. Oxford University Press, New York, 383 pp.

Collinson, M.P., 1987. Farming systems research: procedures for technology development. Experimental Agriculture, 23: 365-386.

Conway, G.R., 1985. Agroecosystem analysis. Agricultural Administration, 20: 31-55.

Conway, G.R., 1987. The properties of agroecosystems. Agricultural Systems, 24: 95-118.

Duchhart, I., 1989. Landscape planning: an approach to local-level planning? In: R. van Haeringen and K.F. Wiersum (Editors), Local Level Planning for Sustainable Land Use. (Bos-document 9) Stichting Bos, Wageningen, pp. 55-72.

Food and Agricultural Organization, 1976. A framework for land evaluation. (ILRI Publication No. 22) ILRI, Wageningen.

Fowler, A.F., 1987. Government decentralization and energy planning in Kenya. Master's thesis, Institute of Development Studies, Brighton, 22 pp.

Fresco, L.O., 1986. The study of farming systems: a perspective on agricultural technology development. In: L.O. Fresco (Editor), Cassava Shifting Cultivation - A Systems Approach to Agricultural Technology Development in Africa. Koninklijk Instituut voor de Tropen, Amsterdam.

Grandin, B.E., 1986. Adding community level variables to farming systems research: a research priority. Paper presented at the IIMI-Rockefeller Foundation Workshop on Social Science Perspectives on Managing Agricultural Technology, Lahore.

Haeringen, R, van, 1988. Approaches to local-level land-use planning, worked out for agroforestry contributions. Master's thesis, Agricultural University Wageningen, 51 pp.

Huizing, H., 1986. Land evaluation - Lecture notes for the N11 specialization. I.T.C., Enschede.

Kleefmann, F., 1984. Planning Als Zoekinstrument. VUGA, s'Gravenhage, 115 pp.

Norman, D. and Collinson, M.P., 1985. Farming systems research in theory and practice. In: J.V. Remenyi (Editor), Agricultural Systems Research for Developing Countries, Proceedings of a ACIAR conference. Canberra, pp. 16-30.

Rocheleau D.E., 1985. Land-use planning with rural farm households and communities. (Participatory Agroforestry Research). International Council for Research in Agroforestry, Nairobi, 76 pp.

Rocheleau, D.E., 1987. The user perspective and the agroforestry research and action agenda. In: Gholtz H.L. (Editor) Agroforestry: Realities, Possibilities and Potentials. M. Nijhoff, The Hague, pp 59-87.

Timberlake, L., 1985. Africa in Crisis, the Causes, the Cures of Environmental Bankruptcy. Earthscan, IIED, Washington DC, 232 pp.

Chapter 1

PLANNING AGROFORESTRY FOR SUSTAINABLE LAND USE

K.F. WIERSUM
Department of Forestry, Agricultural University Wageningen, P.O. Box 342, 6700 AH Wageningen (The Netherlands)

ABSTRACT

The development of agroforestry systems is often proposed as a means to develop sustainable land use. This claim is evaluated in this paper by assessing the meaning of the concept of sustainable land use and by comparing the characteristics of agroforestry to the various properties of this concept. Sustainable land use involves both static and dynamic aspects and includes three categories of major properties: maintenance of production potential, adaptability of management practices, and distribution of inputs and outputs. Although some claims on the possible role of agroforestry cannot (yet) be substantiated, in general agroforestry systems may contribute significantly to sustainable land use once they are well established and provided that they are well managed. In planning new agroforestry systems, it should be remembered that these favorable characteristics have to be built up gradually, and that agroforestry development often takes place within a situation of dynamic rural transformation. Three groups of agroforestry interventions may be distinguished: resource management technology, development implementation tools, and institutional arrangements. If agroforestry is to contribute to sustainable land use, then in planning field implementation projects, attention should not only be given to agroforestry technology development, but also to the need for proper support measures and organization structures, especially at the community level, to stimulate its adoption.

INTRODUCTION

In a recent article (Duchhart et al., 1989), it has been argued that an emergent issue in planning agroforestry systems is the possibility of using agroforestry as a means to develop sustainable land use. Such an assertion may seem paradoxical in view of the fact that the original identification of agroforestry as a form of land use worthy of professional interest was based on the recognition of its potential to contribute to sustainable or even improved use of fragile and wasted lands. Indeed, the first definition for agroforestry stated very specifically that agroforestry is a sustainable land management system (King and Chandler, 1978). But more recently, the rationale of developing agroforestry has been modified to include three alternatives: a higher total, a more diversified and/or a more sustainable production from available resources than is possible with other forms of land use (Lundgren, 1982).

How can this change in thinking be explained? Has agroforestry not lived up to its original expectations or did the concept of sustainable land use prove to be more elusive than originally thought? In this chapter, it will be argued that the answers to these questions may be found through a careful assessment of the meaning of the concept sustainable land use and the general characteristics of agroforestry in relation to the various types of interventions needed for agroforestry development.

2 THE CONCEPT OF SUSTAINABLE LAND USE

There has been considerable discussion in the literature on the interpretation and operationalization of the sustainable land-use concept. A general consensus seems to be emerging that the goal of sustainable agriculture should be to maintain agricultural practices at levels necessary to meet the needs and aspirations of the present population without compromising the ability to meet future needs as a result of excessive environmental degradation (York, 1988). The characteristics of sustainable land use are: conservation of natural resources; sufficient levels of agricultural production; and acceptable livelihood conditions for present and future population.

Although the concept of sustainable land use is sometimes interpreted as a steady-state condition (Tisdell, 1988), the above description indicates that it involves both static and dynamic properties. On the one hand, the level of exploitation of the land resources should remain in long-term equilibrium with their production capability; on the other hand, land use should be amendable to dynamic change under the influence of social and economic tendencies. Indeed, land-use systems are always subject to change in response to various socio-economic developments including: changes in population density and correlated increased demands for food products; development of urban purchasing power and markets; technical progress; changes in institutional conditions including government policies and land tenure conditions; and changing human aspirations.

Because of the various aspects involved in the concept of sustainable land use, operationalization requires that the concept be disaggregated into its various attributes (Oram, 1988). In order to achieve this disaggregation, two basic principles need to be applied: 1) both the static

TABLE 1

Major properties of land-use (sub)systems proposed as diagnostic indicators for land-use performance.

System	Properties	Reference
Land systems	Sensitivity Resilience	Blaikie and Brookfield, 1987
Agro-ecosystems	Productivity Stability Sustainability Equitability Autonomy	Conway, 1985, 1987; Marten, 1988
Farming systems	Ecological stability Management resilience Production sustainability Economic reliability	Wiersum, 1985 adapted from Harwood, 1979).
Agricultural production systems	Ecological stability Technical efficiency Social attractiveness	common usage

and dynamic nature of sustainability and, 2) a recognition that land-use systems are hybrid systems. They involve interactions between the land and its user, and thus encompass numerous interactions between agro-ecosystems and human social systems (Conway, 1985; Marten and Saltman, 1986).

Several authors have tried to identify major properties of land-use (sub)systems to be used as diagnostic criteria for analyzing land use (Table 1). Such properties should serve both as descriptors of the system's behaviour as well as indicators of its performance (Conway, 1985). Although the terms used for the properties sometimes have been used in different contexts (Marten, 1988) or have been defined in various ways, they still may serve as basis for identification of the different attributes of sustainable land use. Obviously, some of the properties shown in Table 1 refer primarily to ecological systems and others to social systems. They may be divided into three basic categories:

-properties related to the maintenance of production potential,
-properties related to the adaptability of management practices,
-properties related to the distribution of inputs and outputs.

2.1 Maintenance of production potential

Several authors indicate productivity as being one of the characteristic properties of land use. A distinction is often made between the short-term consistency or stability of production, and the long-term maintenance or sustainability of production (Conway, 1985, 1987; Marten, 1988). Stability refers to the degree in which productivity is constant in the face of small disturbances caused by normal fluctuations of climate and other environmental variables. Sustainability refers to the ability of a system to maintain productivity in spite of intensive stress or an unpredictable large disturbance (Conway, 1985, 1987). In respect to the ability of land-use systems to withstand such stress or disturbance, Blaikie and Brookfield (1987) distinguish the properties of sensitivity and resilience. Sensitivity refers to the susceptibility of land-use systems to degradational forces, while resilience refers to the ability of a land-use system to recover its original ability after a certain degree of degradation has taken place. Thus, production sustainability may be the result of either a land-use system's resistance, its insensitivity to degradation, or its ability to recover easily from such degradation.

It should be noted that changes in the production level of a certain land-use system do not necessarily result from degradation of the land resources, but may also take place as a result of changed demands for specific products. The term production sustainability should not therefore, be used to refer to the maintenance of the production level (as suggested by some authors), but rather to the maintenance of the production capacity. A clear distinction must be made between the ecological production potential and the actual production level of goods for human beneficiaries.

2.2 Adaptability of management practices

The conclusion that a clear distinction needs to be made between actual production levels of economic goods and biological production capacity, focuses the attention on the fact that the production level of a certain land-use system is not only the outcome of bio-physical properties, but that it depends also on the management abilities of the land manager. Moreover, the production of a land-use system should not be considered as an inherent property of an agro-ecosystem, but rather as an outcome of land-use activities moderated by a farmer. Harwood (1979) described production stability as the outcome of biological stability, the ability to deliver short-term predictable yields, and management stability, the ability and readiness of a farmer to carry out the appropriate management activities to ensure the success of a crop. This distinction is fundamental to sustainable land use, although the term management stability does not seem very appropriate and could better be replaced by the term management resilience (Wiersum, 1985) or management adaptability.

Two different forms of adaptive responses are involved in the property of management adaptability. In the first place, the management techniques should be adaptable in order to resist or overcome year to year environmental fluctuations and long-term degradational forces. Management practices may not only prevent degradation from occurring, but may also be directed at reversing existing degradation processes[1]. In the second place, the management practices should be adapted to changes in the socio-economic conditions, e.g. changes in access to and costs of production factors and price variability of agricultural output. This focuses attention on the economic performance of the land-use system in relation to the risk sensitivity of farmers. In order to deal with the uncertainty created by the lack of control over variable external environmental conditions and market prices, management practices should preferably be amendable to sequential decision making during the production cycle. Flexible adaptations in respect to use of outputs should be possible in response to the conditions evolving during the production cycle (Huijsman, 1986).

2.3 Distribution of inputs and outputs of production

The various properties discussed above may all be identified at the level of the primary land-use unit. It should be kept in mind, however, that land use is a generic term, which refers to various hierarchical levels: field-farm-village-region (Conway, 1985; Marten and Saltman, 1986). The concept of sustainable land use, therefore, may be used not only in relation to the primary land-use units managed by individual land users, but may also refer to higher hierarchical levels. Indeed, as indicated in the discussion on management adaptability, the dynamic nature of land use can often be understood only in relation to change at social levels higher than that of the individual land user.

These attributes of sustainable land use are represented in Table 1 by the properties of equitability (the degree of fair sharing of agricultural production among its human beneficiaries) and autonomy (the degree to which agricultural production can proceed without contacts with systems outside the primary production system) (Conway, 1985, 1987; Marten, 1988). Obviously, autonomy decreases the uncertainty of socio-economic changes influencing the costs and prices of agricultural production. The importance of the property of equitability is best illustrated by the interrelationships between land-use sustainability and poverty (Mellor, 1988). Poverty that results from a lack of equitable distribution of agricultural inputs and outputs has, in many regions, been indicated to have an important bearing on sustainable land use. Poor people are often concentrated in environmentally sensitive areas, where they may not be able to bear the costs of the required management practices adapted to these sites. Furthermore, poor people have a high risk sensitivity. Alternatively, enabling poor people to gain secure rights and means so that they can invest in proper land utilization may be the surest protection for the environment.

In conclusion, it is evident that sustainable land use is dependent upon many variables, several of which are not directly influenced by land-use practices. This means that it is unrealistic to hope that agroforestry systems will be able to address all aspects related to developing sustainable land use. But agroforestry can certainly contribute in attaining such a goal. In the next section, a further assessment will be made of the precise contribution of agroforestry to reaching the goal of sustainable land use.

3 THE ROLE OF AGROFORESTRY IN SUSTAINABLE LAND USE

Three different goals for agroforestry development are now generally identified:

(a) Increasing diversified production, especially under conditions of land shortage. This is not only seen as a means to improve the availability of required agricultural and tree products, but also as a means to relieve pressure on valuable conservation areas.

(b) Contributing to sustained production of crops and livestock either on fragile lands, or in areas of lagging economic development and prevalence of low external input agriculture.

(c) Contribution to land rehabilitation and increased production on degraded lands.

These goals illustrate that agroforestry is expected to contribute to sustained land use by either maintaining or improving the ecological production capacity of lands, or by preventing over-exploitation of sensitive land resources in adjacent areas. But as discussed above, attaining sustainable land use requires that many other aspects be considered as well. Several authors (Budowski, 1981; Wiersum, 1988) have discussed the various potentials and constraints of agroforestry systems. These will be summarized here in relation to the various categories of properties of sustainable land use.

3.1 Maintenance of production potential

As a result of micro-climatic buffering by trees in agroforestry systems, these systems have a relatively high resistance to local fluctuations in weather. The high tree biomass combined with a high humus content in the soil make these systems relatively resistant to erosion and leaching. The productivity of agricultural crops in these systems is enhanced by the large and efficient nutrient cycling system under trees, which includes mechanisms for redistribution of nutrients from sources inaccessible to agricultural crops, to more accessible sources (Young, 1986; Sanchez, 1987). In addition to these ecological service roles of trees, they may also have a direct productive role. The productivity of trees often compares favourably to annual crops because of their photosynthetic activity throughout a longer growing season, coupled with relatively low demands for efficient uptake of nutrients. Together these factors assure improved conditions for maintenance of both short-term and long-term production capacity. It is because of these conditions that agroforestry systems may offer hope for improving yields on degraded lands. However, it should not be taken for granted that the resistance of agroforestry systems against degradation also means that the systems have good resilient properties in respect to land degradation. Little is known about the quantitative role of trees in the recovery of degraded lands (Prinsley and Swift, 1986), especially with respect to the time needed to accumulate nutrients in an enlarged plant-litter-soil nutrient cycle (Kellman, 1979; Kessler and Breman, in preparation). Even on non-degraded lands it may take several years before a newly established agroforestry system reaches the stage of production stability because it takes time for trees to develop their protective properties and to produce harvestable products. As has been discussed in many publications (e.g. Food and Agricultural Organization, 1985) this lack of early tree production is often a limiting factor in establishing new agroforestry systems.

3.2 Adaptability of management practices

The favourable ecological conditions in agroforestry systems with respect to the maintenance of production sustainability should not be interpreted as meaning that all agroforestry systems automatically display all these properties. Rather, they indicate that agroforestry systems have several ecological characteristics which offer good starting points for maintaining productivity. Actual conditions in specific agroforestry systems, such as the composition and configuration of the tree component, the kind of management techniques needed (e.g. soil tillage for annual crops, coppicing of trees), and the amount of harvested products taken out of the system, all influence the real performance of the system. Wiersum (1984) has demonstrated that the success of controlling erosion in agroforestry systems is not the result of the presence of trees themselves, but rather of good management practices that allow these trees to carry out their beneficial influence. The same reasoning can also be applied to production stability and sustainability.

Agroforestry systems display several features, which cause the management of these systems to be relatively adaptable to various changes in the production conditions. For instance, the multi-species composition of agroforestry systems spreads the risk of crop failure, while the multiple roles of trees may allow for adjustments in production output according to actual needs. Furthermore, many tree products are not bound to a specific harvesting time, which may give the tree component of agroforestry systems a reserve function for the agriculture off-season or for periods of special social commitments. Agroforestry systems are, thus, often relatively flexible in the timing of output of tree products in relation to needs or to market prices. Furthermore, the farmer's dependency on external inputs with fluctuating costs is reduced by the service function of trees for crop production.

However, as indicated earlier, the lack of early yields in several newly established agroforestry systems forms a constraint, especially for the most risk sensitive farmers who have either no financial means to bridge this waiting period or no secure rights to yields in the distant future.

3.3 Distribution of inputs and outputs

It is unrealistic to expect that agroforestry systems can make an important contribution to the properties of sustainable land use, which operate at higher organizational levels than that of the primary production unit. Rather, these properties mostly form conditional factors for developing agroforestry. The access to production factors and market relations are therefore important considerations in planning agroforestry systems (Arnold, 1987; Filius, 1988).

Nonetheless, it should also be noted that agroforestry sometimes is used not as a means for production, but rather as a means to gain control over production factors. Such situations may occur because planting trees often establishes a legal claim to the land on which they are planted (Raintree, 1987c). In such a case, the planting of trees in agroforestry systems may not be carried out as a means to contribute to sustainable production, but rather as a means to claim lands. Agroforestry development may result in increased competition or even speculation for land. In such instances, the incorporation of trees in agricultural land use does little to improve equitable access to the production inputs; it even may result in agroforestry systems intruding into sensitive conservation areas (Wiersum, 1985; Van Maaren and Wiersum, 1988). Obviously, such processes do not contribute to sustainable land use.

4 PLANNING AGROFORESTRY SYSTEMS

From the above discussion, it can be concluded that many agroforestry systems may contribute significantly to sustainable land use once they are well established and provided that they are well managed. Based on these favourable properties of agroforestry systems, new systems are being introduced as an attempt to transform unsustainable forms of land use into more sustainable ones. Planning such new land-use systems must take into consideration that the need for these changes results from various processes of rural transformation. Planners of many agroforestry projects in tropical countries are confronted with similar dynamic conditions as described for the semi-arid tropics of West Africa (Norman et al., 1981):

> Many of the changes in the quantities, characteristics, qualities and distribution of production can be explained, partially at least, on the basis of increasing population densities and the gradual emergence of the region into the world economy. The latter has arisen in part due to the continuing development of the communication systems throughout the region. This has had two effects. The first is that it makes the commercialization of agriculture more feasible through the production of export cash crops. The second effect is that there have been changes in the relationships within communities. Ones that were more self-contained and sharing in philosophy have become increasingly dependent upon the outside world and more individualistic in orientations.

In response to such rural transformations, two conflicting tendencies in land use may take place (Ruthenberg, 1980). The growing scarcity of land results in land-use intensification and strengthens the trend toward multiple cropping and interaction between the various production processes. In contrast, economic growth and commercialization tend to promote specialized production processes and the development of open farming systems, which depend to a large extent on the exchange of goods and services. Agroforestry development fits within the first tendency of strengthening interactions between the various farming system components. In areas where the second tendency prevails, formerly well established agroforestry systems may even break down, as illustrated by the history of an Acacia mearnsii agroforestry system in Java (Berenschot et al., 1988).

Planning agroforestry systems as sustainable and adaptable forms of land use should therefore be carried out with due consideration of the dimensions and stages of rural change (Maxwell, 1984). The introduction of a new agroforestry system not only involves a change in land-use practices, but also entails important changes in the nature of needed production factors and in access to these production factors. For instance, new agroforestry practices may involve changing attitudes towards land and labour as productive factors with traditional communal arrangements becoming private and /or subsistence production processes becoming commercial. The development of agroforestry systems should not, therefore, be considered only as a process of technology transfer.

In this respect, a clear distinction should be made between the planning of agroforestry technology development projects and agroforestry implementation projects. In the former, the focus is primarily on the possibility of developing agroforestry as an appropriate technology for improved land use. The latter case deals with the design of the means for supporting adoption of agroforestry technology. Table 2 summarizes the various activities that are needed for agroforestry development. In this table three types of interventions are indicated (Hufschmidt, 1986; Van Maaren and Wiersum, 1988):

-development of resource management technologies,

-development of support measures for implementing agroforestry, and

-arrangements of the proper institutional and organizational setting where implementation can proceed.

As indicated in Table 2, planning for implementation of agroforestry projects attention should be given both to "things to be done" and to "ways of getting things done". In addition, the need for possible reorientation of institutional arrangements should be considered (Van Maaren and Wiersum, 1988).

Until now, most attention in agroforestry development has focused on the applicability of the ICRAF diagnosis and design (D&D) method for planning (Raintree, 1987a, 1987b). The basic features of this method are that it is directed at identifying locally relevant development options for agroforestry technologies and that it takes the farm household with its production processes and socio-economic situation as the central unit of analysis. The approach has also been extended for planning agroforestry extension programmes as well as for use at the community, watershed, or national level (Raintree, 1987a). The merit of this planning methodology is its clear focus on the need to develop agroforestry as an appropriate land-use technology that is adoptable by local farmers. In respect to the adoptability of new agroforestry practices, planners should consider that their objectives for agroforestry development may not be similar to those of the farmers. As a result, it is often necessary to plan agroforestry development according to the "piggyback principle" (Raintree, 1983), by linking the solutions of problems that are not perceived by, or have a high priority for farmers, with solutions of the problems that are of prime importance to farmers. This approach is especially relevant in respect to planning agroforestry systems as a means to contribute to sustainable land use, because many farmers in tropical countries are more concerned with their pressing needs for food or lack of secure access to production factors than with long-term maintenance of productivity.

The D&D method focuses much attention on the planning of agroforestry interventions in the form of resource management technologies, but gives less explicit attention to the need for development of support measures for stimulating the adoption of these techniques. Of particular note, the possible need for planning new institutional arrangements is normally not identified by this method. The need for such rearrangements should not only be considered at the level of government bureaucracies (Wood, 1988), but also at the village level. At the village level, there are a number of important issues of rural development to be considered (Rocheleau, 1987). These issues derive from the fact that normally a variety of land-user groups are present within a rural community with different characteristics in respect to their land-use activities and their rights of access to and ownership of land and/or trees. These different land-user groups may have either a complementary, a compatible, or a competitive relationship to each other. This means that in planning agroforestry, one often has to deal with many land users with different degrees of control, access or responsibility over the agroforestry systems. Such situations are commonly found on various kinds of communal or public lands because on these lands, tree and land tenure is often ambiguous (Raintree, 1987c). In planning agroforestry schemes for these lands (e.g. silvo-pastoral schemes), attention should be

TABLE 2

Interventions needed for agroforestry development (adapted from Hufschmidt, 1986).

Resource management technology	Development implementation measures	Institutional arrangements
Agrosilvicultural practices	Usufruct rights on land	Institutional adjustments
Silvopastoral practices	Extension/education	Local mgt. organization
Agrosilvopastoral practices	Technical help with inputs (seedlings, fertilizer, etc)	Tenure systems
	Financial incentives (Loans, etc.)	Economic policies
	Regulations & Licenses	
	Market cooperation	Reorganization of public agencies (e.g. forestry and agricultural extension
		Liaison with non-governmental organizations

given to the creation of effective local management organizations and to the development of generally accepted rights to agroforestry products. Such institutional arrangements at the community level will often be a key element in planning agroforestry field implementation projects.

Thus, planning agroforestry projects involves various kinds of development interventions. Attention should not only be given to agroforestry technology development, but also to the development of proper tools for effectively implementing this technology. The various approaches for diagnosis and design of the different components of agroforestry development still need to be further unified in an integrated approach, which can serve as a framework for effective planning of new agroforestry systems.

5 CONCLUSIONS

In the introduction to this chapter the question was posed: Why ten years after the identification of agroforestry as a sustainable form of land use, has it recently been indicated that an emergent issue in planning agroforestry systems is the possibility of using this practice as a means to sustainable land use? The answer to this question has been shown to contain three elements:

(a) Although the need for sustainable land use is often mentioned as an essential condition for ensuring maintenance of livelihood conditions and rural development in tropical countries, the operationalization of this concept is still in its infancy. The concept involves both static and dynamic aspects. Planners are faced with the basic problem of designing systems which are stable in the short term, yet are capable of long term changes under influence of economic and social tendencies (Chorley, 1973). Several efforts have been made to identify the emergent properties of sustainable land-use (sub)systems, which may be grouped into three major categories: maintenance of production capacity, adaptability of management practices, and distribution of input and output of agricultural production.

(b) By comparing the characteristics of agroforestry systems with the properties of sustainable land use, it appears that mature agroforestry systems generally display several favourable conditions, which offer good starting points for maintaining productivity. However, the actual performance of agroforestry systems also depends on the effectiveness of management practices. In new agroforestry systems, it may take several years to attain favourable characteristics.

(c) In agroforestry planning, the implementation of projects should take into consideration that the introduction of this "new" agroforestry system will be occurring in a situation of dynamic rural change. Consequently, the planning of agroforestry interventions should not only be directed at the introduction of new agroforestry technologies, but also at the development of effective support measures and the creation of proper institutional arrangements within which implementation can proceed. Only if all these elements are considered during the phase of diagnosis and design of new agroforestry systems, can it be expected that these systems can truly serve as a means to contribute to sustainable land use.

6 FOOTNOTES

1) Blaikie and Brookfield (1987) identified the properties of sensitivity and resilience specifically in relation to human interference, but in reality human activities may take the form of either a stress factor or a factor of resistance or resilience.

7 REFERENCES

Arnold, J.E.M., 1987. Economic considerations in agroforestry. In: H.A. Steppler and P.K.R. Nair (Editors), Agroforestry, A Decade of Development. ICRAF, Nairobi, pp. 173-190.

Berenschot, C.M., Filius, B.M. and Hardjosoediro, S., 1988. Factors determining the occurrence of the agroforestry system with Acacia mearnsii in Central Java. Agroforestry Systems 6: 119-135.

Blaikie, P. and Brookfield, H., 1987. Land degradation and society. Methuen, London.

Budowski, G., 1981. Applicability of agroforestry systems. In: L.H. MacDonald (Editor), Agroforestry in the African humid tropics. United Nations University, Tokyo, pp. 13-16.

Chorley, R.J., 1973. Geography as human ecology. In: R.J. Chorley (Editor), Directions in Geography. Methuen, London, pp. 155-169.

Conway, G.R., 1985. Agro-ecosystem analysis. Agricultural Administration 20:31-55.

Conway, G.R., 1987. The properties of agro-ecosystems. Agricultural Systems 24:95-118.

Duchhart, I., Steiner, F., and Bassman, J.H., 1989. Planning methods for agroforestry. Agroforestry Systems 7:223-258.

Food and Agriculture Organization (FAO), 1985. Tree growing by rural people. FAO Forestry Paper No. 64, Rome, 130 pp.

Filius, A.M., 1988. Economic aspects of agroforestry. In: K.F. Wiersum (Editor), Viewpoints on agroforestry. Agricultural University Wageningen, the Netherlands, 2nd ed., pp. 169-188.

Harwood, R.R., 1979. Small farm development. Understanding and improving farming systems in the humid tropics. Westview Press, Boulder, Colorado.

Hufschmidt, M.M., 1986. A conceptual framework for watershed management. In: K.W. Easter, J.A. Dixon and M.M. Hufschmidt (Editors), Watershed management resources. An integrated framework with studies from Asia and the Pacific. Studies in Water Policy and Management No. 10, Westview Press, Boulder, Colorado, pp. 17-31.

Huijsman, A., 1986. Choice and uncertainty in a semi-subsistence economy. A study of decision making in a Philippine village. KIT Royal Tropical Institute, Amsterdam, the Netherlands.

Kellman, M., 1979. Soil enrichment by neotropical savanna trees. Journal of Ecology 67:565-577.

Kessler, J.J. and Breman, H., in preparation. Agroforestry in the Sahel and the savannah zone of West Africa; its potential to improve plant production.

King, K.F.S. and Chandler, M.T., 1978. The wasted lands; the programme of work of the International Council for Research in Agroforestry. ICRAF, Nairobi.

Lundgren, B., 1982. Introduction. Agroforestry Systems 1:3-6.

Maaren, A. van and Wiersum, K.F., 1988. Policy aspects of agroforestry development. In K.F. Wiersum (Editor), Viewpoints on agroforestry. Agricultural University Wageningen, The Netherlands, 2nd edition, pp. 243-256.

Maxwell, S., 1984. Farming systems research: hitting a moving target.

Discussion Paper DP199, Institute of Development Studies, University of Sussex, Brighton, UK, 24 pp.
Marten, G.G., 1988. Productivity, stability, sustainability, equitability and autonomy as properties for agro-ecosystem assessment. Agricultural Systems 26:291-316.
Marten, G.G., and Saltman, D.M., 1986. The human ecology perspective. In: G.G. Marten (Editor), Traditional agriculture in Southeast Asia, a human ecology perspective. Westview Press, Boulder, Colorado, pp. 20-53.
Mellor, J.W., 1988. The intertwining of environmental problems and poverty. Environment 30(9):8-13, 28-30.
Norman, D.W., Newman, M.D. and Ouedraogo, 1981. Farm and village production systems in the semi-arid tropics of West Africa: an interpretive review of research. Research Bulletin 4(1), ICRISAT, India.
Oram, P.A., 1988. Moving toward sustainability; building the agroecological framework. Enviroment 30(9):8-14-17, 30-36.
Prinsley, R.T. and Swift, M.J., 1986. Amelioration of soils by trees: a review of current concepts and practices. Commonwealth Science Council, London, UK. 181 pp.
Raintree, J.B., 1983. Strategies for enhancing the adoptability of agroforestry innovations. Agroforestry Systems 1:173-187.
Raintree, J.B., 1987a. The state of the art of agroforestry diagnosis and design. Agroforestry Systems 5:219-250.
Raintree, J.B., 1987b. D&D user's manual. An introduction to agroforestry diagnosis and design. ICRAF, Nairobi.
Raintree, J.B., 1987c. Land, trees and tenure. Proceedings of an international workshop on tenure issues in agroforestry. ICRAF & Land Tenure Center, Nairobi & Madison. 412 pp.
Rocheleau, D.E., 1987. The user perspective and the agroforestry research and action agenda. In: H.L. Gholz (Editor), Agroforestry: realities, possibilities and potentials. M. Nijhoff, The Hague, pp. 59-87.
Ruthenberg, H. 1980. Farming systems in the tropics. 3rd ed. Clarendon Press, Oxford, UK. 424 pp.
Sanchez, P.A., 1987. Soil productivity and the agroforestry research and action agendas. In: H.A. Steppler and P.K.R. Nair (Editors), Agroforestry, a decade of development. ICRAF, Nairobi, pp. 205-223.
Tisdell, C., 1988. Sustainable development: differing perspectives of ecologists and economists, and relevance to less developed countries. World Development 16:373-384.
Wiersum, K.F., 1984. Surface erosion in agroforestry systems. In: C.L. O'Loughlin and A.J. Pearce (Editors), Symposium on effects of forest land use on erosion and slope stability. Environment & Policy Institute, East-West Center, Honolulu, Hawaii, pp. 231-239.
Wiersum, K.F., 1985. Forestry aspects of stabilizing shifting cultivation in Africa. Agricultural University Wageningen, the Netherlands. 196 pp.
Wiersum, K.F., 1988. Outline of the agroforestry concept. In: K.F. Wiersum (Editor), Viewpoints on agroforestry. Agricultural University Wageningen, the Netherlands, 2nd ed., pp. 1-26.
Wood, P.J., 1988. Agroforestry and decision-making in rural development. Forest Ecology and Management 24:191-201.
York, E.T., 1988. Improving sustainability with agricultural research. Environment 30(9):19-41, 37-40.
Young, A., 1986. The potential of agroforestry as a practical means of sustaining soil fertility. In: R.T. Prinsley and M.J. Swift (Editors), Amelioration of soils by trees. Commonwealth Science Council, London, UK, pp. 121-144.

Chapter 2

AGROFORESTRY DIAGNOSIS AND DESIGN: OVERVIEW AND UPDATE

J. B. RAINTREE
International Council for Research in Agroforestry, P.O. Box 30677, Nairobi (Kenya)

INTRODUCTION

As a focus of organized scientific activity, agroforestry is relatively new, but as an integrated approach to land use it has existed in traditional farming communities since the dawn of the Neolithic. Indeed, in view of the relatively undomesticated nature of the vast majority of trees used in agroforestry, as compared to agricultural crops, one is tempted to say that agroforestry is still in the Stone Age.

The formal "rediscovery" of agroforestry by the global scientific community can be dated to a point about 12 years ago, when the International Development Research Centre (of Canada) study group led by John Bene published its recommendations on priorities in tropical forestry (Bene et al., 1977) and the International Council for Research in Agroforestry (ICRAF) was established. Despite this global debut, national and regional rediscoveries of agroforestry continue to occur with disarming frequency. At the moment, the most active locus of rediscovery seems to be in the developed countries of the north, where --after a decade of gradually increasing support to agroforestry research and development activities in developing countries of the south-- developed country institutions are coming to the realization that agroforestry may even have a role to play in the rational and sustainable management of their own countrysides.

The opportunity to discuss agroforestry diagnosis and design at this juncture is very welcome, particularly in the context of a book with an explicit focus on planning aspects of agroforestry. Here, at last, it may be possible to give more adequate consideration to some of the fundamental psychological and political dimensions which are part and parcel of the agroforestry planning process. In agricultural, forestry, or agroforestry forums lacking an explicit focus on planning, one is normally constrained to avoid such considerations. One's hope is that this book will be able to deal with the agroforestry planning process in somewhat greater scope than is normally accorded by the agroforestry research and development community.

1.1 The elephant problem in diagnosis and design

In the children's story of the three blind men and the elephant; the mice having gotten hold of different parts of the elephant each came away with a very different idea of what an elephant was. After more than seven years of widely published existence, ICRAF's diagnosis and design (D&D) methodology suffers from similar problems in perception. Part of the confusion stems from the fact that D&D has developed into multiple forms. This paper will attempt to clarify the pattern of development and highlight the most robust and useful of the variations on the D&D theme.

When ICRAF first initiated systematic work on the development of a "diagnostic methodology" for agroforestry (Steppler, 1981), there was a tendency to think of this work as leading eventually, through trial and feedback, toward a single, optimal set of procedures. What has been discovered, however, as experience has been gained with D&D in different settings with different collaborators, was that users had different needs and resources and preferred to use differently adapted forms of the same underlying logic. As a consequence of ICRAF's openness to user modification and feedback, rather than crystallizing into a rigid set of procedures, D&D has preserved an almost colloidal flexibility and, like medical diagnosis and treatment, has evolved into a varied body of literature and practice.[1] Indeed, it might be more accurate to describe D&D as a general paradigm rather than a specific methodology.

Rather than try to impose an artificial standardization on these developments, the emphasis in the D&D project at ICRAF has tended to shift from efforts to perfect the sequence of methodological steps toward clarification of the underlying logic. Specific procedural variants have been elaborated by ICRAF staff, as and when needed to meet the requirements of new types of field applications and collaborative activities. Currently, the most active area of methodological improvement in D&D within ICRAF is in the development of practical methods for activating the "iterative" D&D process throughout the implementation phase of on-farm research and extension projects (Scherr, Chapter 6; also Rocheleau et al., 1989 and Raintree and Hoskins, 1988). From a diffusion standpoint, it is also highly significant that organizations other than ICRAF have adopted the basic approach and developed their own D&D styles and training materials (Buck, Chapter 5; Buck, 1989).

One could, of course, argue that the "elephant" problem in D&D is the result of trying to treat all these "different" approaches as variations on the "same" methodology, and that it would not arise if one would simply call them something different. In answer to this objection, there are at least four compelling reasons for seeking to preserve a unifying perspective on D&D developments. First, in the real-time history of the development of the methodological variations in question, they are in fact elaborations on the original D&D approach. Periodic efforts at reintegration represent, quite simply, an attempt to maximize the scope and coherence of the learning process that has attended the development of these variations and to give the most useful and widely applicable interpretation of D&D results.

There is a common logical core to all of these variations on the D&D theme and that core has a legitimate claim to preservation as a priority perspective because it is, at base, the same fundamental logic on which the human capability for problem solving rests. Another reason for wanting to emphasize a diagnostic approach to agroforestry design is to capitalize on practice in the applied land-management sciences. The medical profession is where the diagnostic-prescriptive paradigm has achieved its highest state of development, but it is a standard of professional practice that can be applied with equal seriousness to human care of the earth. Finally, the integrative urge is further motivated by the need for a common framework, a lingua franca if you will, for the storage, retrieval and communication of information on agroforestry field experience. Ultimately, agroforestry science and practice must come to rest on a firm foundation of empirical evidence. It is hoped that systematic comparative analysis of diagnostic regularities in land use systems and of field experience with various technologies will assist in advancing the state of knowledge on agroforestry in much the same way as the case history approach has contributed to the development of medical science.

2 WHAT IS AGROFORESTRY DIAGNOSIS AND DESIGN?

It may be best to begin the answer to this question with a diagrammatic overview of the various aspects of D&D methodology (Fig. 1) and the operationally significant criteria for deciding between these procedural choices (Table 1).

In the first instance, agroforestry diagnosis and design is a family of procedures for the diagnosis of land-management problems, identification of agroforestry potentials, and design of appropriate agroforestry interventions. It is also the much larger subject of the heuristic knowledge

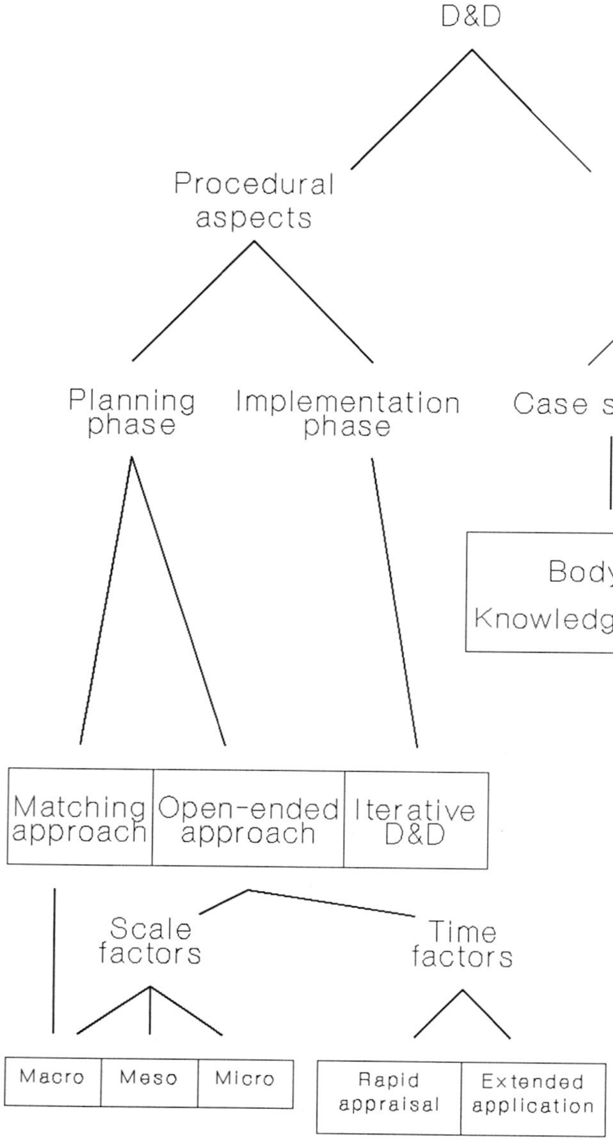

Fig. 1. An overview of agroforestry diagnosis and design.

base on land-management problems and agroforestry potentials, which constitutes the substantive side of D&D and which informs the judgement of the D&D user at each of the decision points within the D&D procedures. Procedural aspects of D&D will help ask the right questions in an efficient way; substantive aspects help give the right answers to those questions. This chapter will concentrate on procedural aspects. On the procedural side, the most important distinction is between the D&D procedures and methodological tools developed for initial planning of agroforestry projects versus those appropriate to the iterative D&D process, which is carried out throughout the project implementation phase (Fig. 2).

2.1 Open-ended discovery procedures versus a matching approach

When ICRAF first started working on the D&D methodology in 1981, the knowledge base on agroforestry was simply too sparse to rely upon and there was little choice but to place one's faith in an open-ended application of the formal discovery procedures on the assumption that the logic of the process itself would lead the D&D team from the description of relevant aspects of the land-use system, through the diagnosis of problems and identification of intervention points, to an appropriate agroforestry design.

In effect, the D&D teams in those days would simply strike out into a new land-use system, armed only with the D&D methodology and a general understanding of agroforestry principles, but no ready stock of "off-the-shelf" technologies and few expectations on what the design outcome would be. This was not only a necessary thing (the shelves were almost bare), but it was also thought to be a good thing, insofar as it kept the D&D team from prejudicing the design result through a preoccupation with "pet technologies". There was a tacit agreement among the group that developed the early D&D methodology at ICRAF not to speak too much at first about what agroforestry was, but rather, to use the D&D studies to discover what it should be.

In the hands of a competent multidisciplinary field team, the open-ended methods have worked well enough, but the transfer of the methodology to collaborating scientists has been somewhat problematical. It was easy enough to show someone how to ask the right questions in the right order --a few days in the field with one or two experienced D&D team members from ICRAF was usually enough to accomplish this-- but, this alone did not ensure that, left to themselves, the trainees would give the right answers to those questions. What was missing was the whole body of heuristic knowledge and decision criteria which an experienced D&D team would intuitively draw upon in coming up with a design that matched the diagnosis at hand.

At this point, it was decided that the effort of trying to make the implicit decision rules and matching criteria explicit would become a priority within the D&D project. By that time, agroforestry research had advanced considerably. As a result, in part, of the experience accumulated by ICRAF in numerous D&D exercises around the world, as well as other sources of information like ICRAF's global agroforestry systems inventory (Nair, 1987), the contents of the agroforester's bag of tricks was no longer such a mystery. It is still the case that not many of the "known" technologies have been thoroughly understood or scientifically confirmed, but it can no longer be said that the shelves are empty. We can at least begin to list and classify the technologies systematically and to develop preliminary recommendation domains and matching criteria for its early phases (see Raintree, 1989; ICRAF, 1989; Rocheleau et al., 1989; Hoekstra, 1989, for the latest developments). Something of a matching approach is now commonly used in ICRAF'S macro level D&D exercises for national agroforestry programmes as an aid to making preliminary assessment of agroforestry potentials over large areas before following up with an in-depth, open-ended micro D&D exercise in the priority areas.

2.2 Scale factors

D&D procedures have been developed and applied at three scales of analysis. Because there is a strong correlation between the type of D&D user and the preferred scale of analysis, these methods will be discussed later in this paper. For the moment, it will suffice merely to register their existence (see Table 1) and note that the micro-scale (household level) methodology was the first to be developed, and that this then gave rise to two separate lines of development, a meso-scale and a macro-scale variant when the need arose to deal with these larger analytical units.

2.3 Time factors

Due to the exigency of the research planning process, most D&D applications, to date, have been carried out within the time constraints of a "rapid appraisal" exercise. Various heuristic techniques have been developed to facilitate a rapid entry into the diagnostically relevant aspects of the land-use system and to sharpen the accuracy of such time-limited applications, but there is nothing within the logic of D&D to discourage longer, more leisurely and thorough-going application of the procedures, if time and resources permit.

TABLE 1
Variable scale D&D methods and applications.

SCALE	FOCAL SYSTEM OR MANAGEMENT UNIT
Micro	Household management unit (e.g. the family farm, household herd, or other elementary land management unit)
Meso	Local community or ecosystem (e.g. neighborhood, village or small watershed)
Macro	Region, country, ecozone

2.4 The iterative nature of D&D

It stands to reason that any rapid appraisal technique used to initiate research or development activities must be backed up by continued monitoring and evaluation of the diagnosed situation as it develops over the course of the project. Indeed, whatever methods are used to identify relevant agroforestry technologies and initiate trials with the intended users, it is almost certain that the initial technology can be improved upon over the course of time. The very act of introducing a new technology may modify the diagnosed situation and this alone is grounds for a repeat diagnosis. Moreover no technology that is selected solely through discussion with the intended users --no matter how extended, varied, community-based and participatory that process might be-- can be regarded as properly evaluated by the users until they have had time to live with it for an adequate trial period. For all these reasons, it is important to insist that diagnosis and design be regarded as an iterative process (Fig. 3).

When properly activated, the iterative D&D process provides a basis for close feedback and complementarity between the different components of a research and development project (R&D). By adjusting the plan of action to new information, the D&D learning process becomes continuous and self-corrective. Ideally, it should not stop until the design is well optimized and/or further refinements are judged not worth the additional effort. Until that point is reached, the D&D logic can be used to identify fresh opportunities for refinement and adaptation of the experimental technologies to meet the needs of the client system. The farmers themselves are often the

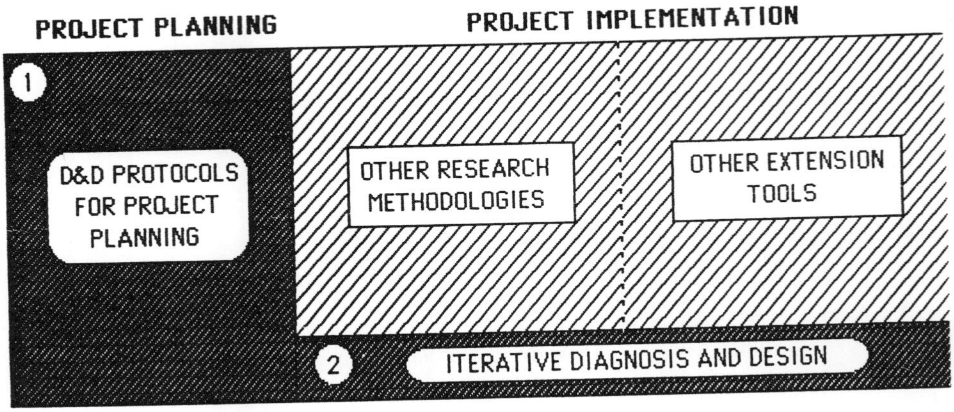

Fig. 2. Two distinct phases of D&D application: the first encompasses a range of procedural options for initial project planning; the second embodies a process of trial and refinement of agroforestry technologies through iterative "rediagnosis" and "redesign" in partnership with the users. This self-corrective process can function as the heart of a project's internal monitoring, evaluation, and planning system.

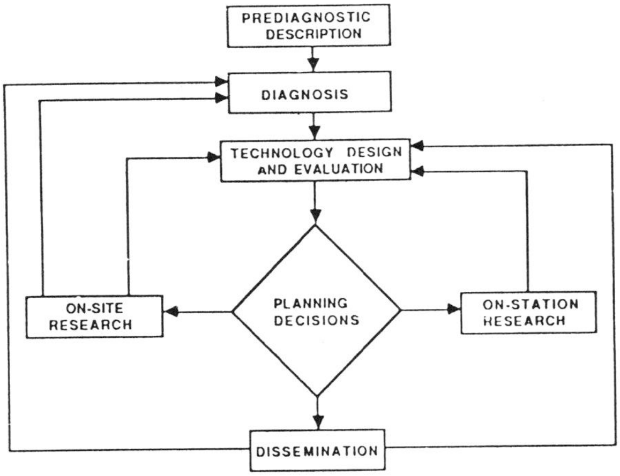

Fig. 3. Feedback between components of a research and development project using iterative D&D for monitoring and evaluation of progress toward project goals. The use of D&D as part of the project's steering system depends upon the feedback linkages shown in this diagram.

best source of relevant (i.e. adoptable) design ideas and improvements. In any case, the iterative D&D throughout the course of an agroforestry project is the heart of the D&D process. In Scherr's chapter in this volume, there is an in-depth, up-to-date exposition of this aspect of the D&D methodology. In the context of this chapter, what is important to emphasize, in the most explicit terms, is the proper relationship between planning and implementation phases of D&D. The initial round of D&D, at the project planning stage, is sufficient to get the R&D or extension trial process moving along more or less the right lines, i.e., toward a generally appropriate agroforestry technology. But no practical person would believe for an instant that the first design will often be the best design or that the D&D process could result in a specifically appropriate technology for the location-specific without several rounds or reiteration. This is only common sense; ultimate fine tuning of the technology is best left to the intended users.

For that reason, it may be counter-productive to belabour the initial D&D planning process too long in the interest of "participation", since the real participation only begins once the preliminary talking is over and the intended user is able to interact with the candidate technology on the ground. While it will never be advisable to rush through the initial planning stage, rapid appraisal techniques may provide just about the right level of resolution to get the process started, without generating too much attachment to plans that are likely to change in the course of time.

3 THE BASIC LOGIC OF THE "OPEN-ENDED" D&D PROCEDURES

Detailed guidelines and reference materials have been developed to assist agroforestry researchers and extension workers with specific types of D&D applications, but in its most basic and robust form the logic underlying the D&D methodology embodies a common-sense problem-solving approach that is applicable not just to agroforestry but to virtually any problem in technology design or, for that matter, any situation requiring a diagnostic approach to planning. Although developed before expert systems came into general usage, the logic of D&D is an example of the most common expert-system, consultation paradigm. It is a basic "open-ended" or "data-driven" inference strategy which seeks to arrive at an appropriate design.

In the medical profession, where the diagnostic approach has reached its highest development, there is a saying that "Diagnosis should precede treatment". It is common experience in all professions that a clear definition of a problem is often all that is needed in order to suggest the

TABLE 2

The basic logic of agroforestry diagnosis and design.

BASIC QUESTIONS	KEY FACTORS TO CONSIDER
Prediagnostic stage	
WHICH LAND-USE SYSTEM?	Distinctive combinations of resources, technology and land user objectives
HOW DOES THE SYSTEM WORK?	Production objectives and strategies, subsystems and components
Diagnostic stage	
HOW WELL DOES THE SYSTEM WORK?	Problems in meeting objectives, causal factors, constraints and intervention points
Design and evaluation stage	
HOW TO IMPROVE THE SYSTEM?	Specifications for problem-solving or performance enhancing interventions
Planning stage	
WHAT TO DO TO DEVELOP AND DISSEMINATE THE IMPROVED TECHNOLOGY?	Research and development needs; extension needs
Implementation stage	
HOW TO ADJUST THE PLAN OF ACTION TO NEW INFORMATION?	Feedback from research and extension trials, independent farmer innovations.

outlines of a solution. This logic is so fundamental to human problem-solving capabilities as to appear, like the human capability to language, to be almost a species' characteristic. As such, it is expressed in many forms. When applied systematically to the diagnosis of land-use problems and designs of agroforestry solutions, it takes the form of the logical framework shown in Table 2.

This logical framework is equally applicable to extension as well as research activities. If the envisaged agroforestry technologies already exist, then the design result can be applied directly as a guide for agroforestry interventions by extension agents and rural development workers. If the envisaged technologies have not yet been developed, or are not yet sufficiently well understood to warrant confidence in extending them to farmers, then the design can serve as a basis for identifying the research gaps that need to be filled in order to develop the "research for development" approach that ICRAF is promoting among agroforestry researchers-- that the research objectives are nested within specific development objectives. There are many different ways of applying the elementary D&D logic. Factors which influence the choice of D&D methods are listed in Table 3.

4 VARIABLE SCALE METHODS

If one were to work out all the theoretical combinations of the different factors shown in the foregoing table, the number of possible methodological variations on the basic D&D approach would be enormous. In practice, however, most of the important variance can be related to the choice of the scale of the D&D methods, as befits the operational scope and purposes of different kinds of D&D users.

For example, if the user is a national research institution in a less developed country and the purpose is to develop a general plan for agroforestry research within the country as a whole, then a large-scale, broad brush, rapid appraisal macro D&D will usually be needed as a first step to land-use systems within the country as a whole in terms of their general agroforestry potential and to identify priorities for more detailed, in-depth micro D&D follow-up exercises in localities that represent the prioritized systems. On the other hand, if the user is an individual or a local organization whose work is already focused within a specific locality, or even on a single farm, then a more complete, intensive and continuous micro/meso D&D approach would be more useful and appropriate.

Within the framework of the basic procedural logic, D&D methods for initial project planning have been developed and applied at three levels in the hierarchy of scale in land-use systems. Some of the most distinctive features of these methods are reviewed briefly in the following sections.

4.1 Micro-scale methods

Initially, the methodology focused on household and farm level diagnosis and design (Raintree, 1981, 1982). This was a logical starting point for a methodology that sought to address farmer decision-making, since

TABLE 3

Factors affecting the choice of procedures for a D&D application.

For whom is the D&D information intended?

- Farmers
- Foresters
- Scientists
- Extensionists
- Planners

Institutional context of the D&D application

- Government research programme
- Government extension programme
- NGO extension or rural development project
- Community-based agroforestry initiatives
- University-based education and research

Scale of concern

- International
- National
- Regional
- Community
- Household
- Individual

Time frame of the application

- Initial project planning
- Iterative D&D throughout project implementation

this is where most of the land-management decisions relevant to agroforestry are made. It is also quite consistent with the general pattern of farming systems research (FSR), of which the micro-D&D methods might be regarded as an agroforestry-specific variant. In comparison with other FSR methodologies, apart from the unique focus on the role of trees within whole land-use systems (and hence a larger diagnostic scope relative to commodity-based methodologies) the most distinctive features of the D&D methodology at this scale are: (1) a basic needs approach to the identification and evaluation of household production subsystems in terms of the land manager's objectives (Raintree, 1983), using a checklist of the system (food, fuel, water, shelter, raw materials for local industry, cash, savings and investment, and social production); (2) a trouble-shooting procedure for identifying critical constraints, leverage points, and functional specifications for agroforestry interventions; and (3) a separate assessment of the sustainability of the land-use system under the most likely future scenario.

4.2 Meso-scale methods

While the household's land-management unit will usually provide one of the key focal points for agroforestry development, D&D procedures at this scale may not in themselves provide a complete description of all diagnostically relevant factors. In the first place, not all of the land-use problems relevant to agroforestry originate within a single farm, nor can they always be solved by action at the individual household level (e.g., erosion problems caused by runoff from farms in the upper watershed, overgrazing the commons, or land for the landless). Also, the household is not always a homogeneous interest group. There are different types of households and different types of producers within the household, particularly where production responsibilities and opportunities are structured by gender roles (Hoskins, 1980; Fortmann and Rocheleau, 1985; Rocheleau, 1987). Larger-scale, inter-household social organizations (e.g. women's self-help groups) may facilitate opportunities for agroforestry-related activities that would be missed by an exclusive focus on households (Rocheleau, 1984; Raintree and Rocheleau, 1987).

For these reasons, meso-scale D&D methods have been developed to address agroforestry-related problems and opportunities at the larger-than-farm, community or local ecosystem level (Rocheleau, 1983a, 1983b, 1984, 1985; Rocheleau and van den Hoek 1984; Duchhart, 1986; Rocheleau et al., 1989). The most distinctive elements of these methods are: (1) the possibility of working with larger-than-farm scale units of landscape analysis and design, (2) identification of agroforestry opportunities through analysis of potential spatial and functional complementarities between different management units within the larger system (Rocheleau, 1983a, 1983b), and (3) local small-group process approaches to the organization of agroforestry activities, e.g. group interview techniques and neighbourhood nurseries based on traditional self-help organizations (Rocheleau, 1984; Rocheleau and Vonk, 1983).

Buck (1989) and Rocheleau et al. (1989) have recently provided new sets of guidelines for extension-oriented, community based approaches to D&D which combine various micro and meso-scale techniques over extended periods of participatory project planning. Both of these are highly recommended for extensionists and community development workers. Among the survey methods recommended by Rocheleau et al. (1989) are:

(a) a review and summary of what is already known by participating research and development workers;

(b) general field visits, or rapid surveys, to describe the landscape, land use and land users;

(c) general and specialized group interviews to identify land-user groups, land-use practices and systems, local knowledge and land-use changes, problems and potential solutions; .

(d) household interviews to discuss the same topics.

(e) walking interviews with individuals and small groups to discuss the landscape, land uses, and activities.

(f) individual interviews focusing on specialized knowledge and skills or discussion.

(g) action interviews (participant observation), which involve working with individuals or small groups to learn from direct experience about their land-use systems and activities.

4.3 Macro-scale methods

The most obvious limitation of micro- and meso-scale approaches is that the scale is simply not big enough for large-scale agroforestry undertakings, like national or regional agroforestry research and development programmes. Even if it were possible to carry out a large number of community D&D studies, that alone would not answer the needs of this level of planning. Something more is needed than just an aggregation of individual site results; sites must be placed within a regional environmental, economic policy, and program planning context. That is the purpose of the macro D&D process.

The impetus to develop macro-level methods came as a direct result of ICRAF's collaborative activities with zonal agroforestry networks in Africa when suddenly, in order to develop plans for needs-based research programs at a much larger scale of operation, ICRAF needed methods for large scale D&D (Lundgren, 1986; Torres 1985, 1986, 1987; Scherr, 1987; Minae, 1989). The general solution was a revised and expanded framework for D&D procedures which incorporates, among other suggested revisions (Huxley and Wood 1984, Young, 1985), a macro-scale framework for agroforestry diagnosis and design, within which both micro- and meso-scale applications can be nested. AFRENA is the name for this continent-wide undertaking. Thus far, only the macro- and micro-level D&D procedures have become institutionalized as standard practice within the multiplicity-country AFRENA networks, but it is still early in the life cycle of these programmes. It is hoped that the larger-than-farm, community-oriented and more participatory methods characteristic of meso-scale applications will come into greater use, once the research projects are more deeply immersed in on-farm research, iterative D&D, and collaboration with extension workers.

From the standpoint not just of the countries involved, but of the global agroforestry community as a whole, one of the most interesting outputs of the AFRENA process is the series of macro D&D reports which are produced by the joint ICRAF/host country D&D task forces, along with the more detailed research plans based on micro D&D exercises in the areas chosen by the national agroforestry steering committees as priority areas (AFRENA reports 1-17, listed in the references). One of the most distinctive features of the macro D&D reports is the now standard feature of using a development strategy based on consideration of the most likely future scenario for the land-use system as a major factor in the assessment of agroforestry potentials and the selection of agroforestry interventions. Of course, the results of the macro-level D&D findings must always be regarded as provisional until confirmed by micro D&D findings, and ultimately by trials with farmers.

5 HOW PARTICIPATORY IS DIAGNOSIS AND DESIGN?

How participatory should D&D be? The answer to this question depends very much on which variant of D&D is being used, the purpose and the user for whom the D&D information is intended, the institutional context of the D&D application, the scale and the time frame of the application. In other words, it depends upon those same factors (Table 3) which influence the choice of D&D procedures in the first place.

"Participation," like "sustainability," has become a shibboleth of project worthiness in all forms of rural research and development activity of late. And well it should be, as a needed corrective to decades of top-down meddling and expert mismanagement in the rural development process. One needs, however, to be a little bit circumspect about this. If the participation criterion is applied in an uncritical manner, then there is a danger of throwing out the baby with the bathwater. As a safeguard against this possibility one needs to modify the basic question to: How effective and participatory is a particular D&D variant? In answering this question, one also needs to consider each of the variants in the light of the criteria by which it was chosen, as well as by the more general criteria of good development practice, and one needs to distinguish between the planning and the implementation phases and assess the degree of participation in each.

5.1 Participatory D&D in extension projects

From the standpoint of maximizing the direct participation of the rural communities which constitute the ultimate intended beneficiaries of agroforestry innovations, the combination of D&D methods used by non-government organization (NGO) extension workers tends to be the most participatory of the D&D variants at the project planning stage, e.g., the CARE methods described by Buck (Chapter 5) and in Buck (1989) as well as those proposed for extensionists and community-based researchers in Rocheleau (1984, 1986) and Rocheleau et al.(1989).

What is it about these types of D&D applications that makes them more participatory in this sense? The ability to maximize local participation in the planning process is due in large part to the scale of the application (a combination of meso- and micro-scale methods emphasizing group discussions and community-based decision making), the time frame (deliberately taking the time to develop a consensus with the people on what needs to be done, through an extended application of participatory D&D procedures), and indeed the deliberate purpose of the project personnel in using the D&D process to provide information for the direct use of the local community and the community-oriented extension workers.

Experience has shown that the use of participatory planning methods at the start of a project is no guarantee that the same level of participation will be carried through the implementation phase of the project. Take the example of the CARE Agroforestry Extension Project in western Kenya, which is the subject of a case study in Scherr (Chapter 6) in this volume. Partly as a reaction of the project manager to his experience with a less participatory research-oriented D&D variant implemented under the aegis of ICRAF in eastern Kenya, when the D&D method was transferred and modified to fit the needs of the more participation-oriented CARE project, there was a conscious effort to avoid being too explicit about the exact nature of the initial diagnostically-based set of agroforestry designs given as "extension messages" to the participating community groups.

As vital as it is to allow scope for people to modify the initial technology recommendations to better fit their own needs and potentials, the lack of explicit documentation on what the project recommendations actually were made it difficult to recognize and monitor a farmer-oriented modification when it did occur. Consequently, the project ended up with a reduced ability to use the iterative D&D process to help the community learn from its own experience with agroforestry innovations by keeping track of and reflecting those experiences back to the community. Thus, community participation in the innovations, which several of its more members were attempting, was not maximized.

Moreover, it was apparent from observing the interaction between extension workers and participating farmers that the extension workers were not adept at the skills of sensitive interview and critical listening needed to get at the farmer's real reaction to the trial technologies. They seemed, instead, content to reinforce the farmers efforts to please the project staff by telling them what they wanted to hear. It was as if the farmers were holding up a mirror to the enthusiastic extension workers, who had not noticed that it was only their own reflection that was being shown to them by the wily farmers. In general, the fidelity of the input from farmers to the project staff seemed to decline over the course of the active extension phase, from its high point at the time of the participatory project planning process until the time when a new ICRAF researcher turned up to collaborate with CARE in reestablishing the iterative D&D learning process.

Make no mistake about it, participation is a two-way street. One does not maximize community participation simply by adopting an attitude of passive acceptance of whatever the community does or says it does. It can be argued that the best efforts of individuals and groups within the local community are most likely to emerge and be recognized by other members of the community when they are actively facilitated by a catalytic agent with the skills to stimulate a genuine two-way flow of information (Raintree, 1978).

5.2 What about research projects: how participatory can they be?

The key to this question is whose participation are we trying to enlist? In most cases, the level of participation characteristic of the planning phase of NGO extension projects is not often achieved in formal research projects. ICRAF's experience has been that the large scale on which the D&D application must be carried out and the severe time constraints which the members of the multidisciplinary team of D&D scientists are subject to, make it very difficult for them to take time to achieve a high degree of community participation in each locality. In principle many of the participation-enhancing methods recommended by Rocheleau can be applied by researchers, but if the team has only a limited time to carry out a macro-D&D survey of the whole country, in practice it will not always be possible to spend the time needed to achieve a community consensus on research priorities in every community at the project planning stage.

This need not be unduly alarming, however. As previously noted, the initial D&D results are only provisional and it may be unwise to place too much emphasis on them in the first place, since they will soon be superseded by the kind of consensus that can only come from real experience with the candidate technologies which occurs during the implementation phase of the project. Since on-farm researchers are, almost by definition, in the community for the purpose of acquiring information, we may expect a higher standard of sensitivity to farmer feedback than from extension workers (unless, of course, the latter have been specially trained for this new "research" type role). In other words, one might have reasons to expect, ceteris paribus, a higher degree of local participation in the iterative D&D process over the course of project implementation from research rather than extension projects, even though the former might have started out on a less participatory basis.

If participation is a two-way street, then what about researcher participation? The main lesson of the whole FSR approach, which over the course of the past decade has changed the face of international agricultural research, is that farmer participation in identification and development of relevant agricultural technology is essential for the success of applied agricultural research efforts, but so also is the participation of researchers. Unless one takes the unsupportable position that formal scientific research has nothing to contribute to the rural development process, then ways need to be found to encourage the participation of individual scientists. As Chambers and Jiggens (1986) have so aptly expressed it, the essential thing in all this is "the clarifying dialogue in which scientists and farmers can match their different systems of knowledge and find out how to get the best of both. The direct meeting is the crux, the occasion and process on which attention must be focussed." It is not good for the scientist to be only "half present" at this meeting. If anything useful is to result from the meeting of minds, then the scientist has to hold up his/her end of this dialogue. If researchers do not have anything to offer the farmers that they could not come up with themselves, then their time is just wasted! In this context, one of the chief values of the whole FSR approach (including D&D in its most common research-oriented variant) is that it has been successful in finding ways to get scientists to be present at this meeting in the first place (by establishing diagnostic surveys and on-farm research as standard scientific practice), and to be there for the farmer in a meaningful way by creating a climate of professional self-interest in hearing what the farmer has to say.

There is no psychologically sound reason to suppose that the participation of the scientists could be improved by expecting them to be utterly self-effacing. In order for most scientists to hear what the farmer is trying to tell them and to take it seriously enough to act upon, they must be able to relate what the farmer is saying to their own style of thinking and system of knowledge. That is why the research-oriented variants of the D&D are deliberately designed to engage the participating scientists in the process in terms that are meaningful to them and why it is all right for the D&D procedures used for this process to be oriented toward getting information from the farmers for the use of the scientist. To imagine that this could or should be otherwise in order to be "more participatory" is to fail to understand the full range of participation that is required.

Having acknowledged the necessary and legitimate role of participation by researchers as well as farmers in agroforestry work with local communities, as with distortions of the participatory process in the extension context, it is also important to be aware of the ways in which an over-emphasis on the research orientation can distort the researcher-farmer dialogue. This can be illustrated by an example from the experience of the ICRAF's Kathama project in eastern Kenya. The conscious purpose of the initial D&D experiences at Kathama was to identify diagnostically relevant and researchable agroforestry technologies for immediate trial within a small, low budget on-farm research project.

All of the identified agroforestry interventions were derived from what the farmers themselves identified as their problems, and all of them were discussed with farmers, who confirmed their interest in trying them out on their own farms. Because of the human-power and budget limitations (a universal constraint at some level in all research projects), we could not do trials on all of the technologies at once; consequently, priorities had to be assigned. The problem arose when it was recognized that the technologies with the greatest need for research (alley cropping, living fences, grazing land improvements), although of considerable expressed interest to farmers, were not the same technologies that the farmers considered top priority (fruit trees, shade trees, boundary plantings).

Although from a research point of view, it appeared to be possible to justify the decision to concentrate the initial farm trials on the technologies of greatest research interest, this proved to be a mistake in retrospect, even from a research point of view. If one of the main objectives in on-farm research is to get the farmers to interact with new technologies in a meaningful way, then it may be self-defeating to try to get farmers to give a fair trial to relatively complex technologies before satisfying their immediate demand for simpler and more immediately adoptable technologies. In other words, given the fairly long-term nature of agroforestry research and the kind of demands it places on the local participants, it might be unwise to attempt to carry out on-farm agroforestry research independent of at least a minimal effort to provide basic extension services.

In any case, it is not until farmers are interacting with the technologies on the ground that the planner finds out (a) what is really of interest to the farmers (an ideal which carries more truth than the words convey), (b) what actually works (it's all hypothetical until you try it), (c) how to make things work (using the iterative D&D discovery procedures), and (d) how farmers evaluate it all (no matter what they or anyone else says at the beginning, they can not be expected to give their best judgement on a new technology until after they have given it a fair trial).

6 COMMUNITY-BASED RESEARCH AND DEVELOPMENT: A RADICAL ALTERNATIVE

The argument has been developed elsewhere (Raintree and Hoskins, 1988; Raintree, 1989) that there are in reality not one but two centres of agroforestry research and development: the formal research centres doing research on agroforestry and the local communities in which the agroforestry innovations must eventually take root. The formal agroforestry research process, as reflected in the operations of the AFRENA network promoted by ICRAF, uses D&D in the context of a large-scale, FSR-type approach, in which D&D is employed in a "consultative" mode to get information from the community for the use of the scientists.

Chambers and Jiggens (1986) have taken the FSR paradigm to task on the grounds that it is not participatory enough. They argue that, rather than just extracting information from farmers to be used by researchers to define the research priorities, farmers themselves should be empowered to articulate their own research priorities to researchers. This critique is not radical enough to justify calling the FSR paradigm into doubt. The FSR paradigm and farmer-first-and-last (FFL) paradigm, which Chambers and Jiggens would like to replace it with, are both oriented toward improving the ability of formal sector researchers to do research for the farmers, rather than assisting the farmers to do their own research for themselves (Raintree, 1989). That would be a sufficiently radical alternative to justify, not the replacement of the FSR approach, which will still be needed as a way of encouraging meaningful participation from researchers, but equal emphasis on providing support to the other centre of agroforestry research and development, i.e. the rural communities themselves.

It comes down to a question of whose thinking process one is trying to catalyze. They may have different rules of evidence from the followers of R.A. Fisher, but farmers have been experimenters since the dawn of the Neolithic. And let us be honest about it: formal science may have some spectacular contributions to make, but it is the community-based research process that has been the backbone of agricultural progress. Scientific agroforestry is a newcomer to the scene. How much more intelligent it would be for us to acknowledge the existence of the community-based agroforestry research process and seek to complement, reinforce and actively catalyze it, rather than supplant it.

There is little reason to doubt the proposition that the way to maximize the rate and quality of agroforestry innovation overall is to support both centres of agroforestry research and development, but the most compelling argument for a dualistic approach to agroforestry research is the argument from fear of failure rather than hope of success. When we look at the way rural communities in the developing world actually integrate agroforestry practices into the rural landscape, we find that it results in exceedingly diverse and complicated patterns on the landscape. The process itself, one suspects, is far too complex, varied, and location-specific for any centralized institution to cope with (Rocheleau and Raintree, 1986).

This is demonstrated quite clearly by the AFRENA experience. It is unreasonable to expect that the macro-D&D reports actually succeed in identifying all of the agroforestry potentials of the diagnosed systems. Typically, they identify eight or ten distinct agroforestry interventions per land-use system.[2] While this may not represent the totality of relevant agroforestry possibilities, even these are far more than the formal research programme can cope with! Typically, given the prevailing constraints of human-power and funding, the national agroforestry steering committee selects two or three technologies as priorities for research. The net result is that many important agroforestry technologies never get chosen for research attention and that certain technologies (e.g. "alley cropping") get too much attention. Even technologies which regularly come up in the number 3 or 4 position on everyone's priority list may fail to become the subject of research, although clearly the results would be widely applicable.

One can hope that some of these anomalies can be dealt with in time by better coordination of the global research effort, but one is left with the intuition that what is discussed here is the kind of problem which no centralized institution can solve because, fundamentally, it requires a decentralized solution. The only ones capable of developing the full, diversified agroforestry potential of the local landscape are the local communities themselves. To persist in promoting only one side of the essential complementarity, i.e. the centralized "official" agroforestry research process which commands the lion's share of available research resources and preempts the potential for support of more decentralized initiatives, is to risk a dangerous over-simplification of rural landscapes and life-support systems.

7 FOOTNOTES

1) Readers already familiar with D&D may recognize material from earlier articles, particularly Raintree (1987). Because of its essential flexibility and because it continues to evolve, there is no single "bible" of D&D. The most extensive overview of the D&D literature, the most readable general introductions to the methodological approach and its major variants are found in Raintree (1987), Rocheleau et al. (1989) and Buck (in press), and the most detailed and extensive collection of useful analytical techniques and agroforestry reference materials are in Raintree (1983) and Rocheleau et al. (1989). Technology and research design are treated in Huxley and Wood (1984). For glimpses of new developments on the procedural side of D&D the interested reader is referred to: Rocheleau et al (1989) for community based approaches to D&D; Buck (1989), Raintree and Hoskins (1988) and Scherr (1987) for D&D approaches in the extension context; Scherr's chapter in this book for "iterative D&D" methods for on-farm research. Those interested in developments on the substantive side of D&D are referred to Hoekstra (1989), Raintree (1989), and Raintree (in press).

2) As this chapter goes to press I have just received a copy of the proceedings of the ILEIA workshop on operational approaches to participatory technology development in sustainable agriculture, which greatly advances the institutionalization of a community based paradigm for research and development. It still remains, however, to strike the right balance between formal and informal sector research efforts and to find a modus operandi for effective collaboration. Perhaps what we need in the research domain is an attitude something like what the Chinese used to call "walking on two legs".

8 REFERENCES

Bene, J.G., Beall, H.W. and Cote, A., 1977. Trees, Food and People: Land Management in the Tropics. IDRC. Ottawa.

Buck, L., 1989. Agroforestry Extension Training Sourcebook. CARE International. New York.

Chambers, R. and Carruthers, I., 1981. Rapid rural appraisal for rural development. Agricultural Administration 8(6): 407-22.

Chambers, R. and Jiggens, J., 1986. Agricultural research for resource poor farmers: a parsimonious paradigm. IDS Discussion Paper. Institute of Development Studies. Brighton.

Depommier, D. (Editor), 1988. Potentiel agroforestier dans les systemes d'utilization des sols des hautes terres d'Afrique de l'Est a regime pluviometrique bimodal Burundi. Report No.2. ICRAF. Nairobi.

Depommier, D. (Editor), 1988. Potentiel agroforestier et propositions de recherche pour le systeme de la crete Zaire-Nil a elevage, Burundi. AFRENA Report No. 14. ICRAF. Nairobi.

Djimde, M. (Editor), 1988. Potentiel agroforestier dans les systems d'utilization des sols des hautes terres d'Afrique de l'Est a regime pluviometrique bimodal Rwanda. AFRENA Report No. 1. ICRAF. Nairobi.

Djimde, M. and Gahamanyi, A. (Editors), 1988. Propositions de recherche agroforestiere pour le systeme de plateaux et collines au Rwanda. AFRENA Report No. 113. ICRAF. Nairobi.

Djimde, M. and Hoekstra, D.A. (Editors), 1988. Agroforestry potentials for land use management in the bimodal highlands of eastern Africa: Uganda. AFRENA Report NO. 4. ICRAF. Nairobi.

Djimde, M. and Raintree, J.B. (Editors). Agroforestry potential in the humid lowlands of Cameroon. AFRENA Report No. 12. ICRAF. Nairobi.

Duchhart, I., 1986. The role of landscape planning in agroforestry projects. Paper presented to the annual meeting of the Council of Educators in Landscape Architecture. University of Georgia. Athens.

Fortmann, L. and Rocheleau, D., 1985. Women and agroforestry: four myths and three case studies. Agroforestry Systems 2(4): 253-272.

Hoekstra, D.A., 1989. Summary of zonal agroforestry potentials and research across land-use systems in the highlands of eastern and central Africa. AFRENA Reports No. 15. ICRAF. Nairobi.

Hoskins, M.W., 1980. Community forestry depends on women. Unasylva 32(130): 27-32.

Huxley, P.A. and Wood, P.J., 1984. Technology and research considerations in ICRAF's diagnosis and design procedures. ICRAF Working Paper No. 26. ICRAF. Nairobi.

ICRAF, 1984. Diagnosis of problems and potentials and design of agroforestry technologies in Batu Arang Forest Reserve, Malaysia. ICRAF. Nairobi.

ICRAF, 1984. Agroforestry alternatives for the compensatory plantation scheme in the forestry reserve at Batu Arang, Malaysia. ICRAF. Nairobi.

ICRAF, 1986. Zonal analysis of country "blueprints" and research implications. Workshop on Agroforestry Research Planning (Unimodal Upland Plateau.) Harare. September 10-12. ICRAF. Nairobi.

ICRAF, 1989. Programme of work for 1989-91(draft). ICRAF. Nairobi.

ICRAF and ICAR, 1984. Report of the joint ICAR/ICRAF diagnostic and design exercise at the Bhaintan watershed in the outer Himalya of Uttar Pradesh. ICRAF. Nairobi.

ICRAF and ICAR, 1986. Diagnosis of land-use systems and designs of agroforestry interventions for the drylands of Karnataka. ICRAF. Nairobi.

ICRAF, INIPA, INFOR, IVITA, CIAT and NCSU, 1982. Research project on developing agroforestry systems for the upper basin of the Peruvian Amazon. ICRAF. Nairobi.

ICRAF, MAG and CATIE, 1984. Development of improved agroforestry systems for seasonally dry uplands of western Costa Rica. ICRAF. Nairobi.

Kamau, I.N. and Odra, J. (Editors), 1988. A blueprint for agroforestry research in the unimodal upland plateau of Tanzania. AFRENA Report No. 6. ICRAF. Nairobi.

Kwesiga, F. and Kamau, I.N. (Editors), 1989. A blueprint for agroforestry research in the unimodal upland plateau of Zambia. AFRENA Report No.7. ICRAF. Nairobi.

Lundgren, B., 1986. ICRAF 1986-1990: report by the director to the board of trustees with proposals on strategy and programme requirements of implementation of the recommendations of the external review panel. ICRAF. Nairobi.

Minae, S. (Editor), 1989. A blueprint for agroforestry research in the unimodal upland plateau of Malawi. AFRENA Report No. 5. ICRAF. Nairobi.

Minae, S. (Editor), 1988. Agroforestry research project proposal for the coffee based system in the bimodal highlands, Central and Eastern

Provinces, Kenya. AFRENA Report No. 16. ICRAF. Nairobi.
Minae, S., 1989. The macro D and D process. Lecture presented to ICRAF course on Agroforestry Research for Development. ICRAF. Nairobi.
Minae, S. and Akyeampong, E. (Editors), 1988. Agroforestry potentials for the land use systems in the bimodal highlands of eastern Africa: Kenya. AFRENA Report No. 3. ICRAF. Nairobi.
Nair, P.K., 1987. Agroforestry systems inventory. Agroforestry Systems 5(3): 301-318.
Okorio, J. and Hoekstra, D. (Editors), 1988. Agroforestry research project proposal for the Kigezi annual montane food crop system in the highlands of Uganda. AFRENA Report No. 11. ICRAF. Nairobi.
Raintree, J.B., 1978. Extension Research and Development in Malandi: Field Test fo a Community-Based Paradigm for Appropriate Technology Innovation among the Tagbanwa of Palawan. University Microfilms, Ann Arbor.
Raintree, J.B., 1981. Diagnostic evaluation of agroforestry-related constraints and potentials. In-house working paper. ICRAF. Nairobi.
Raintree, J.B., 1982. A methodology for diagnosis and design of agroforestry land management systems. In-house working paper. ICRAF. Nairobi. Revised for publication in: National Academy of Sciences. 1983. Agroforestry in the West African Sahel. NAS. Washington.
Raintree, J.B. (Editor), 1983. Resources for Agroforestry Diagnosis and Design. ICRAF Working Paper No. 7. ICRAF. Nairobi.
Raintree, J.B., 1987. D&D User's Manual. ICRAF. Nairobi. (French translation in preparation).
Raintree, J.B., 1989. Information management in agroforestry: by whom and for whom. In: D. Groenfeldt and Moock J. (Editors), Social Science Perspectives in Managing Agricultural Technology. International Irrigation Management Institute and the Rockefeller Foundation. New York.
Raintree, J.B., In Press. Theory and practice of agroforestry diagnosis and design. In: K. MacDicken and Vergarra, N. (Editors). Agroforestry: Classification and Management. John Wiley. New York.
Raintree, J.B. and Hoskins, M., 1988. Appropriate R&B support for forestry extension. Proceedings of the FAO Regional Consultation on the Organization of Forestry Extension. FAO, Bangkok.
Raintree, J.B. and Rocheleau, D.E., 1987. Case study example of the D&D learning process. In: J.B. Raintree (Editor). D&D User's Manual. ICRAF. Nairobi.
Rocheleau, D.E., 1983a. Ecosystems analysis in D&D applications. In: J.B. Raintree (Editor). Resources for Agroforestry Diagnosis and Design. ICRAF Working Paper No. 7. ICRAF. Nairobi.
Rocheleau, D.E., 1983b. Watershed evaluations guidelines. In: J.B. Raintree (Editor). Resources for Agroforestry Diagnosis and Design. ICRAF Working Paper No. 7. ICRAF. Nairobi.
Rocheleau, D.E., 1984. Land-use planning with rural farm households and communities: participatory agroforestry research. Workshop on the Role of Sociologists and Anthropologists in Farming Systems Research. ARPT and CIMMYT. Lusaka.
Rocheleau, D.E., 1985. Criteria for re-appraisal and re-design: intra-household and between-household aspects of FSRE in three Kenyan agroforestry projects. ICRAF Working Paper No. 37. ICRAF. Nairobi.
Rocheleau, D.E., 1987. Women, trees anf tenure: implications agroforestry research and development. In: J.B. Raintree (Editor). Land, Trees and Tenure: Proceedings of an International Workshop on Tenure Issues in Agroforestry. ICRAF and the Land Tenure Center. Nairobi and Madison.
Rocheleau, D.E. and van den Hoek, A., 1984. The application of ecosystems and landscape analysis in agroforestry diagnosis and design. a case study from Kathama Sublocation, Machakos District, Kenya. ICRAF Working Paper No. 11. ICRAF. Nairobi.

Rocheleau, D.E. and Raintree, J.B., 1986. Agroforestry and the future of food production in developing countries. Impact of Science on Society 142: 127-141.
Rocheleau, D.E. and Vonk, R.B., 1983. The role of agroforestry in farming systems research and development. Farming Systems Support Project Newsletter 1(3).
Rocheleau, D., Weber, F. and Field-Juma, A., 1989. Agroforestry in Dryland Africa. ICRAF. Nairobi.
Scherr, S.J., 1987. Planning national agroforestry research: guidelines for land use system description. ICRAF Working Paper No. 48. ICRAF. Nairobi.
Steppler, H.A., 1981. A strategy for the International Council for Research in Agroforestry. ICRAF. Nairobi.
Torres, F., 1985. Networking for the generation of agroforestry technologies in Africa. ICRAF Working Paper No. 31. ICRAF. Nairobi.
Torres, F., 1986. Agroforestry research networks in tropical Africa: an ecozone approach. Presented at the 1st International Conference on Agricultural Research Systems. September 6-11. IFARD. Brasilia.
Torres, F., 1987. The ICRAF approach to international cooperation. Agroforestry systems 5(3): 395-417.
Young, A., 1985. Land evaluation and agroforestry diagnosis and design: towards a reconciliation of procedures. Soil Survey and Land Evaluation 5(3): 61-76.

Chapter 3

ON-FARM AGROFORESTRY RESEARCH PLANNING IN COSTA RICA

J. BEER
Tropical Agricultural Research and Training Center, CATIE-GTZ Agroforestry Project, Turrialba (Costa Rica)

R. BOREL
University for Peace, San Jose, Costa Rica, and

A. BONNEMANN
Tropical Agricultural Research and Training Center, CATIE-GTZ Agroforestry Project, Turrialba (Costa Rica)

ABSTRACT

The Tropical Agricultural Research and Training Center (CATIE) carries out research, training and technical cooperation in Central America and the Caribbean. Since 1976, CATIE's Agroforestry Programme has carried out basic and applied research in Costa Rica, Panama, and Nicaragua. It has also trained many professionals through Master of Science and short course programmes. This chapter analyzes the experiences from different projects in Costa Rica, and proposes some planning principles that have gradually evolved from one project to the next.

In addition to the criteria of "national priority" and potential for extrapolation, the area selection for new applied agroforestry research projects should generally be based on the existence of previous or actual research or extension projects, and the probability that an existing or new extension project will implement the results. The most significant area characteristics to be analyzed before establishing new agroforestry projects will probably be: topography, rainfall regime, land tenure, population density, the actual importance of trees for the farmers in the area, local history (including ethnic considerations), and the presence of other agricultural systems, where agroforestry has no role to play. The mechanisms to set goals have varied greatly between projects but should include 1) consideration from the beginning of the farmers' goals and of the extension service expectations and 2) involvement of the largest possible group of farmers and professionals in the planning stage and during each evaluation. For project implementation, the hiring of local assistants should receive high priority, while the involvement of the farmers (design and evaluation of the research) appears as another essential element. The importance of thorough monitoring of all processes, without forgetting economic and sociological aspects, is also stressed.

INTRODUCTION

1.1 The mission of the Tropical Agricultural Research and Training Centre

The Tropical Agricultural Research and Training Centre (CATIE) is a civil, nonprofit, autonomous association. It carries out, promotes, and stimulates research, training, and technical cooperation (i.e. strengthens the national programmes) in agriculture and natural renewable resources for the benefit of the American tropics, particularly the countries of the Central American Isthmus and the Caribbean.

Guidelines for the CATIE'S operations (see Appendix) are given in a ten-year strategic development plan (CATIE, 1988). CATIE focuses on generating and disseminating knowledge aimed at (a) accelerating agricultural development, (b) developing sustainable agricultural systems for the small farmer, and (c) designing and utilizing integrated natural resource management techniques.

1.2 General remarks on "planning"

Planning of agroforestry systems deals with the management of highly complex natural systems. Components (trees, crops, animals) are used in varying combinations and under different conditions (such as climate, soil, and socio-economic conditions). The theories about the application of scientific methods to managerial functions include two terms: operations research and systems engineering (Encyclopedia Britannica, 1982).

Operation research involves the application of scientific methods to the management of organized systems, in which the components are known and interaction effects calculable, and in which human behaviour may play an important part. Systems engineering is the utilization of scientific and technological knowledge in planning and designing complex systems, in which human behaviour is less important but in which the introduction of new components and their interactions may be essential.

Neither operations research or systems engineering really apply to agroforestry planning. On the one hand, the components are frequently little known and their interactions (theoretically calculable) may take years to be estimated, far beyond the time limit put onto development projects. On the other hand, human behaviour is so important in the design of agroforestry systems that systems engineering techniques are clearly out of place.

Planning in agroforestry, as a process of seeking simultaneous interrelated solutions to a set of inter-dependent problems, appears to be very complicated because of (a) the diversity of expectations and objectives, (b) the large number of possible components and their arrangements, (c) the poorly understood interactions; and (d) the difficulty of quantifying some of the functions (e.g. environmental values and sustainability). As a consequence, the following considerations are especially important for planning an agroforestry project;

-flexibility, allowing for continuous feedback between planning, implementation, and evaluation phases;

-consideration of local human interests from the very beginning;

-avoidance of mathematical models but reliance rather on large planning teams; and

-importance of monitoring, since the significant advances in agroforestry in the next 10-20 years may not come from research but from well documented local development cases.

However, this chapter does not present a theoretical approach to agroforestry planning but rather was prepared to illustrate some of the most important considerations that have been learnt by experience. It could be termed a "bottom-up" rather than a "top-down" presentation which does indeed reflect one of the principal philosophies of CATIE's approach to agroforestry planning!

2 DESCRIPTION OF METHODS USED

The following sections describe the planning and project organizational methods used by two CATIE agroforestry projects, which were established in different priority areas of Costa Rica. The development of planning methods has been mainly an empirical process and lessons learned from a long established project (Acosta-Puriscal, established 1980) were very influential in the choice of methods used in the more recent projects (e.g. Talamanca, 1986).

2.1 Acosta-Puriscal: An empirical planning process

(i) Area selection criteria. The selection was made after field visits and consultations between CATIE staff and personnel of governmental institutions responsible for agriculture, forestry and national planning, with a view to respecting the priorities of all involved. One justification for the choice of Acosta-Puriscal was the possibility for collaboration with an existing CATIE research study of agricultural economics in this area. The specific criteria used were: (a) population density and number of small farmers, (b) standard of living, (c) regional priorities of national institutions, (d) access to the area and infrastructure, and (e) production potential (by means of a subjective evaluation) (Lagemann and Heuveldop, 1983).

(ii) Area description. The Acosta-Puriscal region is situated on the seasonally dry Pacific watershed of Costa Rica, approximately 60 kilometers southwest of the capital San Jose. It falls within the life zones "moist tropical forest" to "premontane rain forest" (Holdridge, 1967) with altitudes ranging from 800 - 1200 meters above sea level. There is little flat land in the area and slopes are moderate (20-30%) to very steep (greater than 100%). Rainfall averages are between 2,100 to 2,500 millimetres per year, but the variation in annual values is 1,300 to 3,400 millimetres per year. There is a marked dry season between November/December and May with only 10% of yearly rainfall in this period. Mean monthly temperatures vary little (20-22°C) (Platen, 1982). Soils, classified as Ustic and Typic Tropohumults, Ustic Humitropepts and Oxic Dystrandepts, are generally clayey, with pH values of from 5-6 and low mineral contents, but topography is a more important limiting factor than soil fertility per se.

Colonization, by Latino farmers from the central valley of Costa Rica, started over 100 years ago and less than 20% of the original forest cover is left. The majority of the farms are small: 0-4 hectares (48%) and 4-10 hectares (31%). The remaining large farms occupy a disproportionate percentage of the area and 65% of farm land is used for extensive cattle grazing (Lagemann and Heuveldop, 1983).

(iii) How goals were established. The farming systems approach (Bertalanffy, 1973; Hart, 1980) was used for analyzing the structure and functioning of the target group farms, in particular the traditional agroforestry systems (Lagemann and Heuveldop, 1983; Borel, 1987). A hypothesis was then formulated: "Traditional agroforestry systems can be used as a starting point to increase production and productivity, by using improved varieties and better management techniques, without a decrease of their ecological stability" (Lagemann and Heuveldop, 1983).

The identification of specific goals (i.e. research topics) and the development of improved or new agroforestry systems was a dynamic process, with modifications being made as new information became available. Initially, the project followed a relatively formal process with questionnaires providing most of the data for identification of development objectives. This was supplemented by case studies designed to give both CATIE and national counterparts a better understanding of traditional agroforestry systems (Beer, 1984a, 1984b).

(iv) Implementation: participation of farmers and national organizations. The socio-economic characterization of randomly selected farms, was based on both static (286 farms) and dynamic surveys (weekly visits to 69 farms over 1 year) (Lagemann, 1982). These on-farm surveys were carried out by specially contracted recent graduates from the local agricultural high school, who were selected because they had the same cultural, linguistic, and educational background as the cooperating farmers (Heuveldop and Espinoza, 1984). This socio-economic study was later supplemented by a survey of traditional agroforestry practices (including for example species and products), carried out mainly by collaborating government extension officers. There was an overlap of the stages of characterization of the farming systems, basic studies of the region, and the on-farm testing of recommendations.

Since bio-physical as well as socio-economic statistics for the project zone were limited, several consultants and post-graduate students (national and foreign) were brought in to work with national staff in order to provide hard data for the stratification of the area and to re-evaluate identified problems as well as proposed solutions. The main activities were a soil survey, evaluation of present land use as well as land-use capacities (Melle, 1984), a survey of firewood production and use (Angern, 1985), and a socio-anthropological study of the farmers' perceptions of agroforestry alternatives which were then being tested by the project (Marmillod, 1987a, 1987b). The project staff carried out a detailed characterization of the traditional coffee-shade systems, identifying components, products, and system architecture (Espinoza, 1985). A complementary watershed management project was also initiated, collecting and publishing useful bio-physical data on the Parrita watershed, which includes most of the Acosta-Puriscal zone (Ministerio de Agricultura y Ganaderia-direccion General Forestal, 1985).

In the final three year phase of the CATIE project, the involvement of the national institutions was greatly increased through their participation in on-farm research (applied research; not technology testing or extension) with improved or new agroforestry systems. At this stage the intention was to investigate which of the proposed improvements of traditional systems or systems developed on experimental stations, were really practical, to provide the foundation for a bilateral agroforestry extension project (i.e. not a CATIE project). The main examples were: dry season forage production from living fence posts or cut-and-carry silvo-pastoral units; total confinement stall-feeding system for dairy goats fed partially with foliage from existing farm trees; management of the naturally regenerated timber tree *Cedrela odorata* in coffee plantations; and the use of the tree *Gliricidia sepium* in an alley-cropping system to provide a nitrogen rich mulch for black bean (*Phaseolus vulgaris*) (Beer and Heuveldop, 1987).

The selection of the farmers for these collaborative trials was from a group of sixty-nine farmers who were involved since the first surveys were made. The individuals therefore were known personally by the project assistants for several years. Selection criteria were not formally organized but included: (1) the availability of a suitable site, (2) the productive potential of the proposed agroforestry system for each candidate farm, (3) accessibility, (4) farm size and structure, (5) the desire to distribute the trial plots over as much of the region as possible, (6) the interest of the farmer, and (7) most importantly, the evaluation of the farmer's potential as a collaborator, which was primarily based on the experience and opinions of the project assistants.

These research trials were managed by the project team (CATIE and government technicians) on land provided by the collaborating farmers. The farmers gave their labour and some inputs which were available from the farm, but the majority of the costs were covered by the project (seed, wire, fertilizers, and even labour). Labour and economic inputs were recorded but have a limited value, since these biological experimental plots are too small, and non-representative of the management intensity, to really judge the profitability of a system (Working Group C, 1987).

The selection of sites took into account possible follow-up extension activities, but only a limited number of experiments could be established for each activity (generally 3-6) due to logistical/financial limitations. Some of the trial plots were subsequently taken over, and the number of sites increased, by the bilateral agroforestry extension project which continues the work in this area (i.e. transfer to national institutions was achieved).

When the on-farm research began, the counterpart contribution changed from being almost exclusively the forestry department to include the regional agricultural centre (Puriscal) of the Ministry of Agriculture. The follow-up agroforestry extension project is, in fact, based in this centre with several full-time team members from both the forestry and agricultural departments working together with two international technicians (an agronomist and a forester). The focus is on promoting self-help agroforestry initiatives through local community organizations.

(v) Evaluation and monitoring. Monitoring of the project was initially only carried out internally as part of the reporting procedure to the funding agency and CATIE at six or twelve month intervals. In the final phase of the project, when consultants and post-graduate students were involved at the same time as on-farm research was being carried out, a more frequent evaluation of progress and distribution of project resources (vehicles, equipment, and field assistants) became necessary. This was achieved through the joint preparation of a one-page summary of a monthly work plan, during a half-day meeting in the regional agricultural centre (not CATIE) of all project personnel, counterparts and students. These meetings had three results: (1) improved coordination of all project activities, (2) consistency in the presentation, to farmers or other technicians, of project goals, methods and organizations, and (3) development of a good understanding with the regional extension officers of this centre, who would eventually be involved in the follow-up project but who had not yet worked directly with the agroforestry project.

The provisional results of the basic studies were used to carry out an interdisciplinary D&D exercise (ICRAF, 1983a, 1983b; Raintree, 1989), wherein the problems and research topics were reviewed by a team of ICRAF, CATIE, and national institutions, that spent two weeks in the area revising available information and interviewing farmers. The project proposal (ICRAF, n.d.) which resulted from this exercise was never implemented as a whole, but it confirmed the choice of existing priorities and introduced some useful modifications into the research programme. A particularly useful concept was the linking of the problems perceived by the farmer to those identified by the technicians, and hence the search for systems which could resolve both at the same time.

The final evaluation of the CATIE project (activities in Nicaragua and Turrialba as well as Acosta-Puriscal) was carried out by an independent consultant contracted by the funding agency, together with their own personnel (Maydell, 1985). The utility of the specific agroforestry systems developed for Acosta-Puriscal, was also evaluated during an "Objectives-Orientated Project Planning" (ZOPP) exercise carried out in 1988 during the implementation of the new extension project (Deutsche Gesellschaft Fur Technische Zusammenarbeit, 1987, 1988). These evaluations emphasized identifying specific problems on the farms, objectives which would lead to the solution of these problems, and an assessment of how far the project activities had contributed to solving the problems. Appraisal of the quality of the research was not the main aim. A change in some agroforestry priorities for on-farm research was recommended in order to emphasize cash generation systems (some on-farm research continues as a component of the new extension project). There were also changes in the list of activities which would be immediately promoted by the extension service. For example, an alternative commercial perennial crop, Bixa orellana (natural food dye), was given high priority.

In summary, it can be stated that the biological feasibility of these agroforestry systems for small-medium sized farms in Acosta-Puriscal was proven, but the economics were often poor and much more attention must be given to the socio-economic limitations. If the farmers are given the appropriate support (for example credit, extension, and germplasm), then a significant number would adopt some of the proposed agroforestry technology. Without support, they will probably continue with many of their destructive methods and additional research will not improve this situation.

It should also be noted that agroforestry cannot cure all the land-use problems of a region. In the case of Acosta-Puriscal, the original goal was to develop appropriate sustainable systems for small to medium farms. In fact the major land-use problem in the region is extensive grazing of beef cattle on steep slopes in large farms, i.e. outside of the target group. Even for the smaller grazing units, it is hard to design an agroforestry system (or any other system) which will reduce erosion from this totally inappropriate land use. The solution must come from another level of agricultural development which involves political and financial decisions (e.g., a critical appraisal of bank loans for pasture establishment) (Borel, 1987).

One general decision which originated from the evaluation of the Acosta-Puriscal project, was to reduce future emphasis on basic regional studies and to increase the on-farm collaborative research. This decision was implemented during the establishment and management of the subsequent CATIE project which began in late 1986 in Talamanca.

2.2 Talamanca: A participative planning process

(i) Area selection criteria. In 1986, new pilot zones were chosen on the Atlantic watersheds of both Costa Rica and Panama, using the following guidelines agreed upon between CATIE and the funding agency:

(a) In contrast to the more common suggestion of choosing the most ecologically damaged areas for new projects, which are usually on the seasonally dry Pacific watershed of Central America where population density is highest, select areas where most new colonization is occurring in order to develop appropriate land use with the settlers before they irrevocably damage their new lands. These new colonization areas are mostly in the low humid tropical zone where ecosystems are especially fragile. This philosophy has been simply expressed as "Run before the problems; not after them".

(b) Test the general research and development philosophy and methodology developed in Acosta-Puriscal, in a completely different ecological zone, with suitable bio-physical and socio-economic conditions for the development of agroforestry systems.

(c) Respect national priorities for forestry and agricultural development.

(d) The final aim should be to establish the foundation for another agroforestry extension project.

Field visits and meetings over a period of nearly six months preceded the final decision by CATIE and the Costa Rican forestry service to select the Canton Talamanca, in the province of Limon. These meetings and field visits involved the Ministry of Agriculture's regional centre (Siquirres); an international private voluntary organization, the Association of New Alchemists Incorporated (ANAI); farmers organizations and cooperatives; local agricultural high school staff; other CATIE projects with experience in the area; and most importantly on-farm interviews to assess, in a general way, the interests of the farmers. The existence of the ANAI community nursery project was quite influential in the choice of Talamanca because it provided a vehicle for gaining access to, and subsequently the confidence of, the farmers. It was also seen as one of the possibilities for concurrent and/or subsequent implementation of the applied research results, through its existing extension organization.

Other specific reasons given for the selection of Talamanca were the following[1]:

(a) Most farms are classified as small to medium.

(b) From the point of view of extrapolation, the Talamanca pilot area is valuable since it falls within the life zones (Holdridge, 1967) "tropical wet forest" and "premontane rain forest" which together occur in 46% of Costa Rica and 49% of Central America (CATIE, 1986).

(c) The existence of several complementary projects in Talamanca, some of which could provide basic data whilst others already carry out agroforestry extension activities.

(d) The importance of agroforestry systems as a traditional practice provided an essential starting point for planning decisions and choosing applied research topics, as well as clearly demonstrating the interest, experience and acceptance of this land-use option by the Talamanca farmers. This was shown by the success of the ANAI community nursery project in which many farmers participated in order to obtain perennial crops (e.g. cacao) and trees (principally little-known fruit species).

(e) The Talamanca farmers associations (reforestation, agriculture), were also seen as useful mediums for contact to the farmers to facilitate on-farm research, and later to promote adoption of the results. The association for reforestation had in fact been pressuring the forestry service for technical assistance.

(ii) Area description. The Canton Talamanca is one of the largest, least developed cantons of Costa Rica. The project has concentrated on the low altitude area ("Baja Talamanca"; 0-300 meters above sea level) which includes coastal plains and hills as well as river valleys. The limits of this area include: on the north side, the Caribbean; and on the west side, the border with Panama. The following description is of the "Baja Talamanca".

"Baja Talamanca" falls within the life zone "tropical wet forest". The temperature varies little throughout the year with an average of 26°C. Annual precipitation is 2,500 millimetres per year and there is no dry season, but soil moisture deficits may occur in September and October. The majority of the alluvial soils have been classified as Typic Troporthent and the soils of the coastal hills as Oxic Palehumults and Aeric Tropaquepts.

Until recently the only areas without the original forest cover were the accessible lands near the coast or the rivers navigable by canoe, and the alluvial soils occupied by a banana company for the last 100 years. The activities of the indigenous people (e.g. Bribri Indians) living inland had had little impact on the forest cover. In the mid 1970s a significant area of relatively flat fertile land, temporarily abandoned by the banana company, was invaded by white settlers who immediately removed all the existing secondary forest in order to claim the land. The coastal area had long been settled (since the last century) by negro fisherman-farmers who established mixed cropping systems in which root crops, fruit trees, and perennial crops (especially cacao) play an important role. Thus, the work zone includes a complex racial mix and settlement history. The infrastructure in this area, including the government's forestry and agricultural extension services, was poorly developed until recently, but is now being improved.

Farm sizes can be quite variable, but the actual area worked on each farm is limited and nearly all farms can be classified as small to medium. In the largest valley, which is occupied by white settlers, the average land holding per community is between 8 and 33 hectares (Segleau and Mora, 1989).

(iii) How goals were established. As mentioned before, the final aim of the CATIE project was to establish the foundation for an agroforestry extension project (i.e. provide applied research results; establish demonstration plots; train national personnel; and establish contacts with national, local, and farmers' organizations). Although this was the same aim as that of the Acosta-Puriscal project, an important difference was the decision to emphasize on-farm research and to de-emphasize basic studies (including the detailed evaluations of traditional agroforestry techniques). This implies the need for some kind of rapid appraisal methods (Chambers, 1981) to identify the main problems which could be at least partially resolved by promoting agroforestry development. It also implies the need to rapidly select farmers and farms for on-farm research, without having the experience and knowledge which come after carrying out detailed dynamic surveys (e.g., the case of Acosta-Puriscal).

The first step, therefore, which was concurrent with the area selection phase, was consultation with all knowledgeable persons and institutions, both governmental and non-governmental. However, the most important step was the development of a simple methodology to determine the farmers' problems, interests, and priorities at the same time as evaluating their potential as possible collaborators for on-farm research. This process should not be confused with a survey!

(iv) Implementation: participation of farmers and national organizations. Interviews of 125 farmers (Segleau and Mora, 1989) were carried out whilst walking around their farms, using a checklist of 12 obligatory open-ended questions, which included: (1) some minimal statistics, e.g. farm size, actual principal farming activities; (2) interests, especially which potential agroforestry activity appealed to him most; (3) experiences and problems encountered when trying to develop his farm; and (4) willingness to collaborate with on-farm research. The interviewer (local project assistant) was by no means limited to these questions, and the second page of a form he had to complete, during or immediately after the visit, was a checklist of observations to be made with respect to (1) the farmer's suitability as a collaborator (such as his age, personality, other commitments) and (2) the suitability of the possible sites on the farm (mainly bio-physical criteria). The final page of this form was designed for the preparation of an approximate map of the farm, showing main activities, and possible sites for on-farm trials (Beer, 1989). The farmers to be interviewed were mostly selected in consultation with the secretary of each community nursery, who also helped present the idea of the project to the community and to programme the on-farm interviews. This implies a bias[2] towards past and present community nursery participants which was considered acceptable since at this early stage there was no intention of trying to identify and collaborate with a representative cross section of all types of farmers within the target group with whom there was the highest probability of demonstrations.

The need to include a more representative group comes later during the phases of technology testing and extension, when hopefully the project and the agroforestry activities it wished to promote have gained credibility. It was also thought better to postpone quantitative survey work until a later phase, in order that the practical problem orientated approach of the project be first demonstrated and thus reduce mis-interpretations (and hence mis-information!) about the real objectives of the project in this socio-economically sensitive region of Costa Rica (mixed cultures, large groups of squatters, and absence of land titles even for long established farms).

The general conclusions from the interviews, about farmers' interests and possible project activities, were discussed with both governmental and non-governmental organizations, as well as with farmers' representatives and groups. A national technician was in charge of the evaluation of the results (Segleau and Mora, 1989). The selection of the agroforestry technologies to be developed and tested by the project were most strongly influenced by the farmers' interests but they were modified according to the technicians' opinions and the project's capabilities, i.e. the selection of agroforestry systems to be developed did not include any which were not frequently requested by farmers, but on the other hand many agroforestry or other systems which were requested had to be left out. This decision-making process does not exclude new agroforestry systems which are not traditional, since these ideas can be introduced and demonstrated through existing extension and community projects, and if considered sufficiently attractive will then be rapidly requested by the farmers. This is exactly what occurred with one of the main lines of applied research chosen by the farmers and hence by the project: testing different living support poles for black pepper vines, an agroforestry system which is certainly not traditional in any part of Costa Rica.

Once a sufficient number of farmers (usually 15-20) had been interviewed in any one community, the interviewer and project coordinator together with each farmer made a review of the results in order to prepare a short list of candidates for follow-up farm visits to discuss concrete proposals for collaborative trials and to select a suitable site. This selection of candidates was made using a list of desirable characteristics for collaborating farmers, based on the criteria used in Acosta-Puriscal but greatly modified and expanded with experience in Talamanca. In particular the importance of socio-economic criteria (farmer), rather than just bio-physical criteria (farm), was emphasized (Beer, 1989).

Logically the project needs more basic data for area characterization and planning than could be gathered through meetings, field reconnaissance trips and simple on-farm interviews. Hence, other activities are supported, such as compilation and synthesis of all relevant published and unpublished information (Kapp, 1989). Basic research activities, usually carried out by CATIE or visiting post-graduate students, are also supported in the Talamanca work zone. A constant effort is needed to evaluate other traditional or new agroforestry systems as sources of new ideas to support existing applied research lines or suggest new possibilities. The support research carried out by the CATIE Nitrogen-Fixing Trees Project is an example of one source of information for Talamanca (CATIE, 1987). As mentioned above, now that the project is established and respected by the target group, quantitative socio-economic studies are being planned, in part to generate the base line data against which future impact can be measured.

(v) Evaluation and monitoring. Evaluation and monitoring activities have already begun even though the project has few quantitative results. Annual review meetings are being planned for each community (not necessarily the nursery group only) in order to present a summary of results and to discuss future priorities. One external evaluation of the impact of the project by a sociologist, supported the selection procedure of farmers and on-farm research activities, but suggested that even more attention needs to be paid to the farmers' interests, limitations, and expectations about the project. In other words, it may not be acceptable to foresee extension activities only in a later phase. There is, for example, an urgent need for explanations of the differences between the management intensity of an experimental plot and of a commercial plantation. Hopefully, such explanations will reduce the expressed doubts of the farmers that these systems are "too expensive for us" (Brenes, 1988). One information gap, which should be closed in 1989, is to record the collaborating farmers' observations and opinions (both criticisms and suggestions) about the management and potential of their plots. A simple form, with a few obligatory questions, will be used as the basis for this evaluation at the same time as some of the monthly bio-physical evaluations are carried out.

The results of these three means of regularly assessing project progress (annual community meetings, external reviews by sociologists, and assessment of on-farm plots by collaborating farmers) will be discussed during an annual meeting of project staff (CATIE personnel and national counterparts) and, together with the medium-term (3-4 years) project action plan (Deutsche Gesellschaft Fur Technische Zusammenarbeit, 1987) will serve as a basis for the measurable indicators of progress.

3 CONCLUSIONS

CATIE's agroforestry research projects must aim at the implementation of results. Thus, the choice of national counterparts and the possibilities for extrapolation of pilot area results are important considerations. Insofar as national resources are available, government technicians are continuously involved (even before research starts) to increase the probability that the results will be accepted, understood, and used. On-the-job training is obviously another benefit. Transfer of the planning methodology, through these technicians to their institutions, is thus a continuous process. After approximately six years, CATIE hopes to hand over all responsibility for each project to an extension organization and then restart the process in a new area (in reality several areas are at different stages of development at the same time). In addition to a transfer of plots and results, there is also a transfer of the planning methodology to each group of national technicians. This same planning methodology has also been transferred during technical assistance (consultancy) missions to rural development projects, both within and outside the region (Beer, 1989).

In order to facilitate extrapolation of the results, the research methodologies and multiple purpose tree germplasm are selected in agreement with other CATIE projects. Results are entered into regional data banks and can be compared with similar trails in other countries. Thus, in addition to their local significance, the results can be used for modelling exercises for areas where trial plot data are not available. However, agroforestry systems are site specific and small changes in environmental variables can significantly affect the interactions between components with unpredictable consequences. Moreover, socio-economic factors can be of overriding importance as determinants for the success of any agroforestry system.

Irrespective of the identity of the cooperating organizations, the final target group is the small to medium farmer (not the farm!). If this aim is really respected at all stages of project planning and development, then many decisions are easily made.

3.1 Area selection

(i) National priority. The criteria of national priority has been mentioned in each project analyzed. It must be stated, however, that in most of the countries, national priorities, as far as the agrarian sector is concerned, are weakly defined and are easily subject to pressures like the expectation of a new (externally funded) project. Although the national priority criteria should in all cases be ascertained, it does not automatically assure, for example, the full support of the central government.

(ii) Extrapolation. The choice of an area has been frequently justified by its ecological resemblance with large tracts of land elsewhere. Again, projects should not be established in unique areas, but to automatically assume that the results can be extrapolated to ecologically similar areas would be an error, particularly when human factors are not taken into consideration.

(iii) Existing projects. The comparative analysis shows that the presence of past or existing projects was always a critical consideration in the selection of new work areas. Of particular significance was the presence of extension projects and of local farmer's organizations.

3.2 Goal setting

(i) Farmers' interests. The interests of the farmers must be ascertained: (a) during project initiation, (b) by frequent monitoring, and (c) during evaluations. At the beginning of a project, rapid rural appraisal methods are more palatable to farmers than detailed surveys but sufficient time and resources must be devoted to developing the relationship and getting more than a superficial impression of the farmers' interests.

(ii) Extension requirements. The probability of successful technology transfer will be greatly increased if extension activities are foreseen and planned from the first day of research. This implies: (a) immediately identifying who will carry out future (or simultaneous) extension activities, (b) selecting the research topics with extension possibilities in mind, (c) selecting collaborating farmers for research who can also act as local extension agents, (d) distributing the long term agroforestry research plots with respect to future extension needs and not just using scientific criteria such as a bio-physical stratification, and (e) include some extension activities in the research project, in particular with a view to developing methodologies for the transfer of the results; i.e., make the researchers think about extension.

(iii) Planning team. In the absence of hard data on agroforestry components and interactions, especially the human factors, the necessity of putting together as many experiences as is practically feasible should be emphasized. This applies to farmers and extensionists as well as to senior professionals. A large group at the planning stage is no guarantee for a successful design, but a small group will most likely be biased and may produce an unbalanced work plan which ignores some critical factors.

3.3 Implementation

(i) Local assistants. It is more important to have assistants who intimately know the target group and their surroundings than those with research experience. The latter can be taught, the former not. Respecting this principal, CATIE transferred its field assistants, who had been contracted and trained in Acosta-Puriscal, to the new extension project. CATIE subsequently had to contract and train new local assistants, from the Talamanca zone, for the next project.

(ii) Farmers as actors. If the farmers are to be true partners in the on-farm research (rather than "sleeping" partners or worst, treated as another farm labourer), great care must be devoted to selecting the most motivated candidates, and secondly by developing the trial together (e.g., by selecting the site together).

3.4 Evaluation and monitoring

(i) Emphasize social and economic aspects. Bio-physical research is not enough. Even if the areas of expertise of the research team do not include economics and sociology, these aspects can not be ignored until a later extension phase. Consultants and local universities should be contacted to provide advise on methodologies for studying these topics.

(ii) On-farm research monitoring. On-farm research has been a constant in all analyzed projects, which is a consequence of the specific orientation of CATIE. There are, however, on-going field projects without formal research components. Nonetheless, these projects are "experimenting" with new species, arrangements, and management methods. The number of formal research activities is grossly insufficient. However, valuable information, some of it quite innovative, could be gained from development projects, provided good quality data are gathered over sufficiently long periods. This consideration calls for a greatly increased effort in monitoring agroforestry development projects. A number of well documented case studies should replace or complement many formal experiments.

4 ACKNOWLEDGMENTS

The authors thank C. Burgos and C. Reiche for their constructive comments on this report. Financial and technical support were provided by the Deutsche Gesellschaft for Technische Zusammenarbeit (GTZ) and by CATIE.

5 FOOTNOTES

1 Letter, 12 December 1986, J. Beer and R. Borel to R. Vargas, Director Forestry Service, Costa Rica,

2 It is assumed that the farmers had automatically carried out a pre-selection since the decision to participate in a nursery suggests a motivated and innovative person with a special interest in perennial crops (particularly cacao) and agroforestry systems.

6 REFERENCES

Angern, M., 1985. Brennholzversorgung in Kleinbaeuerlichen Betrieben in Costa Rica. Diplomarbeit. Forstwissenschaftliche Fakultaet, Universitaet Freiburg, Germany, 62 pp.

Beer, J., 1984a. Estudio y promocion de sistemas agroforestales tradicionales en Centro y Suramerica. In: J.W. Beer y E. Somarriba (Editors), Investigacion de tecnicas agroforestales tradicionales. CATIE, Turrialba, Boletin Tecnico No. 12, pp. 11-19.

Beer, J., 1984b. Introduccion al establecimiento de parcelas permanentes en asociaciones agroforestales tradicionales. In: J.W. Beer y E. Somarriba (Editors), Investigacion de tecnicas agroforestales tradicionales. CATIE, Turrialba, Costa Rica, Boletin Tecnico No. 12, pp. 52-65.

Beer, J. and Heuveldop, J., 1987. A critical analysis of an agroforestry project in Acosta and Puriscal, Costa Rica. In: J.W. Beer, H.W Fassbender and J. Heuveldop (Editors), Advances in Agroforestry Research. CATIE, Turrialba, Costa Rica, Informe Tecnico No. 117, pp. 78-90.

Beer, J., 1989. Proposal for the selection, establishment and maintenance of agroforestry research/demonstration plots in Fiji. In: J. Heuveldop, M. Homola, H.J. von Maydell and C. van Tuyll, (Editors), Proceedings of the Deutsche Gesellshaft Fur Technische Zusammenarbeit Regional Forestry Seminar, Fiji-German Forestry Project, Suva, Fiji, pp. 281-322.

Bertalanffy, L. von, 1973. General Systems Theory: Foundations-Development-Applications, George Braziller, New York, 295 pp.

Borel,, R., 1987. Agroforestry system interactions: man-tree-crop-animal. In: J.W. Beer, H.W. Fassbender and J. Heuveldop (Editors), Advances in Agroforestry Research. CATIE, Turrialba, Coast Rica, Informe Tecnico No. 117, pp. 104-121.

Brenes, C., 1988. Opinion e impacto social del proyecto "Cooperacion Agroforestal CATIE-DGF-GTZ", Canton Talamanca. Instituto Tecnologico de Costa Rica, cartago, Costa Rica, Estudio Preliminar (unpubl.), 33 pp.

Budowski, B., 1987. The development of agroforestry in Central America. In: H.A. Steppler and P.K.R. Nair (Editors), Agroforestry: a decade of development. ICRAF, Nairobi, pp. 69-88.

Centro Agronomico Tropical De Investigacion y Ensenanza. Departamento de Recursos Naturales, 1986. Silvicultura de especies promisorias para produccion de lena en America Central. CATIE, Turrialba, Costa Rica, Informe Tecnico No. 86, pp. 28-32.

Centro Agronomico Tropical de Investigacion y Ensenanza, 1987. Informe Tecnico Anual del Proyecto Erythrina spp. Fase II. CATIE, Turrialba,

Costa Rica, (unpubl.), 123 p.
Centro Agronomico Tropical de Investigacion Y Ensenanza, 1988. Facing the Challenge: A Ten-Year Strategic Plan for 1988 to 1997. CATIE, Turrialba, Costa Rica, 112 pp.
Chambers, R., 1981. Rapid rural appraisal: rationale and repetoire. Public Administration and Development, 1: 95-106.
Deutsche Gesellschaft Fur Technische Zusammenarbeit, 1987. Das GTZ, Interne Planungsverfahren (ZOPP). In: Organisationashandbuch. GTZ. Eschborn, Germany, (unpubl.), 10 pp.
Deutsche Gesellschaft Fur Technische Zusammenarbeit, 1988. Informe sobre el taller "ZOPP" (IV): Proyecto Desarrollo Agricola y Forestal de Acosta y Puriscal (PRODAF). Eschborn, Germany, (unpubl.), 8 pp.
Encyclopaedia Britannica, 1982. Vol. 13, pp. 594-602; vol. 17, pp. 970-976.
Espinoza, L., 1985. Untersuchungen uber die Bedeutung der Baumkomponente bei agroforestwirtschaftlichem Kaffeeanbau an Beispielen aus Costa Rica. Ph.D. Thesis. Gottingen, Germany, 164 pp.
Hart, R.D., 1980. Agroecosistemas: conceptos basicos. CATIE, Turrialba, Costa Rica, Serie Materiales de Ensenanza No. 1, 210 pp.
Heuveldop, J. and Espinoza, L., 1984. El uso de encuestas en la investigacion de tecnicas agroforestales tradicionales. In: J.W. Beer y E. Somarriba (Editors), Investigacion de tecnicas agroforestales tradicionales. CATIE, Turrialba, Costa Rica, Boletin Tecnico No. 12, 42 pp.
Holdridge, L., 1976. Life Zone Ecology. Tropical Science Centre, San Jose, Costa Rica, 207 pp.
International Council for Research in Agroforestry, 1982a. Guidelines for Agroforestry Diagnosis and Design. ICRAF, Nairobi, Kenya, 25 pp.
International Council for Research in Agroforestry, 1983b. Guidelines for Agroforestry Diagnosis and Design. ICRAF, Nairobi, Kenya, 383 pp.
International Council for Research in Agroforestry, No date. Research proposal to improve and develop agroforestry systems for the seasonally dry uplands of Western Costa Rica. ICRAF-CAR-CATIE, Nairobi, (unpubl.), 90 pp.
Kapp, G., 1989. Perfil ambiental de la zona baja de Talamanca, Costa Rica. CATIE, Turrialba, Costa Rica. (In preparation).
Lagemann, J., 1982. Metodologia. In: H. von Platen, G. Rodriguez y J. Lagemann (Editors), sistemas de Finca en Acosta-Puriscal, Costa Rica. CATIE, Turrialba, Costa Rica, Informe Tecnico No. 30. pp. 13-23.
Lagemann, J. and Heuveldop, J., 1983. Characterization and evaluation of agroforestry systems: The case of the Acosta-Puriscal, Costa Rica. Agroforestry Systems, 1: 101-115.
Marmillod, A., 1987a. Farmers attitudes towards trees. In: J.W. Beer, H.W. Fassbender and J. Heuveldop (Editors), Advances in Agroforestry Research. CATIE, Turrialba, Costa Rica, Informe Tecnico No. 117, pp. 259-270.
Marmillod, A., 1987b. Farmers attitudes and behaviour with respect to trees and soil. M.A. Thesis. Reading University, England, 77 pp.
Maydell, H.J. von, 1985. Bericht uber die Evaluierung des TZ-Projektes "Agroforstliche Zusammenarbeit mit dem CATIE" PN 79.2259.4 GTZ, Frankfurt, Germany, (unpubl.), 62 pp.
Melle, G. van, 1984. Estudio sobre la capacidad de uso de la tierra en dos areas de las subregiones Puriscal y Cariagres, Costa Rica. CATIE, Turrialba, Costa Rica, Informe Tecnico No. 40, 40 pp.
Ministerio de Agricultura y Ganaderia-direccion General forestal, 1985. Plan de Manejo de la Cuenca del Rio Parrita: Anexo II, Sector 1-A, Diagnostico. MAG-DGF. Santiago de Puriscal, Costa Rica, 298 pp.
Platen, H. von., 1982. La region de estudio. In: H. von Platen, G. Rodriquez y J. Lagemann (Editors), Sistemas de Finca en Acosta-Puriscal, Costa Rica. CATIE, Turrialba, Costa Rica, Informe Tecnico No. 30, pp. 24-28.

Raintree, J., 1989. Diagnosis and design procedures. Presented at the International Syposium on Planning and Implementation of Agroforestry, Pullman, Washington, April.
Segleau, J. and Mora, F., 1989. Actividades agricolas de finqueros en el Canton Talamanca; sus limitaciones y prioridades. CATIE-DGF, Turrialba, Costa Rica. (In preparation).
Working Group C, 1987. Economics and ergonomics in agroforestry. In: J.W. Beer, H.W. Fassbender and J. Heuveldop (Editors), Advances in Agroforestry Research. CATIE, Turrialba, Costa Rica, Informe Tecnico No. 117, 368 pp.

6 APPENDIX: CATIE RESEARCH AND TRAINING ACTIVITIES

6.1 Research activities

CATIE conducts its research activities through six priority areas:

(a) Perennial Crops (coffee, cacao and plantain). Emphasis on biology and genetics; plant production; tropical soils and climate, and socio-economics.

(b) Annual Food Crops (rice, corn, beans, sorghum). Emphasis on plant protection, plant nutrition and socio-economics.

(c) Promising Tropical Crops. Conservation of germplasm, biology and genetics, emphasis in biotechnological research.

(d) Tropical Livestock (meat, milk, mainly dual-purpose cattle). Research on biology and genetics, emphasis on breeding and nutrition.

(e) Forestry and Agroforestry. Emphasis on biology and genetics, nutrient cycling, socio-economics and crop management.

(f) Production System Development. Integration of production-oriented research activities (crops, tropical livestock, forestry and agroforestry) within the framework: research-education-development.

(g) Integrated Management of Watersheds and Regional Natural Resources. Emphasis on water, soil and vegetation conservation, development of agro-climatological and geographical data bases taking the watershed as the planning unit.

6.2 Scope of CATIE's international development involvement

While Central America and the Caribbean are CATIE's primary areas of influence, the centre serves the needs of other tropical regions around the world. Within the context of horizontal cooperation, universities outside the region play an important role. For example, the Universities of Missouri, Iowa State, Colorado State, Wisconsin, and Wageningen have begun or are planning collaborative activities in some of CATIE's priority areas. Reciprocal cooperation exists with many Latin American institutions and CATIE has provided consultancies in most regions of the world. CATIE is a fundamental part of the most important agricultural research and training network in its working area (Red Regional de Cooperacion en Educacion Superior y Capacitacion Agropecularia y de los Recursos Naturales Renovables; REDCA).

6.3 Financial support

There is a long list of international institutions cooperating through support and providing human resources. In the actual funding structure about 75% of CATIE's annual budget (13 million US dollars estimated 1989) comes from externally financed members' contributions (IICA and member countries), overhead charges,and some productive commercial activities.

6.4 Review of CATIE's involvement in agroforestry

In a history of agroforestry in Central America, Budowski (1987) reviews CATIE's activities and mentions the first descriptive publications (Master of Science thesis and reviews written in the 1950s). Formal research ideas were formulated in the 1970s and the term "agro-silvo-pastoral systems' was introduced in 1976 as part of a programme in CATIE's agroforestry research and education programme. Borel (1987) describes the general research objectives of the centre's agroforestry programme as (1) finding the circumstances in which the use of agroforestry systems is an advantage and (2) developing improved systems for specific areas of the mandate region of CATIE.

The ongoing activities include basic research, e.g. long term tree/crop trials with nutrient cycling modelling, interspecific tree/crop competition, animal nutrition based on tree forage, and selection of germplasm. Emphasis in now given to applied on-farm research, which is carried out through trials in Costa Rica, Panama and Nicaragua.

6.5 Role of education

CATIE has a Latin American mandate for education (Master of Science and other post-graduate training) which includes the option of specializing in agroforestry. Most of the graduate students are qualified professionals working in the agricultural or forestry sectors of their respective countries, to which they return after graduation.

Short courses (2-12 weeks) and seminars on agroforestry are offered by CATIE, several of them annually. Most are conducted by CATIE staff on the centre's campus in Turrialba, others in different countries of the region.

The increasing awareness of the discipline's importance is demonstrated by the rising demand for this kind of continuing education. For a smaller number of professionals, an in-service-training is possible, accompanying the ongoing research work.

Chapter 4

PLANNING OF AGROFORESTRY PROGRAMMES IN JAVA, INDONESIA

ANNET VAN DEN HOEK and TITUS BEKKERING
Kali Konto Project, D.H.V. Consultants, P.O. Box 67, Malang 65112 (Indonesia)

ABSTRACT

The population of Java's uplands is confronted with worsening environmental conditions and decreasing agriculture and forestry production. Present agroforestry programmes developed for the whole of Java, cannot cope with the large diversity of land-use conditions in the uplands. Consequently these agroforestry standard packages are not successful, since they are not adjusted to local conditions and needs. Moreover no link exists between planning of agroforestry programmes on public and on private land. Cooperation between the respective services is poor. This chapter describes a local level planning method to tackle these problems. The planning is based on an agro-ecosystem analysis making use of rapid rural appraisal and a field survey of terrain and land use. Planning on a 1:5,000 scale is carried out per village by local staff of different services and provides a land-use development plan, a land-use development programme, and directives for implementation of agroforestry interventions. This planning facilitates efficient implementation of these interventions.

1 INTRODUCTION

1.1 Background

In Indonesia, and in Java in particular, there is a growing imbalance between population density, and the carrying capacity of agricultural and forest land. The increasing labour force cannot be absorbed by the available opportunities for employment in the manufacturing and trade sector, so the rural population has to depend to a large extent on agricultural and forest land to support themselves. These landless people depend almost exclusively on the forest for their income. The magnitude of the problem is illustrated by the fact that three quarters of the people, who live in villages near the forest, are small or landless farmers with half of the families unemployed (Wirjodarmodjo and Bratamihardja, 1983).

Because of high pressure on land, many areas are confronted with worsening environmental conditions. Large areas of forest land lay fallow and reforestation attempts fail repeatedly (Bekkering and Hartanto, 1987; Nibbering, 1988). Intensive farming on steep slopes has also resulted in the loss of fertile top soils and a reduction in the productivity of many upland areas.

Since the late 1960s and into the early 1980s, the main objective of government land-use policy was to obtain self-sufficiency in food production, especially rice. This goal has been achieved, and present policy is now aimed at increasing upland agricultural and forestry production through sustainable land-management practices for the protection of infrastructures (i.e., dams and irrigation systems) and to protect the natural resources in upper watershed areas. Given the long tradition of agroforestry in Java, and its potentials with respect to productivity and sustainability, these strategies will be one of the main tools to achieve this goal.

The government of Indonesia has a well established institutional structure for tree planting (regreening) and reforestation. Many programmes are being implemented in various regions. This means that a tremendous potential exists for agroforestry in Indonesia. However, the effects of these programmes on halting the problems of dwindling production and environmental degradation of upland areas has so far been limited.

This limited impact can partly be attributed to top-down, centralized planning of standard programmes that cannot cope with the diversity and complexity of problems with respect to land, land use, and land users in the uplands of Java (KEPAS, 1988; Pickering, 1979). Existing programmes of the government are not effective, because they are not adjusted to local opportunities and constraints and often do not reach those people who depend most on the natural resources, namely the small and landless farmers.

In addition, the linkage between agricultural and forest land in the process of planning and implementation of reforestation and regreening programmes is weak. The artificial boundaries between forest and village land have been inherited from the days before independence (Carson, 1987; Nibbering, 1988). Since then, the management of tree and forest resources on both sides of that boundary has been the responsibility of different services, each not feeling the necessity to cooperate with the other. These services are the Directorate General of Reforestation and Land Rehabilitation (RRL) and the Perum Perhutani (State Forest Authority). Although they both fall under the Ministry of Forestry they have their own mandates and their organization is completely separated.

Solutions to these problems of centralized planning and absence of any integration of planning and implementation of policies for forestry and agriculture require the development of a local-level planning methodology which incorporates active involvement of both local population and different government services. By combining efforts, the coordination of development activities on agricultural and forest land may be improved and effective agroforestry programmes may be realized.

1.2 Aim and scope

This chapter describes the development of a local-level planning method for agroforestry programmes in the uplands of Java. This planning method includes the development of a land-use plan, the development of a land-use development programme, and the planning of agroforestry interventions. Agroforestry is a land-use system in which woody perennials are deliberately grown on the same piece of land as agricultural crops either in a spatial arrangement or in sequence. In such systems, the woody components interact with crops. The aim and rationale of agroforestry is to optimize positive interactions to yield a more diversified and/or more sustainable production from available resources than would be possible with other forms of land use (Lundgren, 1982).

2 THE KALI KONTO PROJECT

2.1 Introduction

The land-use planning and implementation efforts described in this chapter were carried out within the framework of the Kali Konto Watershed Management Project. This project is a development cooperation project between the governments of Indonesia and the Netherlands. The project started in 1979 with basic data collection on the forestry situation in the 23,000 hectare Kali Konto watershed. In phase II (1984-1985) the scope of the project was widened and a land-evaluation and land-use plan for forest and village land was developed on a 1:50,000 scale (Thalen and Smiet, 1985). Phase III (1986-present) has focused on the implementation of watershed development measures involving village and forest land. A range of programmes directed toward sustainable land use, increased local income, and physical and social infrastructure enhancement are being developed. Other activities in this third phase are research and monitoring to support other planning activities, training of staff of the cooperating Indonesian agencies, and the further development of a methodology for watershed planning. This chapter is part of the latter activity.

2.2 Project area

The Kali Konto project area is located in the upper watershed of the Konto River, a tributary of the Brantas River, in the province of East Java (see Fig. 1). Soils are of young volcanic origin and consist of deep, highly permeable and relatively fertile andisols and cambisols. Average annual rainfall is 2400 millimetres. The highland varies in elevation from 620 metres to 2868 metres above sea level. Two thirds of the area is clearly demarcated state-forest land managed by Perum Perhutani. The area covers two administrative sub-districts: Ngantang and Pujon. Within this region there are approximately 100,000 people. Agriculture is the major source of income. The regions at higher elevations are characterized by intensive vegetable cultivation and dairy farming. The lower elevations are dominated by rice and coffee cultivation and mixed gardens. Dairy cattle are a key to the agricultural productivity of Pujon and are becoming increasingly important in Ngantang.

2.3 Forestry

The forest land is divided into different forest types (RIN, 1986) as is shown in Fig. 2. Trees are grown on village land in a multitude of land-use types such as road sides, home gardens, dry arable lands, and mixed gardens (i.e. coffee gardens or citrus gardens). On forest lands, shrubs have often replaced the former vegetation that consisted either of forest (natural and plantation) or of state-owned coffee gardens abandoned in the beginning of this century and subsequently turned over to the forest authority (Nibbering, 1988). The long-term goal of Perum Perhutani is to reforest all shrub areas.

The forests provide fodder, fuel, and timber for the local population. Fodder and fuel are supplied from all mentioned forest types, whereas the source of timber is illegal cutting in the natural forest. The existence of a ready market for these forest products combined with large numbers of landless people has led to forest degradation (Hadipoernomo, 1980; PKK, 1986; Bekkering and Hartanto, 1987; Nibbering, 1987). The proximity of an abundance of woody biomass to the villages (maximum 4-5 kilometers), the cool climate and the extra fuel wood used to heat and prepare cattle feed has resulted in high per capita fuel wood consumption (0.93-1.43 cu/capt/yr) (Rusmanhadi, 1987).

The commonly used method for reforestation of the shrub areas is the _taungya_ (locally called _tumpangsari_) system. Landless farmers receive 0.25 hectares of forest land on which they have to plant trees. These new plantations may be intercropped with annuals because labour is paid in-kind with the yields of the annual crops. The trees benefit from soil tillage,

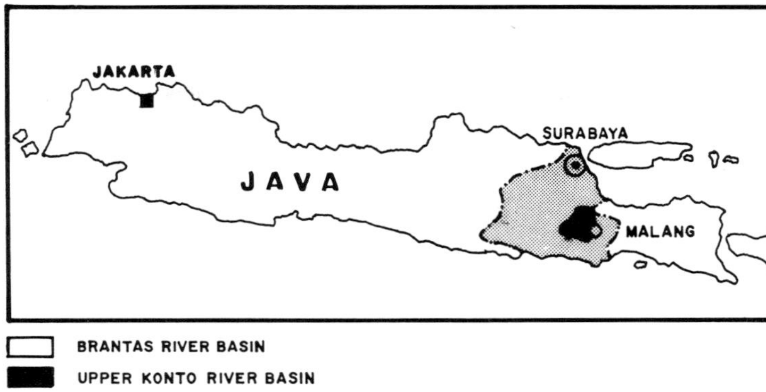

Fig. 1. Location of the project area.

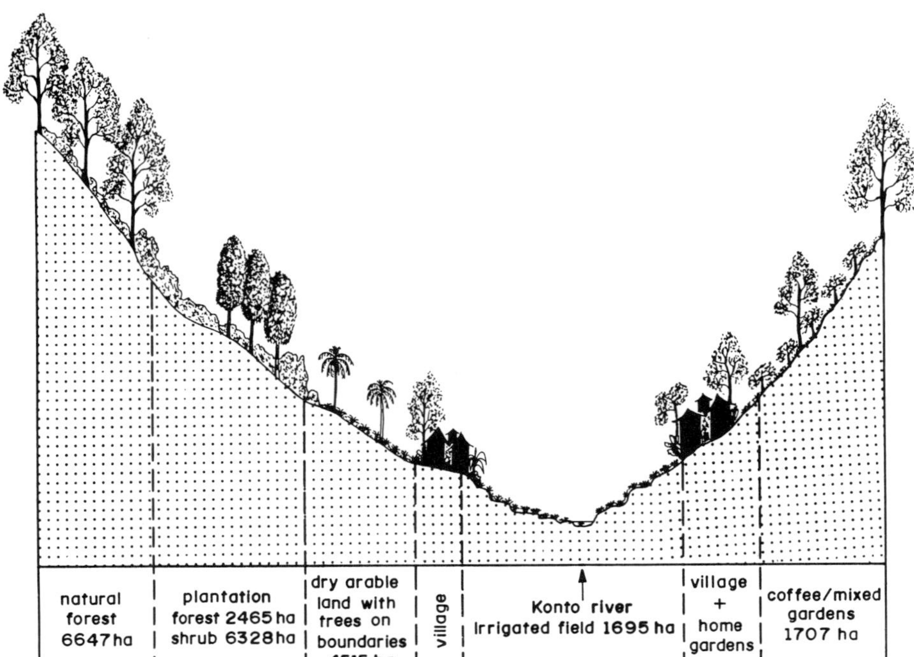

Fig. 2. Cross section landscape of the project area.

added nutrients, and tending in these arrangements (Wiersum, 1987). Similar arrangements exist for interplanting plantations with grass.

The results of the taungya system are not always optimal when pressure on land is high. Because the contract between the farmers and Perum Perhutani is usually short term, farmers do not have a vested interest in the long-term well-being of the trees during the contract period. Additionally, the fact that the annual cropping contract sometimes can be extended if the trees fail to become established can lead to sabotage. Successful plantations can still fall victim to illegal cutting after the period of intercropping is over, because nobody feels responsible (Perum Perhutani, 1985; Bekkering and Hartanto, 1987; Nibbering, 1987). To date, stricter enforcement of management requirements through increased forest police patrols has proven to be ineffective.

Given these problems, Perum Perhutani initiated a social forestry programme (in cooperation with the Ford Foundation) which aimed at improved forest management through increased farmer responsibility for the state forest. In this programme, the emphasis is on the introduction of new agroforestry systems. To stimulate farmer involvement, the newly established tumpangsari sites have a wider spacing of timber trees. Between the timber trees, farmers are allowed to plant fuel wood, fodder, fruit trees, and other perennial cash crops. The duration of the contract between the farmers and the Perhutani is extended beyond two years, and an effort is made to actively involve the local population in planning, implementation, and management (Bratamihardja, 1987).

Other trials at developing new approaches to reforestation are those carried out by the Kali Konto project. In these approaches, farmers receive a small plot of forest land where they can grow annual crops. In return, they have to plant and maintain a neighbouring plot of forest. This forest part is often underplanted with grass (Nibbering, 1987; Bekkering, 1988).

3 AGROFORESTRY LAND-USE PLANNING

3.1 Existing land-use planning procedures

(i) Planning procedures of RRL (the service for land rehabilitation, soil conservation, and tree planting). The Indonesian 4th Five-Year Development Plan (REPELITA 4 1984-1989) identifies 36 priority watershed areas in Indonesia. Of this total, 22 have been assigned super-priority status because of the large investments in infrastructure development. RRL is responsible for the planning and implementation of land-rehabilitation and soil-conservation measures in these watersheds on privately-owned land.

The provincial offices of RRL have the responsibility of preparing master plans (scale 1:250,000) and five-year plans (scale 1:50,000). Sub-centres are responsible for the design of technical implementations plans (1:1,000 and 1:1,500). There appears to be a gap between the five-year planning and the yearly technical implementation plans. The former are too general and unadjusted to micro-conditions to form a basis for the preparation of detailed, local-level plans. In addition, planning often does not reach beyond the analysis and problem identification phases. Consequently, the process does not provide recommendations for the organization and implementation of programmes. An extra step in this planning process on the local level is necessary to fill this gap.

(ii) Planning procedures of Perum Perhutani. Perum Perhutani is a semi-autonomous sub-unit of the Ministry of Forestry. Their mandate is to manage approximately 2.9 million hectares of forest divided over 57 forest districts on the island of Java. In managing these forests, the classified mission of Perum Perhutani is profit earning and public service (Bratamiharja, 1987). The planning of Perum Perhutani is independent of other government organizations and directed specifically at the management of state forest areas. Planning officials are active on central, provincial, sub-provincial, and forest district levels. A key role is played by the planning section, where a ten-year management plan is prepared for the forest district. This plan is based on a forest inventory made every ten years. Based on this ten-year plan, annual technical plans are made. These annual plans are then used to develop a sub-forest district work-plan.

During the second phase of the Kali Konto project, efforts have been undertaken to develop a new methodology for identifying the location of various, forest-land utilization types by application of a land-evaluation method (Thalen and Smiet, 1985). However, this method has yet to be adopted into standard planning practice.

Traditional planning practices focus primarily on timber production. These techniques are not equipped to deal with problems which result from land-use pressure exerted by landless people who rely on the forest for obtaining fuel wood, fodder, and other forest related products. Recently, within the social forestry programme efforts have been undertaken to adapt the Perum Perhutani planning procedure to address these needs by developing planning methods which take such local village conditions into account.

(iii) Other developments. Recently, the group for agro-ecosystem analysis (KEPAS) developed a new approach for evaluating the large diversity of land-use systems in the uplands using rapid rural appraisal (Conway,

1985). This methodology has drawn the interest of Perhutani and to a lesser extent of RRL. Perhutani has started to introduce this approach in their planning procedures.

3.2 Local-level, land-use planning: method and tools

The local level, land-use planning method makes use of agro-ecosystem analysis and rapid rural appraisal for basic data collection. Aerial photos, sketch maps, cross sections, and maps are used as tools for the analysis and design.

(i) Agro-ecosystem analysis. The diversity of upland conditions means that land-use development programmes cannot be standardized. The large number of services involved in the development of uplands, combined with the complexity of the problems faced by upland farmers, supports the need for a conceptual framework that can be used for both problem identification and planning of development activities. Agro-ecosystems may serve this purpose (KEPAS, 1988).

The aim of agro-ecosystem analysis is the investigation of the relationships between the bio-physical characteristics of upland regions. The major upland agro-ecological zones are based on elevation, terrain, and climatological characteristics of eastern Java. Each zone can be further defined by key social and economic factors affecting their management (Konto River Project, 1988).

Analysis is carried out on the basis of the following system properties: productivity, stability, sustainability and equitability (Conway, 1985; KEPAS, 1988). The analysis provides a description of terrain, land use, socio-economic conditions, and the identification of land management problems from both farmer and researcher perspectives. Recommendations for agricultural extension, research programmes, and land-use management within each zone are made. However, the local-level planning method goes beyond analysis and the formulation of recommendations for agricultural development. It aims at developing a land-use plan, a land-use development programme and directives for implementation of agroforestry interventions for agricultural and forest lands.

(ii) Rapid rural appraisal (RRA). Rapid rural appraisal is a process of learning about rural conditions in an intensive, iterative fashion. Multi-disciplinary research teams employ a range of methods, tools, and techniques specifically selected to enhance understanding of rural conditions. A major emphasis in this approach is the use and incorporation of indigenous knowledge into the analysis (Grandstaff, 1987). These characteristics make the RRA very suitable for local-level, land-use planning. However, wherever skillful researchers are lacking, RRA may be counter productive and result in questionable findings (Grandstaff, 1987).

Many local field workers do not yet have the skills to carry out the interviews or to adopt the techniques of RRA. On the other hand, field workers who have long-term experience in the area know the major development constraints and historical changes which can facilitate the use of RRA. In order to guarantee a certain uniform level of planning results, part of the planning process carried out with the help of a RRA should be standardized by the use of field-survey and observation forms, as well as questionnaires.

A variety of different elements of the RRA are used in local-level planning. The terrain and land-use survey and questionnaire support the other RRA techniques with objective data. These data are valuable for the development of a land-use plan and for the planning of interventions, as well as for the development of a monitoring and evaluation system. Existing maps (secondary data), the result of the informal interviews, observations, and the existence of knowledge among field staff on land use and terrain, provide sufficient information to enable a quick field survey.

(iii) Planning tools. In Indonesia the use of aerial photographs for planning has considerable potential in planning. While aerial photographs are not always easily accessible because of government security rules, the number of people within RRL who have the skills in aerial photographic interpretation is growing rapidly. Consequently, there is a large body of technical expertise which can potentially be quite useful in planning.

Aerial photographs with a scale of 1: 20,000 that can be enlarged to a 1: 5000 scale are of great value for local-level planning. These photos are ideal for rapid assessments of natural and human resources in villages or small watersheds (Carson, 1985; Van den Hoek *et al*., 1988). Unlike existing maps in Indonesia, aerial photos cover both village and forest land and show agricultural, social, and economic influences on the landscape. In planning, such maps can assist in the localization of problems and linkages in land use and terrain conditions can be clarified. Another benefit from the use of maps is that programmes can be localized in a final plan, that facilitates the planning and implementation of interventions (1: 1000 scale). Additional sketches, diagrams and cross sections are necessary for analysis of the local situation and understanding of the agro-ecosystem.

3.3 Experiences with local-level, land-use planning

The local-level, land-use planning method is being applied and refined in the Kali Konto watershed management project. The objective is to prepare both a local land-use development programme as well as directives for actual agroforestry interventions. Several aspects of this planning method will be illustrated with examples derived from a study carried out in Pagersari village in the project area.

The planning method has the following characteristics:

-The planners form a multidisciplinary team (maximum of five individuals) of local staff, who live in the area and who have considerable work experience in that area.

-The planning is carried out within four weeks time.

-The planning process is easy and feasible for the local field staff.

-The planning team is organized within an existing government structure to guarantee that the team's expertise can be used repeatedly. Proposals for development programmes will only be accepted if they are forwarded through official lines.

-The planning takes place with active involvement of the local population.

-The results of the planning process are accessible to policy makers at district and provincial level, as well as to field workers.

(i) Composition of planning team. The local extension centre, in which the services for annual crops, estate crops, livestock, and soil conservation are represented, was responsible for local-level planning. The local field staff of Perhutani (Mantri), the village official and a representative of the government at sub-district level were added to this team.

(ii) Selection of the planning area. Based on the results of the five-year plans for watershed rehabilitation of RRL and the ten-year forest management plans of Perum Perhutani, an area was selected for local-level planning by the local government. The area was representative of the agro-ecological zone in which it is situated. Because agroforestry programmes on forest land were to be included, the following criteria were met:

(a) a high level of interaction between the community and the forest,

(b) a clear conflict over land management between the community and the government, and

(c) the possibility of compromise between these two parties (Seymour and Fisher, 1987).

The field survey on terrain characteristics and land-use patterns was carried out within the boundaries of a sub-watershed. This physiographic unit provides the strongest indication of the interaction between land and water uses (Hamilton and King, 1984). However, administrative boundaries of the forest and village were used for the planning. Since development plans have to be forwarded for each village to the planning agency and for each sub-compartment to Perum Perhutani, the use of administrative boundaries facilitated the implementation of programmes considerably.

(iii) Preparation and secondary data collection. Secondary data collection was the responsibility of district-level services. An agro-ecosystem analysis carried out by KEPAS (1988) in various agro-ecological zones of east Java was a major source of secondary information. This analysis provided recommendations for research and extension programmes and indicated the key issues for development of soil conservation, and livestock, fruit tree, and food crop production in these zones.

Additional secondary data were collected, including maps and aerial photos, by representatives of the district offices assigned to the local planning team at the start of the planning exercise. The planning team also collected additional data on historical land-use changes and village demographics.

(iv) Definition of key issues. Through a brainstorming session of the planning team, a list of key issues in the planning area was developed. By focusing on such key issues the planning process could proceed quickly and efficiently. Key issues in the case study were: (a) deforestation, high fuel wood needs; (b) low productivity in agriculture, especially coffee gardens; (c) large number of landless farmers (high unemployment); (d) drinking water availability; and (e) accessibility to the hamlet of Gombong. A first orientation in the field was made to discuss these topics and prepare field work. Discussions with the village officials, key informants, and personal observations provided information necessary to further specify or adjust the key issues. From this process a checklist, questionnaire, and field programme were prepared.

(v) Field work. During field work, the aerial photo interpretation was checked and additional information was gathered on land-use and terrain characteristics. Observations were carried out on terrain (slope, soil texture, and depth) and major land-use forms; cropping patterns; planting systems; and the agricultural, social, and economic situation in the area.

A check list of key issues and aerial photographs were used for discussions with individual farmers in the field, farmers groups, and key informants of village institutions and organizations. Participatory observations were carried out and sketch maps were made. Village meetings

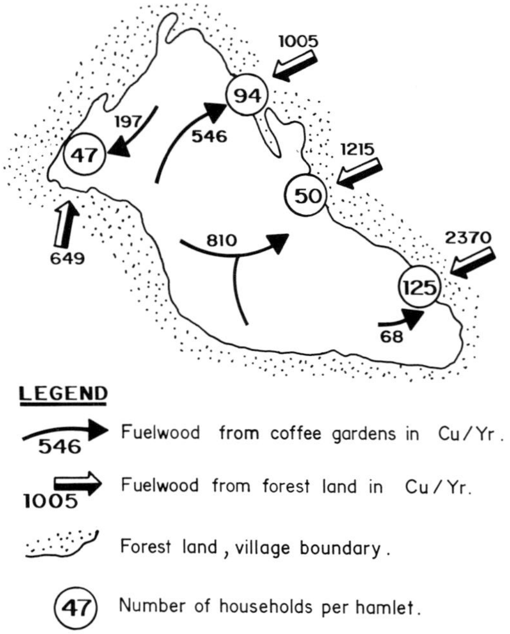

Hamlet	Total no. of house holds (h.h)	Own cons.	Sales
		(cu/hh/yr)	
Sumber-sari	125	4	15
Claket	94	4	12
Gombeng	47	4	14
Pager-sari	150	4	4

Fig. 3. Analysis map of fuelwood problem in Pagersari village.

were also organized in which all groups of the village society were represented (women, landless and rich farmers, traders, and farmers involved in reforestation). In these meetings, the villagers could voice their problems and wishes. Previously collected information was checked during these meetings. Formal interviews were held to obtain comparable data on employment, education, and the farming system.

(vi) Data processing. The data collected in the field were processed so as to be understood by both the villagers and local field officers of the government services involved. Consequently, considerable attention was given to visual representation of results in the form of maps, tables, cross sections, and diagrams. Fig. 3 illustrates the results of the analysis of the fuel wood problem in Pagersari village. Similarly, maps showing priority areas for soil conservation measures or proposals for changes in land use were prepared as well as a map illustrating the key issues for land-use development (see Table 1). Visual representation was an important tool for stimulating discussions among the planning team and village representatives concerning the plan.

(vi) Preparation of plan and programme. The various thematic data were integrated in a proposed land-use plan on a 1: 5000 scale (Fig. 4) and a land-use development programme. A priority ranking of the proposed programmes was made and presented to the head of the sub-district. This land-use plan and development programme were first discussed by the local government at sub-district level, which sent it to the district government and planning agency for approval and budget allocation.

4 LESSONS LEARNED FOR PLANNING OF INTERVENTIONS ON FOREST LAND

4.1 Introduction

A land-use development programme is not operational unless concrete proposals for the implementation of actual interventions have been made. This planning of interventions was not part of the initial planning procedure followed in the Pagersari village. However, experiences with the implementation of various separate activities, notably agroforestry on forest land, are available from the Kali Konto area. These experiences are used here to illustrate various aspects which need to be considered in the planning of agroforestry interventions.

The planning of agroforestry interventions should include and focus on the following items (derived from Hufschmidt, 1986; Wiersum, 1987): (a) forest and tree management practices choices of species, arrangements of trees (spacing), tending and exploitation; (b) provision of tools to local

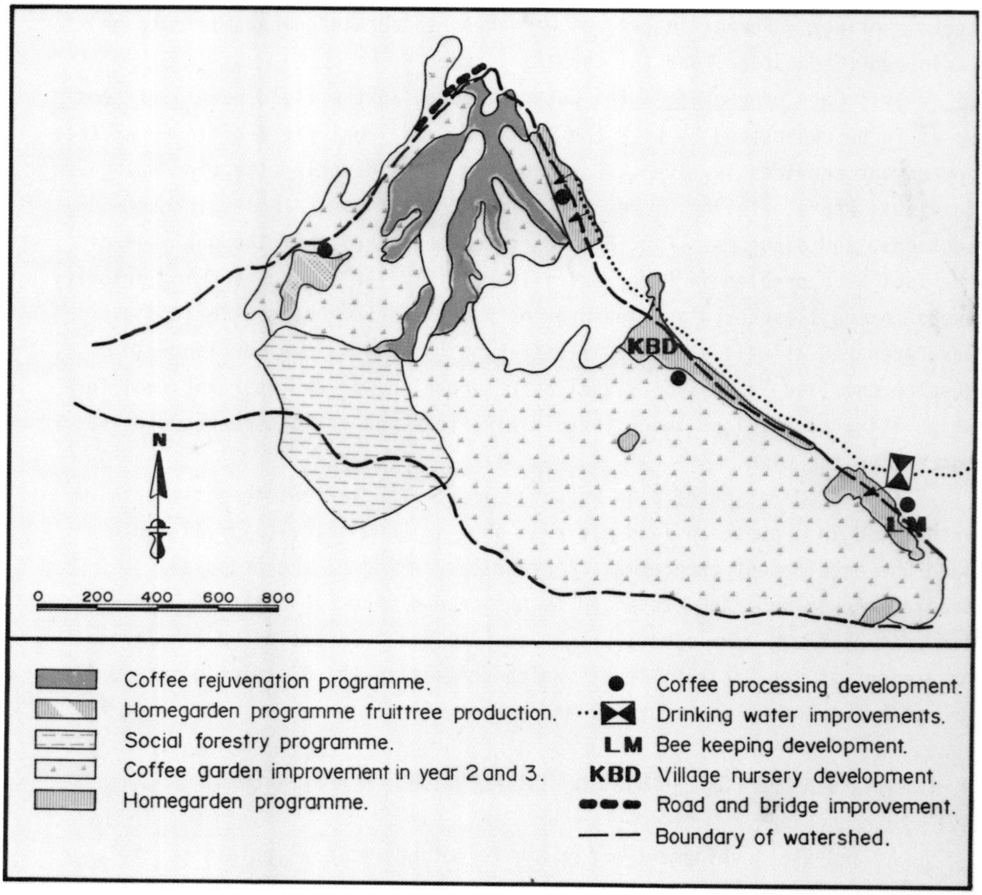

Fig. 4. Land-use plan for Pagersari village. Agroforestry has the potential to solve important key problems encountered in Pagersari village.

TABLE 1

Key features for land-use development.

Low productivity of mixed and coffee gardens due to:

- -low quality trees (too old and/or unselected clone)
- -bad spacing of coffee trees
- -poor maintenance of coffee trees
- -poor choice and maintenance of shade trees
- -low inputs of fertilizers or seedlings

High percentage of landless farmers especially in Sumbersari

High dependence on the forest to find an income resulting in failure of reforestation and deterioration of the natural system. Involvement of local population in reforestation so far has been limited. Mutual distrust exists between the local population and Perum Perhutani personnel. The hamlets of Sumbersari and Gombong sell the largest quantity of fuel wood per capita.

Potential erosion on steep slopes of the river valley because of land-use changes from perennial gardens to annual cropping systems.

Poor agricultural productivity because of limited inputs (credits) and limited knowledge of farming technology (poor education).

Limited access to water in the villages which restricts seedling production and dairy cattle development, and which creates additional burdens for village women.

Weak village institutions and organization which influences the vitality of development activities.

---

people for carrying out actual field implementation of tree growing incentives, such as training and technical support; and (c) arrangements of a proper institutional and organizational structure within which implementation can proceed.

4.2 Planning of tree-management practices

At this writing, tree management practices in agroforestry on forest land (taungya system) are largely prescribed by Perum Perhutani. The main interest of the farmer is to grow food and cash crops. Perum Perhutani wants to preserve forests and produce fuel wood, pulp wood, and timber. Consequently, there is a permanent tension between the two parties, under which farmers continuously try to stretch the regulations and apply their preferred tree management practices until they are corrected by Perum Perhutani. An example is the continuous lopping of trees for fodder. Fodder is of more interest to these farmers than fuel wood, especially in the dry season when grass production on agricultural fields is low (Nibbering, 1988).

In order to minimize the conflict over optimal tree-management practices, it has been suggested that in the design of more appropriate agroforestry systems on forest land (consisting of a mix of timber, fuel wood, fruit/other perennial trees) existing agroforestry systems should be used as a starting point for development (Oldeman, 1983; Wiersum, 1982). Because of its successful history in the area, the smallholders' mixed coffee gardens on village land might serve such purpose. However, it is doubtful whether with the present knowledge of the composition, structure, and interrelationships within such agroforestry systems is sufficient to permit designing an intervention which would address the whole rotation. Such data are needed to fit the agroforestry plans within standard forest management plans.

Some of the complications caused by competing interests in various agroforestry products are minimized in systems with a permanent annual crop component that is spatially separated from the forest component. Although the basic conflict between Perum Perhutani and the farmer also exists in such systems (farmers have a tendency to try to enlarge the area under annual crops at cost of the forest area), it is much easier to control. Furthermore, the annual fluctuation in benefits for the farmer are lowered (because the annual component is constant), and therefore it is easier for the farmer to decide whether the incentives are sufficient to participate in such a programme.

A clear but flexible framework should be made for deciding on tree-management practices. It should indicate who has responsibility over which tree-management practices. It should maintain the balance between uniformity with respect to the needs for effective administration by Perum Perhutani and options for sufficient diversification in adaptation to local conditions. Some freedom should be given to the local field staff and participants to make decisions on tree-management practices. Such a framework will probably need to be further operationalized for each agro-ecological zone.

4.3 Planning of implementation tools

(i) Incentives. The local population does not perceive the problem of forest degradation as a serious issue. Therefore, local inhabitants will need incentives in order to participate in solving the forest degradation problem. Experiences with projects in which incentives were provided in the form of wage payments have had poor results. Local motivation in these cases was dependent upon funds being available. When these funds ceased, so too did participation in these programmes (Nibbering, 1988). Allowances to collect fodder and the promise of future fuel wood collection have also proven to be ineffective incentives. The allowances used were too meagre an incentive for farmers to participate in reforestation activities, and farmers have been skeptical as to whether they would actually get the promised future permission to harvest fuel wood.

A much better incentive appears to be the right to grow annual crops (food/cash) and fruit trees on forest land for a certain period of time. With this incentive, however, care needs to be taken to ensure, once annual cropping is no longer possible because of a closed tree canopy, that the revenues from fruit trees form a large enough incentive to maintain the motivation of the participants.

(ii) Selection of participants. In various activities, involvement of the poorest members of society in agroforestry activities proved to be difficult. This was particularly so for forest lands where large profits could be earned by annual cropping. In such cases, richer farmers may pay the forest guard or village head for participation in the scheme or they may buy the forest-use permits from poor farmers. These are strong forces because forest guards do sometimes welcome the addition to their salary. Additionally, there is a widespread perception that wealthier farmers, because of their experience in agriculture, make better participants in reforestation efforts than landless farmers. While there are some benefits to this scheme it is the landless or land-poor farmers that depend on the forest (Peluso, 1986). It is of critical importance to accommodate such landless farmers with income earning opportunities because they are the ones who put the burden on the forest.

(iii) Training. Two aspects, which are often overlooked, have been found to be important in planning training activities. First, it should be recognized that reforestation activities are normally carried out by both husband and wife, and consequently training in planting techniques, vegetative propagation, and other tree management techniques has to be targeted to both. Second, both local villagers and Perum Perhutani field staff must be involved in additional training. It is the Perum Perhutani field staff who usually understand the problems of conflict between the farmer and Perum Perhutani. Unfortunately, field staff have been trained to deal with conflicts paternalistically rather than through a process of mutual agreement or consensus. A greater emphasis on the value of the latter approach will enhance the possibilities of success. This social learning approach is still alien to many foresters who have no experience with social forestry and who have worked their whole career in a strict hierarchy. A system where field staff are regularly transferred to other areas might assist in this process of education.

(iv) Marketing. The planning of agroforestry systems for large-scale production of fuel wood, but even more so for fruit production, requires information on the marketing prospects of these products. For instance, fruits are bulky and perishable and marketing of large quantities might be difficult (Dijk, 1987). Access to various sources of information and insight on marketing needs on a sub-district level is required. This should become the responsibility of Perum Perhutani staff at the sub-district level.

Experience has shown that when thinning was carried out on a sub-compartment basis in accordance with the rules of Perum Perhutani, the local village markets are quickly saturated. Transport of the fuel wood to neighbouring villages becomes unattractive because of the relatively low value per volume. The local field staff of Perum Perhutani together with the farmers groups need more freedom in the planning of thinning and sale of fuel wood.

4.4 Planning of institutional arrangements

Forest management by Perum Perhutani is, like in many developing countries, based on a European forestry tradition. During the long history of plantation forestry in Indonesia, rules and procedures have become well established. Some of these rules and procedures need to be reviewed because of difficulties which have arisen during their application to agroforestry planning.

(i) Decision-making. With the inclusion of fuel wood and perennials for the local population in forest plantations, more management decisions need to be made than was originally the case for traditional plantations. No guidelines exist yet on what these decisions should be (e.g., selection of fruit trees in combination with other trees and their respective spacing), how the decisions should be made (e.g. what the selection criteria are for choices of fruit trees), and who should make the decision. At present, most decisions are still made on a higher level than field staff, which makes it difficult to plan agroforestry interventions that are well adapted to local conditions and needs. A planning framework needs to identify decision-making responsibilities.

(ii) Forest types. It is expected that fuel wood will remain the major energy source into the next century and that the production of fuel wood on village land will not rise substantially (Gier, 1987). There is a need for the establishment of exclusive fuel wood plantations instead of mixed timber/fuel plantations. Under such a system, present administrative arrangements will have to be changed. Fuel wood plantations have the advantage that the planning of a balance between local supply and demand is much easier than with mixed plantations. Fuel wood plantations can also be exploited in a rotational arrangement of cutting blocks (followed by regeneration) so that the amount of fuel wood produced is constant over the years.

(iii) Contracts. Arrangements about duration of the contract are often unclear to the local farmer (Munneke and Vorsterman van Oyen, 1987). Perum Perhutani keeps contract arrangements vague because of a fear that farmers will claim ownership over land once they have been allowed to cultivate it for several years (Bekkering, 1988). However, farmers tend to be interested only in the production of fruit trees and fuel wood, provided they have a contract of at least 8 to 10 years in which it is clearly stated that the produce accrues to them. Clear contracts with rights and obligations are a prerequisite for the success of these agroforestry programmes.

(iv) Organization of farmer groups. Several indications exist that it may be easier for Perum Perhutani to deal with specific farmer groups rather than with individual farmers. The heads of the farmer groups play a key role in the organization. These individuals have the difficult task of maintaining the full trust of participants and of keeping close ties to the field staff of Perum Perhutani.

5 CONCLUDING REMARKS

Agro-ecosystem analysis is a useful alternative to standardized national implementation programmes and can form a framework for planning at the local level. Such a framework would be based on agro-ecological zones and would give guidelines for research, extension, and development programmes.

In addition to this technical framework, administrative procedures should be developed to make local-level planning feasible. For realistic local-level planning, the perspectives of both local villagers and government services should be taken into account. This requires a merging of top-down and bottom-up approaches.

The local-level planning method presented is adjusted to the capabilities of government staff. Present planning procedures can be incorporated into this method, consequently this local-level planning method is already being implemented in east Java.

A multi-disciplinary planning team of field staff of the local extension centre will have to be able to carry out local-level planning since they have experience in the area and know the key issues for development. At the local level, cooperation between services and the formation of a multi-disciplinary team has been possible.

Rapid rural appraisal (RRA) is used as a major data collection technique. A terrain and land-use survey as well as formal interviews are incorporated in the RRA to make it less dependent on personal capacities of the planners. By focusing on key issues for development, the planning process can proceed quickly and efficiently.

The result of local-level planning is a land-use development plan, a land-use development programme, and the planning of interventions. Since the proposed land-use development programmes are interrelated and are only effective if a package of integrated interventions is carried out, the development programme should be implemented as a whole.

Some of the proposed activities in the programme are specific for each village. Other activities are representative for the agro-ecological zone as a whole. These general programmes become part of the conceptual framework for planning.

6 REFERENCES

Bekkering, T.D., 1988. Agroforestry trials on forest land: experiences and options. Konto River Project, Working Paper No. 20. Malang, Indonesia.

Bekkering, T.D. and Hartanto, H., 1987. Forest establishment through agroforestry: some suggestions for improved tumpangsari. Konto River Project, Working Paper No. 8. Malang, Indonesia.

Bratamihardja, M., 1987. Social forestry on state forest land. Paper Workshop Planning and implementation of social forestry programmes in Indonesia. Faculty of Forestry Gadjah Mada University Yogyakarta.

Carson B.R., 1985. Assessment of a small watershed using aerial photography: an example form a remote hill region in Nepal. Paper presented to the international workshop on watershed management in the Hindukush-Himalay Region, 14-19, October 1985. Chengdu, China.

Carson B.R., 1987. Agro-ecosystems; a comparative study of agro-ecosystems in East Java and Nusa Tenggara Timur, report KEPAS, Malang.

Chambers R., 1983. Rural Development: Putting the Last First.

Conway, G.R., 1985. Agro-ecosystem analysis. Agricultural administration 20: 31-55.

Conway, G.R., (in press). The properties of agroecosystems. Agricultural Systems.

Dijk, M.P. van, 1987. Perennials in reforestation: the economic aspects of some alternative agroforestry systems. (Konto River Project, Working Paper No. 15). Malang, Indonesia.

Gier, A., 1987. The energy situation in the Konto River Project area.

(Konto River Project, Project Communication No. 6). Malang, Indonesia
Grandstaff, T., 1987. Narrative report to the Ford Foundation, on international conference on rapid rural appraisal. In: C.P. Veer, The Other Half of Social Forestry: Small-scale Forest Based Enterprises. Paper Workshop Planning and implementation of social forestry programmes in Indonesia. Faculty of Forestry Gadjah Mada University, Yogyakarta.
Hadipoernomo, 1986. Dukuh Cukal, profil tanah sebuah desa di tepi hutan pegunungan. Duta Rimba, Jakarta.
Hamilton, S.M. and King, P.N., 1984. Watersheds and rural development planning. East-West Centre, Honolulu.
Hufschmidt, M.M., 1986. A conceptual framework for watershed management. In: K.W. Easter, J.A. Dixon, and M.M. Hufschmidt, (Editors), Watershed Resources Management. Westview Press Studies in Water Policy and Management No. 10, Boulder, Colorado, pp. 17-31.
KEPAS, 1985. The critical uplands of eastern Java; an agroecosystems analysis. East Java Agro-ecosystem Working Group, Malang, Indonesia.
KEPAS, 1988. Pendekatan Agro-ekosistem pada pola pertanian lahan kering: hasil penelitian di empat zone agro-ekosistem Jawa Timur. Kelompok Penelitian Agro-ekosistem, Malang, Indonesia.
Konto River Project, 1988. Screening Study Brantas Watershed: executive summery (draft). Malang, Indonesia.
Lundgren, B., 1982. Introduction. Agroforestry systems, 1: 3-6.
Maaren, A.van and Wiersum, K.F., 1988. Policy aspects of agroforestry development. In: Viewpoints on agroforestry, 2nd ed.
Munneke, H.H. and Vorsterman van Oyen, E., 1987. Socio-economic conditions in the hamlet Cukal: a study on the impact of the Bendosari agroforestry schemes. (Konto River Project, Working Paper No. 11). Malang,Indonesia.
Nibbering, J.W., 1988. Forest degradation in a highland area in Java Paper Presented at the IUFRO tropical forest history conference, 16-18 May 1988, Canberra, Australia.
Oldeman, R.A.A., 1983. The design of ecologically sound agroforests. In: P.A. Huxley (Editor), Plant Research and Agroforestry. ICRAF, Nairobi, pp. 173-206.
Peluso, N.L., 1986. Report on social forestry field research in West and Central Java, October 1984-October 1985. Perum Perhutani and Ford Foundation, Jakarta.
Perum Perhutani, 1985. Studi Kasus social forestry beberapa deluarga di Jawa Tengah jan Jawa Barat (II). Jakarta, Indonesia.
Pickering. A.K., 1979. Soil conservation and rural institutions on Java. IDS Bulletin 10(4): 60-66.
PPK, 1986. Final report phase II Kali Konto Project. Kali Konto Project, Malang, Indonesia.
RIN, 1986. Evaluation of forest land Kali Konto upper watershed, East Java. Kali Konto Project, Malang, Indonesia.
Rusmanhadi, 1987. Review of Some fuel wood consumption standards. In: Bekkering and Rusmanhadi (Editors), Issues on fuel wood. (Konto River Project, Working Paper No. 12). Malang, Indonesia.
Thalen, D.C.P. and Smiet, A.C., 1985. Quantified land-use policy options in forest land evaluation for watershed management. Journal Agricultural Science 33(2): 89-103.
Van den Hoek, A.I., Schomaker, M.S. and Zonneveld, J.M., 1988. Local land development planning: a methodology for integrated village development. Konto River Project, (Working Paper No. 22). Malang, Indonesia.
Wiersum, K.F., 1982. Tree gardening and taungya on Java: examples of agroforestry techniques in the humid tropics. Agroforestry Systems, 1:53-70.
Wiersum, K.F., 1987. International experience in social forestry and implications for research support. Paper workshop planning and implementation of social forestry programmes in Indonesia. Faculty of

Forestry, Gadjah Mada University, Yagyakarta.
Wirjodarmodjo, H. and Bratamihardja, M., 1983. Policies, strategies, and design of forest development on the island of Java. In: K.F. Wiersum (Editor), Strategies and designs for afforestation and tree planting. Proceedings of an international symposium 19-23 September, 1983, Wageningen, The Netherlands, pp. 363-374..

Chapter 5

PLANNING AGROFORESTRY EXTENSION PROJECTS, THE CARE INTERNATIONAL APPROACH IN KENYA

L. BUCK
Agriculture and Natural Resources Specialist, P.O. Box 256, Bogor (Indonesia)

ABSTRACT

The Agroforestry Extension Project (AEP) supported by CARE International in Kenya was designed to direct the promised benefit of agroforestry land use to subsistence or near-subsistence level farmers through a pilot technology-generating extension scheme. Initiated in 1983, AEP aimed to test the adoptability of prototype agroforestry interventions, and to select and modify the most appropriate for improved performance, before widespread extension. A uniquely trained team of project-supported field staff would facilitate the technology improvement and extension process by influencing indigenous organizations of farmers, and school children to design, establish and monitor agroforestry interventions on their farms and school compounds.

Planning, training and field monitoring and evaluation activities are guided by ICRAF's diagnosis and design (D&D) methodology, adapted for extension purposes by AEP and its predecessor projects. This case study describes the project design process and key characteristics of AEP's implementation methods and experience. Strengths and limitations are identified and prospects for improvement considered. Implications of AEP's experience are suggested for advancing an increasingly promising extension and research linked project approach to agroforestry technology and rural development.

1 INTRODUCTION

CARE International in Kenya has supported the Agroforestry Extension Project (AEP) since 1983. In AEP's five year history it has provided technical and material assistance to approximately 600 women's groups and primary schools in Nyanza Province of western Kenya to establish and manage their own agroforestry tree nurseries (see Fig. 1). These groups of women, men and school children represent approximately 100,000 households who have been directly or indirectly assisted in designing and establishing some 23 different agroforestry interventions and at least 36 tree species on their farms and school compounds for a variety of purposes. Of the approximately 9 million trees planted by April, 1988 over 50% were surviving by June, 1988. Periodic formal and informal external evaluations throughout the project

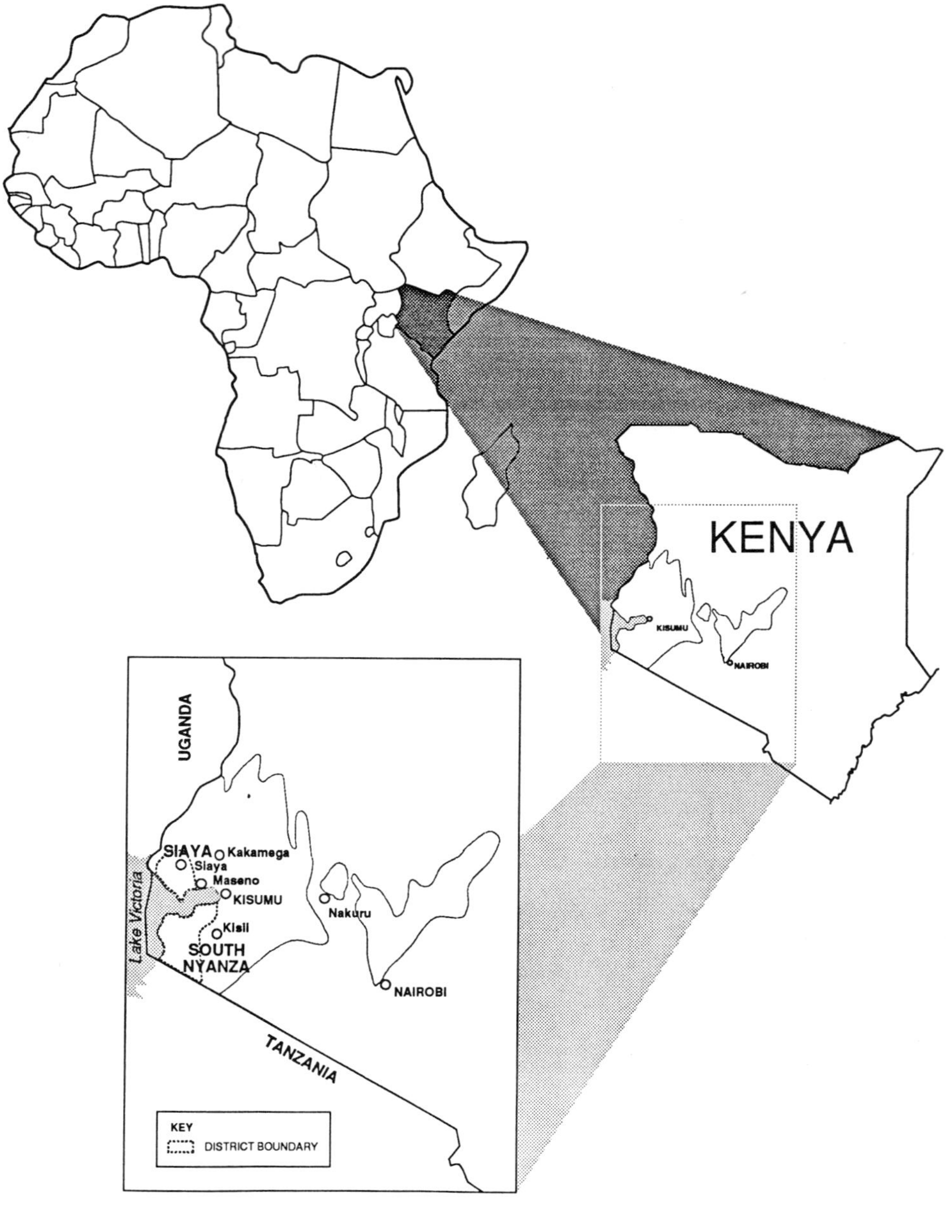

Fig. 1. Nyanza Province in western Kenya.

history indicate significant farmer satisfaction with agroforestry assistance provided by CARE.

AEP was unique among CARE and other non-government organizations (NGO) programming initiatives in the agriculture and natural resources sector in 1983. Designed as a pilot project, AEP aimed to generate information on agroforestry technology performance and impact, resulting in a set of "farmer-tested prototype designs" for a variety of subsistence and near subsistence farming systems, which extension agency field staff throughout the pilot district could recommend with confidence. An internal monitoring and evaluation system was initiated for this purpose, to complement a limited formal research component. Simultaneously, AEP sought to influence district extension agencies, community organizations and households to participate in an ongoing district level agroforestry development process.

The heart of this process was field-staff consultation with farmers to design and improve agroforestry techniques for a variety of farming system types, and systematic information collection from the on-farm activities. Conceived as neither a classic research nor traditional extension project the AEP design embodied elements of each in a hybrid technology-generating and organizational development approach.

This chapter reviews the agroforestry extension and research project experience in Kenya which influenced the design of AEP. After describing the project planning context, methods which guided the development of AEP are discussed. Constraints in meeting its agroforestry technology generation aims to date are identified. A summary of project strengths, weaknesses, and potential for fulfilling an extension and research-linking mandate suggest characteristics of project design which could support this innovative and promising type of agroforestry development activity.

2 PROJECT PLANNING CONTEXT

CARE's Agroforestry Extension Project was born of a youthful, but promising and rapidly advancing state-of-the-art of agroforestry research and development in Eastern Africa, and a sponsoring agency newly committed to principles of sustainable community development, eager to apply agroforestry to this purpose. These conditions allowed for a project design which would establish CARE in partnership with ICRAF to develop research methods and agroforestry technology recommendations in an extension project context. AEP would also establish collaborative relationships between CARE, Government of Kenya (GOK), and non-governmental organization (NGO) development agencies to help secure the presumed benefit of agroforestry for rural people.

2.1 Agroforestry in Kenya

In 1983, when CARE-Kenya's Agroforestry Extension project was initiated, agroforestry extension activity in Kenya was in its infancy (Getahun, 1988). ICRAF's first national seminar in agroforestry two years prior had revealed a budding interest in the science and practice of agroforestry, but little research or extension experience (Buck, 1981a). Farmer-focused tree planting assistance at the time was limited primarily to the sale of subsidized seedlings by the Forest Department of the Ministry of Environment and Natural Resources (MENR). Species were of limited variety, selected mainly for plantation reforestation.

Institutionally, three GOK ministries claimed leadership interest in the sector; Environment and Natural Resources (Forest Department), Energy and Regional Development, and Agriculture (Soil Conservation Unit). However, the most innovative and aggressive agroforestry programming was lodged in the NGO sector, whose efforts were supported by national development policy at the time (Van Domelyn, 1986; Buck, 1984, 1986; Ndegwa, 1982).

CARE's Agroforestry Extension Project grew from a succession of two previous NGO-initiated agroforestry projects in Kenya aimed at assisting farmers to establish agroforestry interventions on their own land. Mazingira Institute's Agroforestry Plots for Rural Kenya project was a pilot research and development effort funded by the Ford Foundation and the Dutch government in 1981 to help generate, monitor and evaluate local organizational capacity to develop and manage common and private agroforestry demonstration plots (Buck, 1981b).

The Mazingira project collaborated with ICRAF in the development of its agroforestry diagnosis and design (D&D) methodology (Raintree, 1983). A variety of planning, monitoring and evaluation tools for agroforestry extension were developed in the course of the project (Buck, 1983a).

ICRAF's Kenya Tree Seed Project was an outgrowth and complementary effort to the Mazingira rural plots project. Through the Tree Seed Project and in collaboration with the Mennonite Central Committee (MCC) and the Kenya Energy Non-governmental Organization (KENGO), two other leading NGOs in rural tree development, an extensive inventory was undertaken of indigenous, naturalized and exotic tree species in Kenya's farming landscape (Buck, 1983b; Teel, 1984). In 1982 and 1983, CARE-Kenya assisted the joint Mazingira/ICRAF agroforestry extension effort with a full-time field assistant and material support for small-scale nursery water supply at several sites. CARE sought agroforestry project experience in anticipation of eventually initiating its own project.

The design of AEP was significantly influenced by the methods and findings of these pioneering research and development efforts. It provided an opportunity to apply and further develop several innovative land-use assessment, agroforestry design, monitoring, and evaluation methods initiated by AEP's "mother projects" in the context of a larger scale rural development activity.

Key lessons were drawn from the rural plots and tree seed projects which influenced the design of CARE's AEP. Although comparatively commonplace perhaps, some five years later, in 1983 the findings represented a relatively advanced outlook regarding rural forestry (Buck and Alitzi, 1983; Buck, 1983a, 1983b).

2.2. CARE-International in Kenya

Prior to 1983, CARE's 15 year development assistance programme in Kenya consisted primarily of material and product delivery to rural communities. Aid was provided on an ad hoc basis throughout the country. In 1983, CARE reorganized its programme to cluster area and sector-specific initiatives and increase training and technical assistance to communities in order to achieve more sustainable, developmental effects. Ten of Kenya's 42 administrative districts were selected for cluster programming following an assessment of needs and opportunities in each district (CARE-Kenya, 1983).

From the problem areas identified by the national level needs assessment, five project sectors were chosen, one of which was agroforestry extension. On the basis of experience with the Mazingira Institute and ICRAF projects, CARE-Kenya hired a manager to design and launch the agroforestry extension project within its ten districts focus area. Design guidelines provided by CARE were essentially limited to a funding estimate (US $1.2 million) and specification of an approach similar to the rural plots project.

3 PLANNING METHODS

AEP was designed as a pilot project to test an extension approach, as well as prototype agroforestry interventions for their long-term viability in addressing land productivity problems as they affect rural small holders. The cornerstone of the project design was an ongoing, internal planning process for guiding project implementation and evaluation (Buck, 1983b). AEP's planning process was guided by the following principles. Frequent planning workshops were to take place between project management and staff, the project unit and agency collaborators and the project unit and local organizations. An ongoing, iterative cycle of planning events to guide changeable training, implementation, monitoring, and evaluation activities was needed. An open planning environment in which experience and ideas are freely shared and critically examined was desired. Well informed planning participants reliant on project-generated information with shared decision-making among parties involved regarding objectives, activity targets, and resource commitments were sought. Finally, the process needed a flexible project view toward output targets, guided by the project's strategy and approach for meeting project objectives.

These guidelines were to apply similarly to project planning with staff, government, and NGO collaborators, as well as local participants. In addition to the mutual learning expected to occur, the aim of planning with agency collaborators was commitment of joint responsibility for technical, administrative, or policy support. The aim of planning with community organizations and farmers was to establish objectives and respective responsibilities for site-specific agroforestry development schemes, including technical, training, material, management, and labor inputs (Buck, 1983b).

In accordance with CARE's multi-year project planning and funding cycle, the AEP project document covered an initial three-year implementation period. However, the project design was based on a minimum six-year funding horizon and recommendation that the planning process set in motion be maintained and monitored for 10 years.

3.1 Establishing a strategy and approach

The Agroforestry Extension Project's original rationale was to relieve a chronic and worsening household level wood fuel supply shortage identified by the Ministry of Energy and the Beijer Institute in their 1981-1982 national level energy study. The study indicated that certain areas of Kenya's Western and Nyanza provinces faced extreme supply shortfalls in the coming decade, and agroforestry was the most appropriate land-use strategy for improving household access to wood fuels (Beijer Institute, 1982). Thus, a project strategy was conceived to address institutional and land-use constraints resulting in the wood fuel availability problem by initiating an agroforestry development process at household, community, and district or sub-district government levels.

The project had to develop its own internal unit of expertise, which in turn would lend support to existing organizations who had responsibility and/or capability for respective elements of agroforestry development including technical assistance, material acquisition or delivery, and administrative and policy support. Project assistance would be primarily in the form of knowledge and institutional capability building through training and information networking. Minor material assistance would be provided initially to local farmer organizations and schools as incentive to raise agroforestry tree seedlings for their own use.

To maximize the value of project experience for evaluating agroforestry potential and performance, a project area was sought which would represent a range of agro-climatic conditions and farming systems types. To improve the potential for integration of agroforestry concepts and practices into agency and rural environments, new initiatives would be based on existing interest, knowledge, and capability at each level (Buck, 1983b).

The strategy called further for the training and field support of a dual level extension team to catalyze change in the rural environment. The project field team would embody analytical proficiency and personal familiarity with local farming systems, and would be trained to work as complementary partners in assessing problems and opportunities, as well as formulating and communicating potential agroforestry solutions to address them. The field team in turn would train and encourage local organizations, namely women's groups and primary schools, to work on a "self-help" basis to design, select, and adapt appropriate agroforestry interventions, raise agroforestry tree seedlings for their own plots, plant the trees, monitor their performance, and adapt these interventions accordingly. The field based technical assistance team would encourage local activity by suggestion and demonstration, but would not assume responsibility for the activities.

Fundamental to the project's approach was a view that CARE as the implementing agency should participate in the local developmental process, versus the more common development view of engaging local people to participate in project activities. This implied an understanding of local development dynamics and pursuit of project activities which would enhance these.

3.2 Identifying a project cycle

AEP's ongoing and iterative planning process is reflected in the model of its project cycle (Table 1). The model includes five basic project activities; planning, training, implementation, monitoring, and evaluation. These are repeated throughout the later three phases of the cycle, with varying degrees of emphasis in each. The content of the five activity categories is further specified through the project's internal planning process.

TABLE 1

Agroforestry extension project cycles (Adapted from Agroforestry Monitoring and Evaluation Methodology Programme [AFMEMP], in Buck <u>et al.</u>, 1988.

<u>Phase I:</u> <u>Project Conception</u>

Identify rationale
Assess historical, institutional experience.
Assess institutional potential
Prepare concept paper

<u>Phase II:</u> <u>Project Design</u>

Assess agroforestry need: national/regional levels
Assess agroforestry potential: national/regional levels
Identify collaborators and initial project area
Establish approach
Develop hierarchy of objectives
Identify activity types to meet objectives
Develop staffing structure and management strategy
Design preliminary monitoring and evaluation plan

<u>Phase III: Negotiation and Commitment</u>

Select and train staff
Identify/select local sites and participants
Assess local agroforestry need and potential
Design area-specific interventions
Negotiate staff/farmer responsibilities (men's and women's)
Develop extension activity schedule

<u>Phase IV:</u> <u>Establishment</u>

Site and develop nurseries
Design/select site/farm-specific interventions
Raise and plant trees
Monitor configurations and species planted

<u>Phase V:</u> <u>Ongoing Management</u>

Management and use trees and other agroforestry system components
Monitor tree management objectives and techniques
Assess participant satisfaction/adoption
Assess technology performance
Assess organizational capacity to manage activities
Assess potential for expansion; repeat Phases III and IV at new sites and adjust management practices accordingly

Assess potential for sustainability (and further evolution) of respective activities and intended outcomes

The following activities are repeated in Phases III-V, with varying degrees of emphasis:

Planning
Training
Implementation
Monitoring
Evaluation

The first one to three iterations of Phases III-V constitute the project's pilot stage.

The AEP design specified that the first iteration of the project cycle following the design phase (Phases III-V) constituted a pilot state. This was for testing the viability of AEP's strategy and approach as well as methods for planning and implementing realistic activity targets (outputs) for a longer term, full-scale effort to meet project objectives. During the pilot stage, AEP evaluations would tolerate significant deviation between planned and actual activity outputs, given reasonable assumptions and explanations.

The dynamic of repeating the cycle of (post-funding) project activity each season or year, depending on local planting and harvesting cycles, was termed AEP's rolling planning process (Buck, 1983b). This process was to guide the simultaneous implementation of various phases of activity at different sites, externally assisted reviews for assessing the effectiveness of the process, and adequacy of the outcome.

Iterative, internal, participatory planning has gained increased acceptance by development practitioners during the 1980's. In 1983 however, the approach departed from most leading development agency planning practices, including CARE's. Rather, classic, comprehensive, rational planning theory prevailed as the predominant paradigm for rural development project design. The later approach situates planning at the beginning of the project cycle, under the control of the expert professional planner. The success of the project or activity is judged at its pre-determined end point on the basis of correct adherence to the plan. The approach assumes nearly perfect <u>a priori</u> knowledge of the planning environment, the intervention, and the dynamic between the two. Development experience however, has confronted practitioners repeatedly with the folly of this assumption in the context of cultural discontinuity, unpredictability, and rapid change which characterize most development situations.

Agroforestry sector planning for rural development provided a unique opportunity to confront and challenge the viability of the rational, comprehensive planning approach. Agroforestry's youthful stage of development meant far from perfect knowledge. Although the field was too young to claim experts, it was too promising to ignore until scientists were convinced of its validity. Thus, a process was required in which research (knowledge-generating) and extension (promotion) objectives could be pursued simultaneously. Creativity and compromise were required which encouraged an alternative development paradigm altogether. AEP's challenge was to fashion methods which could link these distinct sets of activities into a development project framework.

3.3 Identifying collaborators and project area

With a project strategy and approach defined, meetings and interviews were conducted early in the design process to identify potential collaborators and consult on further aspects of project design. This effort revealed ten organizations, listed in Table 2, with a direct interest and role to play in the type of project CARE envisioned.

All agencies endorsed the forest department, particularly its Rural Afforestation Extension Scheme (RAES), for the role of primary government collaborator. Assistance from the Swiss Development Corporation's Rural Tree Development Project enabled RAES to advance its farm forestry extension activity throughout the country. RAES encouraged CARE to lend technical field support to this effort within a selected area of the country in exchange for forest department policy and programme support. AEP met frequently with national and district forest department officials throughout the project's design phase and beyond. Other agencies were consulted as needs for information and opportunities for joint activity arose.

The collaborators' first major decision was the selection of a district focus area for the project. The forest department was sympathetic to national wood fuel study findings that Nyanza province was a wood fuel supply deficit area. Accordingly, this was a RAES priority area for intensive farm forestry extension assistance. RAES viewed the AEP project initiative as a timely complementary effort.

Key among AEP project site selection criteria were (a) diversity of climatic and farming systems types to maximize learning about agroforestry, and (b) significant representation of poorer sectors of the rural population. Visits by the project designer to Siaya district's forest officer (DFO) quickly confirmed compatibility between AEP's and RAES's development

TABLE 2

AEP's collaborating organizations (aside from the Forest Department and Agricultural Research Institute, direct government agency collaborators were represented at the district level only).

Ministry of Agriculture (MOA), Soil Conservation Unit
Kenya Agricultural Research Institute (MOA).
Ministry of Energy, Renewable Energy Division
Ministry of Environment and Natural Resources, Forest Department's Rural Afforestation Extension Scheme
Ministry of Culture and Social Services
Ministry of Education
University of Nairobi Forestry Department
ICRAF (International Council for Research in Agroforestry)
KENGO
Mennonite Central Committee (MCC)

strategies and approaches. The climatic diversity CARE sought in a pilot project site was present in Siaya district (Fig. 2). Furthermore, the area was characterized as among the least developed in Kenya. An agreement was reached to base the first phase of the project in Siaya district with the possibility of expansion into neighbouring South Nyanza district.

The second major decision concerned project collaboration and influence with other district agencies. An AEP/DFO initiative was accepted that an agroforestry subcommittee of the district development committee be established to guide agroforestry policy and activity. CARE was invited to be a voting member of the subcommittee which was active throughout the first three years of the project. This placed AEP in an excellent position to share influence in the district's formal planning process.

3.4 Establishing goals, objectives, and activities

AEP's original goals and objectives stemmed from the project's strategy and approach, and were designed also to support them. CARE-Kenya's acceptance of AEP's pilot project status allowed for departure from the organization's standard project planning guidelines in the design of goals and objectives.

CARE's status as a human welfare organization dedicated to improving the conditions of people's lives, specifies its highest level or "final goal" be expressed in terms of "human welfare" improvement. Second level or "intermediate goals" are concerned with the land, institutional, infrastructure and/or human development changes required to bring about the

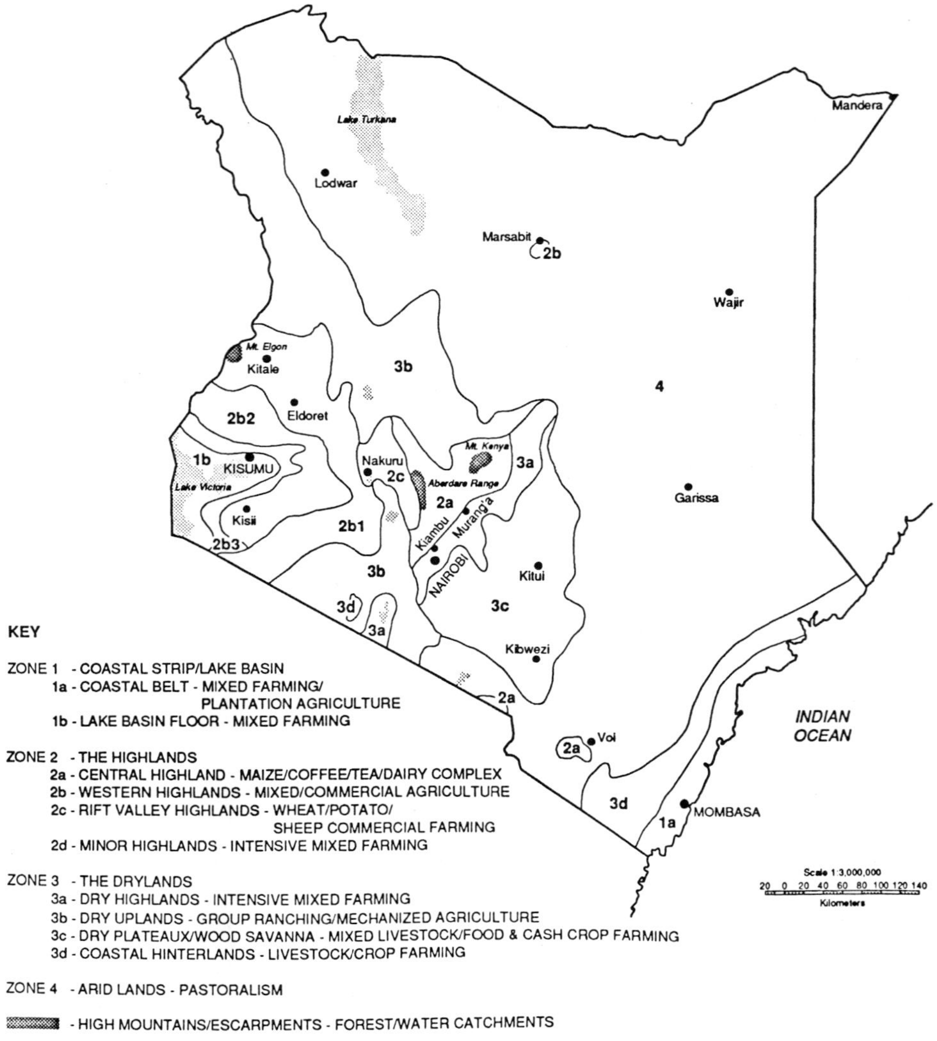

Fig. 2. Climatic regions of Kenya.

desired improvement in people's lives. For agriculture and natural resource projects, intermediate level changes, in recent years, have come to be expressed in terms of the land-use systems of which people are a part; extension or service delivery systems for upgrading the level of technical and material assistance to communities; and the change in knowledge, attitudes and practices which communities and farmers themselves must exhibit to ensure lasting improvements.

When AEP was designed, CARE's project proposal guidelines reflected programming emphasis on material delivery to rural communities. The logic of linking such activity and its presumed output to intermediate and final goal achievement was assumed, although rarely tested. In the intervening five years, CARE's approach to project design has evolved in response to its adopted programming principles. Significantly more attention now is given to planning, monitoring, and evaluating goals, as well as activity achievement, and to validating assumptions which link these, based on knowledge of development dynamics and the project environment (Buck, 1988).

In designing AEP however, it was inappropriate to assume causal linkages between agroforestry development and improved human welfare, or within three years to test it. Too much remained unknown about agroforestry's technical, institutional, or socio-economic viability. Instead, the project attempted to create and institutionalize a learning process which eventually would result in better understanding of these linkages, so higher order effects could become predictable. In effect, causal linkages in AEP constituted "hypotheses" rather than assumptions, to be tested during the pilot stage of the project and probably beyond.

The project's initial intermediate goals reflected essential components of the agency and household agroforestry institutionalization process. Another departure from conventional CARE project design practice was specification of activity types, but omission of quantitative targets or outputs in the project implementation document. These were to be developed with collaborators and participants through the project's internal planning process.

The project's second multi-year plan (1987-1989 MYP) included a more specific and limited hierarchy of goals and quantified activity outputs. Intermediate goals were reduced from eleven to three and quantified by certain indicators represented originally as goal statements. In the revised 1987-1989 MYP also, fuel wood production ceased to be a specific objective as AEP found that farmers in the project area did not consider wood fuel shortages a serious problem!

AEP's modified goal and activity hierarchy reflected better informed expectations about the significance of intermediate goal achievement on realization of the final goal, as well as quantities of activities and outputs achievable and their effect. However, the second MYP also increased output targets well beyond original numbers, thus formalizing the preceding year's trend to emphasize nursery establishment, seedling production and tree planting at the expense of on-farm technology optimization activities. Anticipated advances in technology improvement were to drive the project's limited, formal on-station research component and individual farmer judgement.

3.5 Staff training

AEP's staff development plan stemmed from the assumption that both the farmer and the technical specialist had knowledge needed to transform the promise of agroforestry into practice by significant numbers of households. Although each had a role to play in improving agroforestry practices, a communication gap existed between scientists/technicians and farmers.

The project design thus called for a dual-level field team to perform these two-way information and communication functions. The role of each level included knowledge building with varying degrees of emphasis throughout the project cycle. To accomplish the knowledge-building role, more educated field staff (field officers) were responsible initially for ensuring project use of the technical information base on agroforestry. This was derived primarily from investigating agroforestry need, potential, and practices in the project environment, supplemented by scientific and popular literature.

High-school level field staff (extension workers) recruited from the project area were responsible initially for motivating farmers to participate in the project's communication process, and for representing local farming-system experience and attitudes in the project's staff planning process. Later in the project, extension workers assisted and monitored farmer adaptations of agroforestry ideas and practices while field officers coordinated and supervised these activities. Over time, farmers assumed a larger share of the agroforestry promotion function through encouragement by project field staff. The aim of the AEP's staff training component thus was to prepare the dual level extension team to play their respective and overlapping communications and information roles proficiently.

AEP's agroforestry extension training curriculum evolved from training activities initiated during Mazingira's Rural Plots project. A communication component was added by the RAES rural forestry advisor. Diagnosis and design training activities were enhanced by AEP's first field-based manager during a Kathama training exercise with initial AEP field officers. Training in agroforestry seed supply and propagation methods was enriched by MCC's seed supply investigator on the project. In each case training activities were planned and conducted on a "trial and error" basis as most material was untested, but fashioned specifically for project purposes. Based on staff feedback and evaluation by the various trainers, a staff training curriculum was prepared. The curriculum is detailed in CARE's Agroforestry Extension Training Sourcebook (Buck, 1989).

The training programme has been considered highly successful in meeting its skill development objectives by appropriate means (CDP Associates, 1985). AEP field staff continue through the present to exhibit extraordinary planning, analysis, and communication skills relative to common expectations for extension staff performance (Scherr, 1988b; Davis-Case, 1988).

3.6 Land use diagnosis for agroforestry

Assessing the need and potential for agroforestry interventions in Siaya farming systems was a predominant project activity during AEP's first year. Agroforestry diagnosis provided excellent staff training, project learning and farmer motivation values, and laid groundwork for agroforestry design, monitoring, and evaluation activities. The centrality of land-use diagnosis to AEP's extension methodology was a function of senior staff use and adaptation of ICRAF's research-focussed agroforestry diagnosis and design (D&D) methodology.

The D&D was based on state-of-the-art farming systems research experience which departed from previous agricultural research approaches by focusing analysis on households rather than crops. ICRAF's approach went further than farming systems approaches of the day however, in specifying components of the farming system in terms of their role in satisfying basic household level needs. Farming systems and sub-systems were identified in terms of outputs (products) needed to sustain the family and the farm, including food, energy, shelter, cash, water, and raw materials for home industry (Raintree, 1987b).

AEP's project strategy and approach however, called for modifications in ICRAF's diagnostic approach to enhance compatibility and improve its utility for extension project purposes (Buck and Leteipan, 1986; Vonk, 1987). First, AEP placed increased emphasis on identifying existing agroforestry activity and potential in the project environment. Its strategy was to improve existing land-use and agroforestry systems incrementally on the basis of traditional knowledge and practices. This required project information about how trees contribute to farming systems (and subsystems) performance. Whereas ICRAF's method focused the user almost entirely on problems, AEP was similarly concerned with opportunities.

The design and use of diagnostic survey procedures provided excellent staff training value. The logical, step-by-step, problem-focused, and "common-sense" framework of investigation and analysis was conducive to trainee self-discovery about the project environment and to teaching skills in observation, survey design, interviewing and other research methods for project knowledge-building (Buck, 1989). The D&D method used as a training tool resulted in several minor modifications, mainly concerning the sequence of questioning for better farmer response and the rigor of the investigation according to educational and skill levels of the user. The effect was to reduce the formality and length of ICRAF's household problem identification survey (Vonk, 1987).

The scale of analysis was expanded to include landscape (or larger-than-farm level) land units and, within the household, to include gender-specific perspective on land use problems and agroforestry potential, including specific tree uses. The methodology then accounted for differences in women's views regarding household access to on and off-farm resources (Rocheleau, 1985; Feldstein et al., 1987).

Data collection methods were adapted to the sources available. AEP emphasized the use of qualified informants as data sources, through facilitated group interviews. By creating "expert" roles for agency personnel and farmers, they were given a larger stake in project decision-making about agroforestry. Group interviews were used for both pre-diagnosis and diagnosis activities. Well facilitated group interviews produced quality data efficiently, due to the breadth of knowledge and range of experience represented in a single forum. With farmers, the motivation and extension value of diagnostic group interviews was apparent to field and management staff.

Data collection methods were also adapted to the educational and skill levels of project "researchers". Although the number of variables and depth of analysis were reduced from ICRAF's D&D research guidelines for both pre-diagnosis and diagnosis activity, acceptable confidence in the data was assured through (a) long-term staff presence in the project environment, (b) repetition of the group survey with most participation farmers groups (primarily for motivation/extension purposes), (c) follow-up with a sample of on-farm problem identification surveys, and (d) occasional use of technical consultants to examine causes of identified supply problems.

Diagnostic summaries were presented in simple format as a basis for joint staff and farmer participation in designing agroforestry interventions/modifications. A planned adaptation of the diagnostic method for extension project purposes of farmer group interview and household survey data was to be used as baseline information for measuring extension progress (adoption), technology performance and impact. For reasons discussed later however, this use has not materialized. The following outlines specific steps and uses of AEP's adapted version of ICRAF's agroforestry diagnosis Methodology.

(i) Pre-diagnosis (step 1). AEP used secondary data sources and qualified informants to classify land-use and farming systems types within the district; trained field officer level staff in research skills and land use analysis; and identified initial, representative project sites. This activity occurred over a two month period during the project design stage. Information obtained was incorporated into a project area profile for use in later site selection and staff training activities. The profile however, was lost before put to this use.

(ii) Farmer group interview (step 2). This step was not initially part of ICRAF's D&D methodology. Early in the project AEP collaborated with ICRAF to develop a group interview process for community level assessment of land-use problems, production constraints, and landscape and gender specific roles of trees in local farming systems. Staff were trained to facilitate the group interview for maximum participation by all members.

During the first cycle of site specific implementation activity AEP implemented a "Stage 1" group interview, whose primary purpose was for project information and agroforestry design. Later in the project, AEP implemented "Stage 2" group interviews primarily to stimulate farmer awareness of the existence and potential for agroforestry in local farming systems. Stage 2 interviews were used also to identify site-specific variation in land-use and social conditions relevant to agroforestry interventions. The Stage 2 group interview has become the project's principle and probably most effective extension tool.

(iii) Problem identification survey. Several farmers were selected from each group following the group interview, as respondents to the on-farm problem identification survey. Field staff walk farm plots with farmers to observe practices and discuss problems. The result is a household level assessment of most issues addressed during the group interview. The exercise provides information, training, and motivation value in varying proportions, depending on the stage of project development.

(iv) Agroforestry tree inventory. AEP added this step to the diagnostic methodology. An inventory and assessment of local trees and shrubs is conducted for their agroforestry potential. The exercise also includes description of predominant, traditional agroforestry systems. It has highly significant training value and provides excellent information for agroforestry design. As farmers are the primary data sources for the exercise, it also has an awareness-building and motivating effect.

(v) Causal analysis. ICRAF's analysis of on-farm problem identification survey data through "causal diagramming" was not substantially modified by AEP except to increase emphasis on the perceived and potential effect of existing agroforestry practices in linking sub-system supply problems, causes or constraints, and agroforestry opportunities. Again, the training value of this exercise is high, thus it is repeated frequently.

3.7 Agroforestry technology design

AEP's agroforestry technology design process was planned to involve project staff and farmers in overcoming the constraints of limited scientific knowledge about traditional and induced agroforestry systems. The aim of design was not to establish technically optimal interventions within the short term, but to generate "prototype" interventions farmers would find attractive enough to try, and to establish an ongoing monitoring and evaluation system for their gradual improvement and "optimization" based on practicing farmer's performance criteria.

This process was patterned on ICRAF's iterative D&D approach to technology design (Raintree, 1987). Its adaptation for extension project purposes however, involved greater reliance on qualitative data sources, particularly staff and farmer judgement. In the context of AEP, the process was more reliant on information about traditional agroforestry systems in the project environment as a basis for technology design and how these could be fine-tuned for increased productivity.

Technology design was a strategic component of AEP's extension staff training programme, thus training activities contributed substantially to early stages of the design process and outcome. AEP's technology design process included (a) specifying design principles, (b) identifying elements or components of an agroforestry technology, (c) identifying and developing appropriate information sources about agroforestry technologies and potential, (d) developing and using design tools, and (e) testing prototype designs, prior to and after initial establishment.

(i) Design principles. Principles used to guide the AEP design process stemmed from the state-of-the-art of agroforestry, and AEP's approach of participating with local people and organizations to improve land-use systems. Although the concept of agroforestry was substantiated by little scientific evidence, theory and traditional practice were promising. Significant short-term pay-offs of certain interventions and enthusiasm for trees by many Kenyan farmers indicated the likelihood of their cautious involvement in technology adoption and testing. These characteristics of the planning environment suggested several design principles.

First, interventions need to be based on what farmers already know. This does not preclude the option of introducing new plants or practices, but suggests that at least certain elements of an intervention --species, configurations or managements practices-- need to be familiar to promote adoption and facilitate evaluation of performance and effect.

Related to familiarity is the principle of adaptive or incremental change. Land-use changes are likely to be more adoptable and sustainable if made incrementally instead of abruptly. In introducing an unfamiliar feature, such as a hedgerow of an exotic trees species, requiring intensive management, it is wise to establish one row at a time and assess the affect before adopting more.

Third, most project-area farming systems are diverse. This characteristic should be preserved and when possible enhanced to limit financial and nutritional risk, especially to subsistence or near-subsistence farmers. Interventions should include various components and management objectives to address the farmers' several needs.

Lastly, agroforestry most often involves land-use intensification, as pressures or incentives dictate. Intensification may be achieved by creating a more linear pattern to the farming system with rows or hedges, particularly in field crop areas. In garden areas near the home compound, it may involve establishing additional niches through species diversification.

(ii) Design elements. Design elements address what trees and other system components are intended to do or produce for the farming system and what household-objectives they are intended to satisfy (function). This includes questions concerning species of trees/shrubs, crops and/or livestock, and their adaptations to climate, soil, landscape, and other site conditions, as well as regeneration, termite resistance, form, shade, rooting habit, and other physical characteristics (components). What are the spatial arrangements in the farming landscape in relation to other components? How should these species be managed (management)?

(iii) Design information. Information available to the project from formal technical/scientific sources was limited. In particular, few findings were available from on-station or on-farm agroforestry experimentation. Accordingly, the project relied primarily on AEP staff evaluation of existing agroforestry practices and components, and review of popular literature on the theory and potential of agroforestry.

Prototype trials involving farmer testing and evaluation at AEP's predecessor Mazingira and Kathama project sites were another important information source on technology design. Monitoring and evaluation of AEP design prototypes was planned to provide more robust information for future design activities. Farmer trials with prototype designs were to be complemented by a more formal research component. This was to be used to measure selected performance variables on some of the most promising yet potentially risky interventions.

(iv) Design tools. A number of design tools were used by project staff to organize and evaluate agroforestry information for designing AEP interventions. These were borrowed or adapted from the Mazingira's rural plots project. They are illustrated in CARE's Agroforestry Extension Training Sourcebook (Buck, 1989).

(v) AEP prototype designs. Using the guidelines and information, AEP field and management staff created an initial set of prototype, or "best-bet" designs for nine initial project sites representing Siaya's four agro-climatic zones. Farm-specific intervention plans were evaluated and adapted with participating groups and individual farmers prior to planting.

Following planting, designs were evaluated by extension staff. Unfortunately, files containing the lists and descriptions of these initial prototypes, as well as the post-planting evaluation data, were lost from project records during the office move which accompanied AEP's first major expansion. Furthermore, this process was not systematically continued beyond establishment of AEP's initial nine project sites.

In later iterations of AEP agroforestry site development, the more common approach to design involved field staff assistance to farmers in considering design elements and characteristics of the various species available. Farmers were then encouraged to "design" and adapt interventions by their own methods. This non-prescriptive approach undoubtedly resulted in more variable configurations and practices, probably too variable to be efficiently managed for learning purposes.

AEP's approach to technology design and promotion following the first iteration conducted as part of initial staff training activity is indicated in the project literature's progression of intervention descriptions. Initially, six basic features were used to define the project's prototype designs. Each was variable according to species and management recommendations for meeting an inventory management recommendation (Buck, 1985).

AEP's mid-term external evaluation team, six months later, felt the preceding inventory did not adequately represent the enormous diversity of actual practices. Rather than classify these into specific design types, the evaluators prepared a matrix of design elements in an effort to catalogue their respective frequencies (CDP Associates, 1986). A half year later AEP shared its technology design experience in a training forum on the basis of a traditional versus modern classification scheme. Here the number of prototypes described was substantially collapsed relative to preceding records (Vonk and Chitechi, 1986).

During case-study research by the CARE/Food and Agriculture Organization Agroforestry Monitoring and Evaluation Methodology Programme (AFMEMP) in 1987, the investigator was unable to trace a progression of learning or experience with prototype designs due partially to inconsistent classification schemes. Iterative consultation with field staff finally resulted in a list of 19 interventions ranging from highly specific to general regarding specification of species and configurations. Soon afterward ICRAF scientists assisting AEP conducted a field staff workshop and pilot survey to classify agroforestry interventions (Scherr, 1988a). Thirty-three were identified. The exercise resulted in a project commitment to conduct a comprehensive inventory of interventions as a necessary starting point for assessing technology performance, and ultimately impact (AFMEMP, 1988).

The foregoing may be an inevitable outcome of AEP's deference to farmer judgement in the agroforestry "design" process. Although responsive to farmer enthusiasm, this laissez faire approach to date has limited project-level learning from the design process and probably the rate of technology advancement. The approach also raises the issue of "risk" impact on farmers.

3.8 Monitoring and evaluation

AEP's monitoring and evaluation component was planned to support the pilot, technology, and extension-testing characteristics of the project. Its primary purpose was to focus the learning required to improve staff and farmer confidence in recommending and implementing new practices, and in planning and implementing project activities. A second purpose was to generate information needed for periodic internal and external assessments of progress toward project goal achievement.

Evaluating goal achievement was done through an iterative, three-stage process to (a) measure programme impact (social, institutional and technical change) based on attainment of final goals, (b) measure programme progress based on attainment of project activity targets and progress indicators for project goals, and (c) measure satisfaction of programme expectations by collaborators and participants.

Impact and progress indicators were required to obtain and interpret these measures. Development of the project's monitoring and evaluation plan constituted a three-step process. First, impact indicators and means of measurement were identified. Second, progress indicators and means of measurement were identified. Finally, measurement tools and training or advising appropriate for participants were created.

The project learning objective for AEP's monitoring and evaluation system required staff skill and involvement in documenting and analyzing project activity achievement. It also required an understanding of the relationship between activity achievement and progress toward project objectives. Thus, initial staff training for internal monitoring and evaluation focused on the logic of the linkages between AEP's final goal, intermediate goals and activities; the value of staff participation in project monitoring and evaluation activity, particularly the potential for improved extension activities and recommendations; and what to measure, and how to identify appropriate indicators and tools.

To answer the latter question, a monitoring and evaluation planning workshop was undertaken by the project manager, two ICRAF scientific (D&D) advisors, and the project designer to consider appropriate indicators and measurement methods for assessing progress and impact. The key challenge faced by this monitoring and evaluation methodology planning group was the virtual absence of technology and extension performance standards.

The pilot nature of AEP extension activities and technical recommendations meant that established or standard performance and achievement measures were not available. Therefore, the adequacy of desired social, institutional, and technical change resulting from project influence was based largely on the judgement of those concerned in response to questions such as: Are things getting better or worse? How do you know, or what convinces you? What role has the project played in bringing about these changes?

A complementary strategy involved formal research to reduce uncertainty about performance standards and expectations. Given its relatively high cost, questions answerable by this mode needed to be carefully selected for their strategic utility to project learning and accountability.

Preparation of the plan was followed by recruitment of a social-research supervisor for the project staff, a social-research consultant to design the baseline survey, and a technical-research advisor to design the research plots. In addition, a monitoring and evaluation consultant was hired to review and improve the use of ongoing monitoring and evaluation tools, particularly field notebooks, nursery and seed supply manuals, and prototype design follow-up forms. This activity indicated AEP's initially strong commitment to fulfilling its knowledge-building aims.

The project's monitoring and evaluation plan however, has resulted in minimal output, or project learning about technology adoption, performance, or impact. Several factors were responsible. First, AEP's ambitious research and evaluation plan was not adequately incorporated into the project's implementation plan in the form of activity targets or goal indicators. Thus, monitoring and evaluation activities fell outside of day-to-day project management concerns. Neither in the monitoring and evaluation plan nor in the project's trimester and annual implementation plan, CARE's formal internal planning documents, were linkages specified between outputs of the monitoring component and ongoing planning, training, or field implementation events. Thus there was little incentive to link decisions about expansion, intensification, extension strategies, or technology recommendations to existing experience. Instead, the significant work load specified by the plan assumed more an accountability than a learning function, which in itself limited staff motivation for the activities.

Secondly, large quantities of data which began to be generated could not be assimilated by project information management devices. AEP first acquired a computer soon after the research and evaluation plan was developed. However, staff were not adequately familiar with appropriate programs nor was data management activity allocated sufficient time to produce useful information on a timely basis.

Thirdly, AEP's choice of baseline survey consultant was unfortunate. Lacking training in rapid rural appraisal methods or the use of minimal data sets to complement staff knowledge about the project environment, the consultant used a classical, social anthropological approach to developing a baseline questionnaire. Further, the consultant did not link the baseline survey directly to D&D findings from group interviews and on-farm surveys. In reality, this material itself may have provided sufficient baseline information (Gibson and Mueller, 1987).

Because the survey approach was inappropriate for intended uses, as well as the manpower and data handling tools available, usable information never materialized. Instead, a large proportion of project resources was devoted to the consultant and his survey, diverting these from more productive uses. In addition, the project experienced a computer problem during baseline survey data entry. The demoralization which followed prevented re-entry of the data.

Fourthly, as monitoring and evaluation activity intensified from various consultant and staff inputs, AEP undertook a major expansion of project extension activity. Farmer enthusiasm and demand provided the rationale for dramatically increasing the number of schools and women's groups the project would assist in establishing nurseries and planting trees during AEP's pilot stage. Farm-focused technology monitoring and evaluation planned for a project assisting up to 33 nursery groups became unmanageable following a sudden increase to 87 and then 351 groups, spread over two large districts. Nursery establishment and tree production suddenly became the overwhelming project priority and success indicator.

A fifth factor affecting project learning was that AEP's manager was transferred prior to reconciliation of these competing project orientations. He was not replaced with a manager of similar professional interest and background. Instead CARE divided management responsibility for AEP between the previous assistant manager and a newly appointed community development programme coordinator. The revised senior staffing strategy was part of CARE-Kenya's structural reorganization aimed to de-emphasize sector programming. As AEP's research and evaluation component had not become well established prior to this influential change, it was fated to a minimal priority status.

Finally, researcher-managed experimental agroforestry plots, aimed to examine the effects of hedgerow intercropping (alley-cropping) with selected species on soil fertility in respective agro-climatic zones, was the element of AEP's original research and evaluation plan maintained throughout the project. The plots are managed by the Kenya Forestry Research Institute (KEFRI) through an agreement with CARE, which provides for CARE financial assistance and specifies the production of periodic reports on research findings (KEFRI, 1987). For practical purposes, this activity became virtually distinct from extension activity following departure of AEP's agroforestry-trained manager.

Lacking regular input from technology-monitoring activities, field staff base their technology recommendations to farmers on initial project training combined with their personal experience and judgement. The specified project strategy for prototype design improvement remains "let the farmer take the lead". Consequently, many field staff feel inadequately prepared to respond to farmers' questions and concerns regarding technology management and performance (Scherr, 1988c; Davis-Case, 1988).

Despite minimal performance to date in the use of ongoing, project monitoring, evaluation, and research tools, sight has not been lost of their potential value. AEP staff involvement in AFMEMP has raised interest in reviving original monitoring and evaluation tools and initiating new ones (AFMEMP, 1987).

The strongest element remaining of the project's early commitment to improving agroforestry technologies and disseminating this information through a sustainable agroforestry extension network are AEP field staff trained in this approach. AFMEMP's and ICRAF's recent efforts to assess the status and potential for improving AEP's monitoring, evaluation, and research component reveals that staff are keen observers, critical thinkers, good two-way communicators eager to share their ideas and experience and genuinely interested in their work (Scherr, 1988c; Davis-Case, 1988).

Accordingly, staff are themselves good sources of data, provided appropriate methods of extracting it are employed. With minor additional training, they can obtain reliable data on technology performance and impact from farmers. With this asset, the gap between AEP's potential to assume an agroforestry technology generation and institutionalization role and its comparatively limited accomplishments to date could be readily overcome given the institutional commitment and management expertise. On the other hand, AEP may choose to continue on its present course. Although this limits full performance, it would not negate important lessons learned about planning a viable agroforestry extension project. On the basis of adoption criteria, fundamental to extension success, the project works.

Furthermore, continued collaboration with ICRAF in its year-old impact-assessment activity can result in vastly improved knowledge about the technology and farming system effects of project extension activity to date. Although this activity is externally-guided, its basis on familiar D&D methods should preserve AEP's option to incorporate ICRAF input into a viable internal monitoring and evaluation system. With AFMEMP assistance, and depending on its interpretation of AEP's recent donor-sponsored external evaluation, this option may soon materialize as the project focus returns from growth targets to a more informed approach to agroforestry extension.

4 SUMMARY OF STRENGTHS, LIMITATIONS, AND FUTURE IMPLICATIONS

CARE's Agroforestry Extension Project was designed with "a leg up" toward ensuring satisfactory agroforestry adoption rates. It benefited from the experience of two predecessor projects which suggested a project strategy and approach for influencing Kenyan farmers to establish innovative agroforestry prototype interventions. AEP's design process focused on meeting these conditions.

The underlying premise of AEP's design was that satisfaction of these conditions could help achieve a higher order extension activity in which uniquely trained field staff would participate with self-motivated farmers to improve agroforestry technologies using a prototype testing approach. Through a system of ongoing, internal monitoring and evaluation among farmers and AEP's dual-level field staff, a set of agroforestry technology recommendations would develop which were understood and promoted by a district-wide formal and informal agroforestry extension network. Joint training and planning among organizations involved in technical assistance to rural communities would provide the basis for this agroforestry institutionalization process.

The methodological framework for guiding the agroforestry extension and testing process was an adapted version of ICRAF's D&D approach to agroforestry research. Initial AEP training and planning activities, as well as its predecessor projects, provided the context for modifying D&D for extension project purposes. The resulting extension modified D&D methods have been used to promote well-informed staff and farmer involvement in the design of adoptable agroforestry interventions.

The method of project design itself was instrumental to AEP's early success. Influential conditions and characteristics included:

(i) Adequate time. Many aspects of AEP's design were innovative to CARE programming. The lack of precedent, plus the need to collaborate with potential project partners about needs and opportunities in the project environment, and an appropriate strategy and approach for addressing them required an unprecedented project design period. CARE leadership in Kenya during AEP's design phase provided such support.

(ii) Pilot phase and rolling planning process. Project design activities continued following project funding, as a function of the rolling planning process established during the design phase. This process allowed for continual testing of assumptions with collaborators, staff, and participants about keys to project effectiveness prior to committing to specific objectives and activities. A well defined strategy and approach based on prior project experience guided the rolling planning process and ongoing "re-design". Designation of AEP's first full funding cycle (three years) as a pilot stage provided flexibility in planning and accounting for project activity needed to innovate and adapt as conditions in the project environment warranted.

(iii) Flexible management style AEP's first mid-term evaluation one and one half years after project funding attributed AEP's early success in large part to a flexible leadership style exhibited by project management. This characteristic combined with unique technical training in agroforestry (for that time) were instrumental in advancing AEP's ambitious methodological and local impact objectives.

(iv) Extension-modified D&D. This provided an appropriate analytical framework for assessing problems and opportunities in the project environment, beginning with the use of pre-diagnosis information gathering guidelines for project site selection.

Were adoptability the sole criteria for agroforestry success, the agroforestry extension programming challenge would have been virtually met by AEP. At least 23 types of interventions and 36 different species have been adopted by some 100,000 farming households and probably many more in five years. Farmer demand for additional assistance remains beyond project capability to respond and the Kenya Forest Department attempts to address the need.

The productivity of agroforestry interventions, their immediate and long-term effect on the larger farming systems and the potential for district-level institutional effectiveness and sustainability in extending agroforestry knowledge and inputs remain untested. AEP was designed to create reliable information on these aspects of agroforestry as well and to use it to establish a prototype district-level agroforestry extension scheme for repeatability by the collaborating Forest Department.

Several related factors have so far limited realization of these aims. First, AEP's initial mid-term evaluation applauded the project's achievements for creating an appropriately guided participatory development process resulting in impressive adoption rates. Rather than exploiting this achievement as a springboard for agroforestry technology testing and improvement as planned, however, evaluators recommended reduced sector emphasis in favour of better integration with other programming sectors. They also recommended expansion to meet project-induced farmer demand for tree-planting assistance. Compared with other CARE projects in Kenya, as well as agroforestry initiatives by other organizations, AEP was already overwhelmingly successful, using extension effectiveness as the criteria. AEP's knowledge-generating aims were not considered equally important.

Second, AEP's pilot phase activities and leadership were not continued long enough to develop a convincing record about the value of internal monitoring and evaluation. Thus predominant institutional biases toward extension activity and evaluation criteria were strongly favored by AEP's second stage of organization and leadership.

Third, the monitoring, evaluation, and research plan for meeting AEP's technology generation and institutionalization objectives was too ambitious for project resources to manage. The use of tested and promising monitoring and evaluation tools for AEP's extension context was overshadowed by the inappropriate information gathering and analysis techniques.

Fourth, the project decision to expand its nursery and tree planting extension services resulted in much reduced interaction among project staff due to logistical and associated cost factors. Thus internal staff planning and training activities were substantially reduced, limiting the communication and learning so fundamental to the project design.

Finally, in the absence of leadership and support for systematic, prototypic agroforestry technology testing, field staff relied on their own best judgement to guide their recommendations. By admission, certain field staff recommended their favorite prototypes with the confidence that they are tested. Others are less prescriptive, and some leave all choice to the farmer. The highly variable effect is predictable, but the responsibility of such an approach is debatable.

Conditions appear to be ripening for a restoration of AEP's technology improvement and internal impact assessment objectives. There are indications from the agroforestry research community of an emerging new focus on research and extension linkages for technology improvement and impact. Not only may AEP provide the context for pioneering and refining methods which extension staff can manage for these purposes, but AEP adoption of these should limit the problems it experienced during its initial attempts to collect and manage appropriate information affordably.

In the meantime, CARE-supported projects throughout eastern Africa have adopted agroforestry planning methods developed by the Kenya Agroforestry Extension Project. AEP's strategy and approach as well as its training, diagnosis and design methods have been shared through senior staff training workshops, CARE's Agroforestry Extension Training Sourcebook, its agriculture and natural resources sector strategy document and project visitation by CARE's regional technical advisor in Eastern Africa. Increasingly the linkage of extension and research, which AEP was designed to achieve, needs to represent state-of-the-art agroforestry programming within CARE's expanding sphere of influence in Eastern Africa.

6 REFERENCES

AFMEMP, 1987. Report on Kenya Country Workshop, December 9-12, CARE/FAO, Nairobi.

AFMEMP, 1988. Correspondence between the research and training team of the Agroforestry Monitoring and Evaluation Methodology Programme and CARE's Agroforestry Extension Project, December, Nairobi.

Beijer Institute, 1982. Energy development in Kenya: problems and opportunities. Draft Discussion Document for consideration by the Steering Committee for the Kenyan Fuel wood Project, Stockholm.

Buck, L.E. 1981a. (Editor) Proceeding of the Kenya National Seminar on Agroforestry. University of Nairobi and ICRAF, ICRAF, Nairobi.

Buck, L.E. 1981b. Agroforestry plots for rural Kenya project: preparation phase report and modified work plan for phases In Burley, J. and Von Carlowitz, P. (Editors.) Multipurpose Tree Germplasm, Proceedings of a Planning Workshop to Discuss International Cooperation, ICRAF, Nairobi.

Buck, L.E., 1983a. Kenya agroforestry tree seed project report, ICRAF (Working Paper No. 4). ICRAF, Nairobi.

Buck, L.E., 1983b. Agroforestry extension project multi-year plan, FY 1984-86, CARE, Nairobi.

Buck, L.E., 1985. Tree management guidelines for Siaya agroforestry interventions: CARE-Kenya Agroforestry Extension Project, Consultant's Report for CARE-Kenya, August, Nairobi.

Buck, L.E., 1986. Toward a Kenya national on-farm forestry extension strategy, The World Bank. Kenya Forestry Sub-Sector Review, Nairobi. In J. Burley and P. Von Carlowitz (Editors), Multipurpose Tree Germplasm, Proceedings of a Planning Workshop to Discuss International Cooperation, ICRAF, Nairobi.

Buck, L.E., 1988. Personal observations as Agroforestry Advisor for CARE in Eastern Africa, 1985-1988.

Buck, L.E., 1989. Agroforestry Extension Training Sourcebook, CARE International, Nairobi, Kenya, 350 pp.

Buck, L.E., and Alitizi, A., 1983. Agroforestry plots for rural Kenya project, progress report on phase II. Mazingira Institute, Nairobi.

Buck, L.E. and Leteipan, D. 1986. Agroforestry problem diagnosis. Paper presented at the Training Workshop of Ugandan District Forest Officers in Siaya, Kenya July 27-August 8, 1986, CARE International in Kenya, Kisumu.

Buck, L.E., Davis-Case, D., Mueller, E., Ngugi, A., 1988. A synthesis of experience from the Agroforestry Monitoring and Evaluation Methodology Programme. Paper presented at AFMEMP Regional Workshop, Kisumu, 25 pp.

CARE-International, 1984. Draft programming principles presented for discussion at CARE's Eastern Africa Conference for Country Directors, Nairobi.

CARE-International, 1988. Agriculture and natural resources sector strategy, New York, draft, 22 pp.

CARE-Kenya, 1983. Multi-Year Mission Planning Document, 1983-1988, Nairobi.

CDP Associates, 1986. An evaluation of CARE International in Kenya's Agroforestry and Energy Project, Final Report, September, Nairobi.
Davis-Case, D., 1988. A Case Study of CARE Agroforestry Extension Project. CARE/FAO AFMEMP Programme, Nairobi, draft, 70 pp.
Feldstein, H., Rocheleau, D.E., and Buck, L.E., 1987. Intra-household decision-making in agroforestry diagnosis and design: the role of gender in planning interventions for CARE-Kenya's Agroforestry Extension Project. Population Resources Council, New York.
Getahun, A., 1988. A review of agroforestry development in Kenya, ICRAF, Nairobi.
Gibson, D.C. and Mueller, E.U., 1987. Diagnostic surveys and managements information systems in agroforestry project implementation: a case study from Rwanda. (Working Paper No. 49). ICRAF, Nairobi.
Huxley, P.A., 1987. Recorded comments during an informal ICRAF evaluation of AEP's on-station and on-farm agroforestry research potential for ICRAF's Agroforestry Development in Kenya Project, Kisumu.
KEFRI, 1987. A report on the alley cropping trials, prepared for the CARE-Kenya Agroforestry Extension Project, Forestry Research Institute, Nairobi.
Ndegwa, P., 1982. Report and recommendations of the Working Party on Government Expenditures Appointed by His Excellency the President, Presented July, 1982, Nairobi.
Nyamai, D., 1988. Tree survival count-CARE Agroforestry Extension Project (a survey conducted in Siaya and South Nyanza Districts), KEFRI, Nairobi.
Raintree, J.B., 1981. Preliminary diagnosis of agroforestry related land use constraints and potentials in the eastern Kanzalu range of Machakos. In-house working paper. ICRAF, Nairobi.
Raintree, J.B., 1982. A methodology for diagnosis and design of agroforestry land management systems. In-house working paper, ICRAF, Nairobi.
Raintree, J.B., 1983. Preliminary diagnosis of land use problems and agroforestry potentials in northern Mbere Division, Embu District, Kenya. (Working Paper No. 1). ICRAF, Nairobi
Raintree, J.B., 1987a. A user-friendly information management system for participatory agroforestry research and extension projects. In-house working paper and proposal. ICRAF, Nairobi, 11 pp.
Raintree, J.B., 1987b. The state of the art of agroforestry diagnosis and design. Agroforestry Systems, Vol.5, NO.39: 219-251.
Raintree, J.B., 1988a. Appropriate R&D support for forestry extension. Paper presented to the FAO Expert Consultation on Organization of Forestry Extension, March 7-11, Bangkok, Thailand, 28 pp.
Raintree, J.B., 1988b. Factors to consider in developing an information system for agroforestry extension projects. Paper presented at the AFMEMP Regional Workshop, May 15-20, Kisumu, 11 pp.
Rocheleau, D.E., 1985. Criteria for re-appraisal and redesign: intrahousehold and between household aspects of FSRE in three Kenyan agroforestry projects. (Working Paper No. 37). ICRAF. Nairobi.
Scherr, S.J., 1988a. Pilot survey of adopted agroforestry practices in the CARE Agroforestry Extension Project. (ICRAF-CARE Project Report No. 6). ICRAF, Nairobi, 34 pp.
Scherr, S.J., 1988b. Current agroforestry practices and extension recommendations of the CARE Agroforestry Extension Project. (ICRAF-CARE Project Report No. 8). ICRAF, Nairobi, 58 pp.
Scherr, S.J., 1988c. Personal communication in relation to planning for an ICRAF/CARE Workshop, ICRAF, Nairobi.
Scherr, S.J., 1989. The "D&D" approach to agroforestry project planning and implementation: examples from western Kenya, Paper for the Washington State University/University of Wageningen Symposium on Planning Agroforestry Projects, April 26-30, Pullman, Washington.

Teel, W., 1984. Tree Seeds in Kenya, KENGO Nairobi.
Van Domelyn, J. 1986. NGO's in tree planting: background paper for Kenya Forestry Subsector Review, The World Bank, Nairobi.
Van Schaik, M. 1986. The role of Markamia platycalyx in the farming system in Siaya District, Nyanza Province, Kenya: a qualitative analysis. Unpublished draft report, Wageningen, The Netherlands.
Vonk, R.B., 1984a. Thesis presented in partial fulfillment of the requirements for a MS in tropical agriculture, Wageningen University, Wageningen.
Vonk, R.B., 1984b. Research and Evaluation Proposal for the CARE Kenya Agroforestry Extension Project, Nairobi.
Vonk, R.B., 1986. Agroforestry Extension Project Multi-Year Plan, FY 1987-FY 1989, CARE, Nairobi.
Vonk, R.B. and Koech, L., 1986. Background Information of Project Area, Paper presented at the East Africa Regional Agroforestry Training and Planning Workshop held in Kisumu-Kenya February 3-12, 1986.
Vonk, R.B. and Chitechi, P., 1986. Participatory Agroforestry Design, Paper presented at the Training Workshop of Ugandan District Forest Officers in Siaya, Kenya July 27-August 8, 1986, CARE International in Kenya, Kisumu.
Vonk, R.B., 1987. Indigenous agroforestry systems: a case study from the CARE-Kenya Agroforestry Extension Programme, Text of a lecture presented at the Department of Forestry and Resource Management, University of California, Berkeley, 20 pp.

Chapter 6

THE DIAGNOSIS AND DESIGN APPROACH TO AGROFORESTRY PROJECT PLANNING AND IMPLEMENTATION: EXAMPLES FROM WESTERN KENYA

S. J. SCHERR
Economist (Senior Scientist), International Council for Research in Agroforestry (ICRAF), Nairobi (Kenya)

ABSTRACT

The diagnosis and design (D&D) approach to developing appropriate agroforestry technologies begins with land-user knowledge, practice, and objectives. Through "diagnosis" of the land-use system, agroforestry options are identified. Development of suitable technology designs--which specify components, arrangements, and management for particular users, sites, and functions--involves review of existing information, then experimental research, and field monitoring of farmer trials to generate "missing" design information. Proposed technology designs are validated through on-farm testing. Once validated designs are disseminated, farmers adopt and adapt them, leading to new problems and opportunities for agroforestry technology re-diagnosis and re-design.

The process is illustrated for two field projects in western Kenya working to improve designs for farmer alley-cropping technologies. The ICRAF-Government of Kenya AFRENA project emphasizes experimental research. The ICRAF-CARE/Kenya project emphasizes field monitoring and evaluation methods. A model for technology development is proposed which involves integrated inputs from farmers, extensionists, and researchers. New criteria for agroforestry project planning are recommended, based on the D&D approach.

INTRODUCTION

The objective of agroforestry diagnosis and design (D&D) is to develop agroforestry technologies whose design is tailored to the needs and conditions of the land users (Raintree, 1987). Raintree (Chapter 2) emphasized the initial process of "diagnosis", or identification of promising agroforestry technologies for a particular land-use system. "Technology" refers here to a type of agroforestry practice, such as fodder banks, living fences, tree lines in cropland, alley-cropping, or multistrata home gardens. This chapter emphasizes the subsequent process of agroforestry technology "design" and the different but complementary roles of land users, extensionists, and researchers in continually improving designs from the users' perspective.

"Design" corresponds here to a specific type of "intervention" in extensionists' terms. A complete agroforestry design identifies multipurpose trees and shrub (MPT) varieties and associated non-woody components (crops/grasses/animals), arranged and managed in specific ways to meet defined land-user objectives in defined types of sites. A design embodies an integrated set of principles about tree and crop growth patterns and land management in a particular environment. It is probably neither feasible nor desirable in agroforestry to encourage adoption of a set of standard designs, "extension packages", to farmers, because of the extreme variability of site, objective, component combinations, and management criteria. Rather, extension can provide information about these principles to farmers, by demonstrating and explaining how such prototype designs can work, and under what range of design variation the principles will continue to hold. The farmers can then custom-design their own agroforestry practices.

The major technical challenge in agroforestry development today is the lack of ready, validated designs which can be disseminated with confidence to land users, and the lack of reliable technical information on components and component management with which to develop more general agroforestry design principles.

This situation has not prevented --and should not prevent-- continued international encouragement of agroforestry in rural landscapes, to meet pressing needs for tree products (e.g., fuel wood, building materials, leaf fodder) and services (e.g., shade, wind protection, soil protection). It does, however, pose a major challenge to the institutions involved. Agroforestry researchers <u>must</u> focus their modest resources on technical research which addresses priority information gaps in technology design for defined land-use systems--gaps identified by farmers and extensionists, rather than scientific curiosity alone. Agroforestry extensionists <u>must</u> take responsibility for the quality of the information they disseminate, critically evaluating the performance of their recommendations and exploiting their close contact with farmers to encourage, develop, test, and document innovations.

Such a mandate has important implications for planning agroforestry extension or research projects, for it implies a blurring of the commonly accepted division of labor, and a modification of conventional project planning. In this chapter, the author addresses extensionists and researchers together, encouraging them to consider their potentially complementary and preferably linked roles in the techology design process. A strategic framework for technology design is discussed, and then two case studies from western Kenya--one an extension project and the other a research project--are presented to illustrate the application of the model, and some lessons for agroforestry project planning.

2 A STRATEGY FOR AGROFORESTRY TECHNOLOGY GENERATION

The proposed model for agroforestry technology generation has the following sequence: (a) land-user knowledge and practices, (b) a common definition of land-use system needs and promising agroforestry interventions, (c) development of technology designs, and (d) validation of technology designs. With large-scale farmer adoption and adaptation of new technologies, new problems and opportunities arise, leading to a new cycle of re-diagnosis and re-design (see Fig. 1).

2.1 Land-user knowledge, practices, and objectives

The starting point of the process is the land-use system--a farming system, but more broadly defined to include rangeland, forest, etc., and the possibility of multiple users (e.g., herders, farmers, landless). Any technology design activity should start with existing local knowledge and practice in land husbandry and agroforestry, and with land users' own objectives in mind.

2.2 Identifying land-user needs and possible agroforestry systems

The land-use system is evaluated systematically through a diagnostic procedure whereby project planners join their outside technical knowledge with farmers' local knowledge to identify priority problems and development opportunities which the introduction of new tree components or management systems could help to address. Priority functions for trees or shrubs for land users are identified, for specific landscape niches. These determine the range of agroforestry technologies which would be applicable to that land-use system. Other bio-physical, socio-economic, and organizational factors further narrow the list of relevant agroforestry technologies (Raintree, 1989).

2.3 Developing technology designs

After diagnosis and selection of relevant technologies, the process of developing specific technology designs begins. Design information which should be specified includes: site suitability, functions which the technology can fulfill, woody and non-woody components, spatial and temporal arrangements of components, management practices for woody and non-woody components, expected performance in terms of production and service

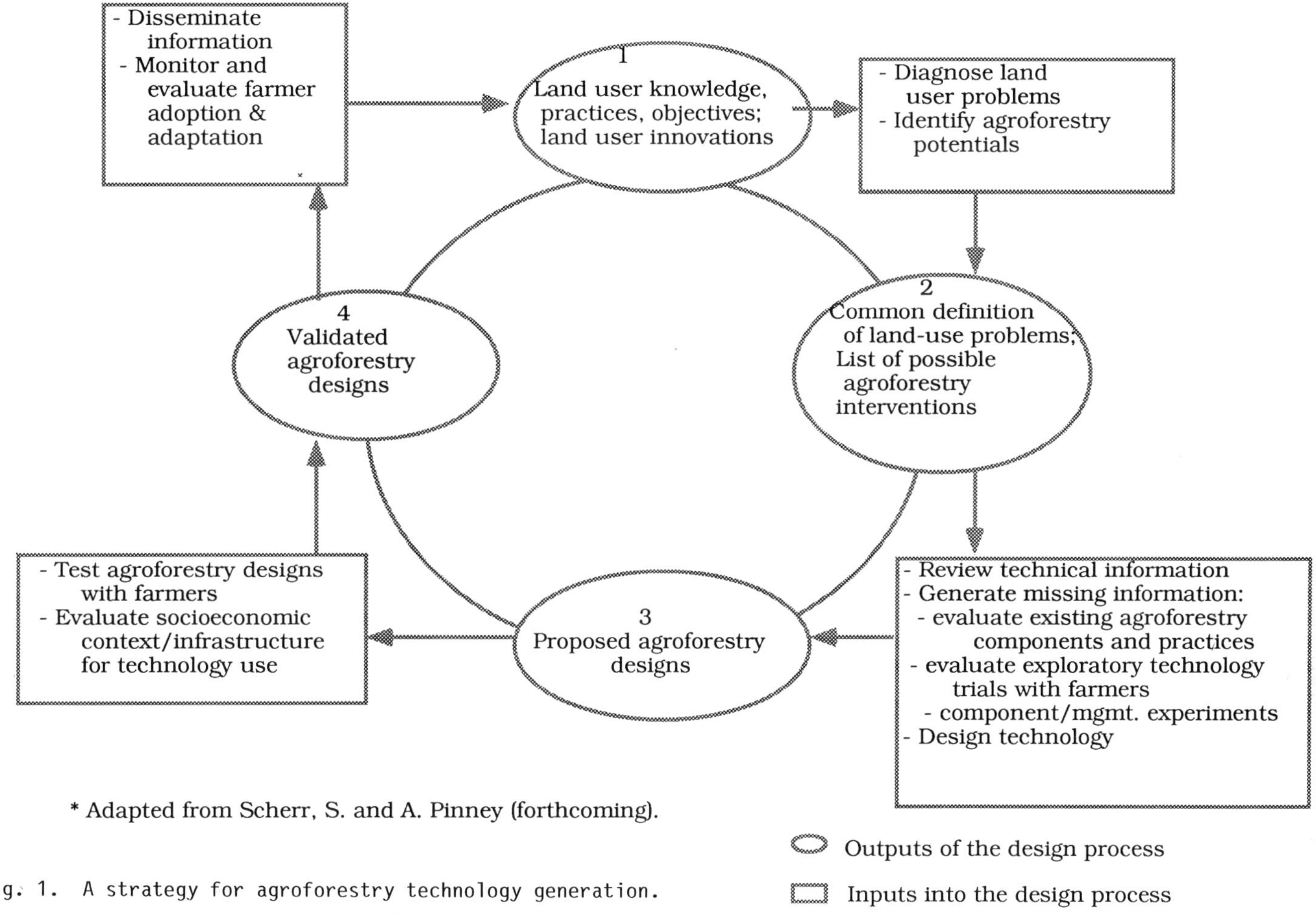

* Adapted from Scherr, S. and A. Pinney (forthcoming).

Fig. 1. A strategy for agroforestry technology generation.

functions, and required inputs for effective management. Such complete information is available today only for a few well-studied agroforestry systems, such as coffee under shade trees.

An attempt is first made to develop an appropriate design from existing scientific and technical information about components and designs used in similar sites for similar functions. Three basic outcomes are possible from this review.

(i) Type 1 technology. The review may identify designs "off the shelf" which are proven or likely to be biologically and economically viable under local land use conditions. Recommendations based on these designs could be initiated with farmers as soon as their relevance was confirmed by local validation trials. Not many technologies fall into this category.

(ii) Type 2 technology. The review may provide substantial design information, but with certain key information missing or clearly not adapted to local specifications. Some adaptive research activities will be necessary prior to large-scale dissemination to farmers, but these would commonly involve substantial collaboration and trials with a pilot group of land users.

(iii) Type 3 technology. The review may identify so many design gaps that project planners would not feel confident introducing the technology to farmers on more than a limited trial basis. In such cases, a substantial research effort may be needed prior to extension, involving a smaller number of land users.

Even an agroforestry design "proven" to be effective (i.e., to work as well or better than alternative systems available to land users), however, is almost certainly performing under potential in terms of meeting users' objectives. The dearth of past indigenous or scientific efforts to enhance tree productivity means that there is significant potential for improvement with new or genetically improved MPT species or varieties, or through more intensive or innovative management regimes. Thus, even in the case of Type 1 or Type 2 technologies, there can be a valuable role played by systematic research.

Where key design information gaps are identified, a number of different methods can be used to obtain it, depending on the type and quality of information needed.

(i) Basic/applied research on components and management. These are experimental trials or bio-physical evaluation of existing agroforestry plots carried out under controlled conditions, to explore the relationships between different variables in plant growth and productivity. The objective is to generate basic information relevant over a relatively wide range of land-user conditions. It is most commonly undertaken by scientists.

(ii) Monitoring and evaluation of on-farm agroforestry. This involves studies of farmer use, adoption and adaptation of agroforestry designs. The objective is to capture information generated by farmers about agroforestry design, including performance, preferences, and innovations. It is most commonly undertaken by extensionists.

(iii) Adaptive component/management research. These are experimental trials or bio-economic evaluations of existing agroforestry plots undertaken to find out how certain components and designs can be expected to perform under location-specific conditions. The objective is to develop design recommendations for a specific set of land users. Adaptive research may be undertaken in an adaptive research project, or by researchers or technically competent extensionists in an extension project.

Information of all three types is valuable in developing a specific set of agroforestry technology designs. These "best bets" should integrate available information to identify the combinations of components, arrangements and management practices most relevant for the land users concerned. This role requires effective skill in integrating technical and land-use system information in a practical way.

2.4 Validating agroforestry designs

Many extension and research projects release their recommendations once specific technology designs have been proposed, without explicit testing of the integrated designs. Validation is commonly ignored in agroforestry extension programmes, due to a perception of formal testing as a "research" activity not appropriate to extensionists, project constraints which encourage early large-scale extension, and/or a presumption that the mere establishment of new trees is necessarily beneficial to farmers. Validation is commonly ignored in agroforestry research programmes, because the question "does this design work better here than the other one?" is seen to generate too little extrapolable scientific information at too great a cost (particularly if done on-farm) and/or because researchers see this as the responsibility of extension.

In fact, however, validation is an essential part of the technology design process, and the step most likely to encourage good initial design. Validation consists of evaluating designs objectively in terms of their biophysical and socio-economic performance in contributing to farmers' objectives. Interventions should be clearly beneficial to farmers, and offer clear advantages over existing alternatives.

Thus, technology testing is complementary to experimentation. Discrete and significantly different alternative designs and principles for land management are compared in technology testing. Experiments, by contrast, permit evaluation of yield responses to varying levels of particular parameters and can be used to identify the most effective range of parameters for technology design.

Proposed designs can be tested on research or demonstration plots, but direct testing by land users is preferred, as this offers greater opportunity for farmer judgement and a more realistic situation. Tests should allow a critical comparison between the proposed design, and control or comparison plots using existing farmer practice or alternative agroforestry or non-agroforestry practices. For reliable validation, they need to be tested over a number of sites, and over a period of years. One of the priority tasks for researchers is the development of streamlined methods and assessment criteria for technology validation trials, appropriate for use in adaptive research and extension programmes.

Some level of validation should precede large-scale dissemination, but for low-risk interventions, well-monitored pilot dissemination can help both to refine designs and also extension materials and strategies. Unlike annual crops technologies, validation can be a lengthy process, and it is unreasonable to postpone all action until complete. Another part of the validation process is exploration of key factors in the socio-economic and institutional environment without which widespread farmer adoption would be unlikely. Examples might be distribution networks for tree seed and other key inputs, resources for farmer training in the case of more complex technologies, or markets for tree products in the case of commercially oriented technologies.

2.5 Re-diagnosis and re-design

Once a new agroforestry design has been disseminated or spontaneously diffused through a group of land users, the natural process of farmer adaptation begins. At the same time, new technical problems (e.g., pest damage or poor performance in particular sites) or new opportunities (a new market for leaf fodder or growing demand for poles) may emerge which require or suggest modifications in the design. This can in turn initiate a new cycle of "re-diagnosis and re-design", which further strengthens the knowledge base of land users to achieve important objectives through agroforestry. The application of this process in research and extension will be demonstrated in the following two case studies.

3 D&D IN AGROFORESTRY RESEARCH: THE KENYA-AFRENA PROJECT

3.1 D&D in national agroforestry research planning: AFRENA

The D&D approach has a number of applications in research: in regional and national research programmes to allocate scarce resources to topics and technologies of relevance and priority for development, and in specific research projects to provide a framework within which to evaluate research findings and effectiveness.

In 1986, the International Council for Research in Agroforestry (ICRAF) began collaborative agroforestry research activities with national institutions in Africa based on these principles (Torres, 1987). Agroforestry Research Networks for Africa (AFRENAs) were initiated in four major ecological zones: the humid highlands of east Africa, the upland plateau of southern Africa, the humid lowlands of west Africa, and the semi-arid lowlands of west Africa. Each AFRENA network includes country-specific projects for farming systems of high national priority and zonal projects for research on multipurpose trees and shrubs and technologies of priority interest across the zone.

The selection of research projects is made systematically, on the basis of a zonal D&D summarized in Raintree (Chapter 2). In brief, it involves the identification of major land-use systems in the ecozone and their problems and potentials, identification of promising agroforestry technologies, and selection of priority systems and technologies for network research. Decision-making on the research agendas for AFRENA lies with national policy-makers, through the national agroforestry steering committee whose chairmen together form a zonal agroforestry steering committee. ICRAF and national scientists work together in all projects, with ICRAF zonal scientists serving as resource persons in their field of expertise across the zone.

The priority ecozone selected for AFRENA research in east Africa was the humid and sub-humid highlands above 1000 meters with at least 1000 mm of rain, where most of the population is settled (see Fig. 2). In Kenya, agroforestry research and development efforts have been underway for more than decade, through an unusually broad range of research and development institutions. The national steering committee for AFRENA in Kenya includes four research institutions, three university faculties, five ministries, a presidential commission, and a coordinating agency for non-governmental organizations (NGOs) working in the field of energy. Most agroforestry development and research projects in Kenya are linked to one of these organizations.[1] The principal implementing agencies are the Kenya Forestry Research Institute (KEFRI) and the Kenya Agricultural Research Institute (KARI).

The identification of land-use systems and agroforestry potentials for the humid and sub-humid highlands of Kenya was carried out in 1987, by a multidisciplinary team from ICRAF, KEFRI, KARI, and the Kenya Ministry of Agriculture. Fourteen land-use systems were identified. Initial diagnostic appraisals were carried out for the six systems of priority concern from a development perspective (Minae and Akeampong, 1988). Of these, the steering committee selected the food crops system of western Kenya as the focus for the AFRENA zonal research project (Hoekstra, 1988).

3.2 Identifying agroforestry options for the food crops system

The food crops system is a mixed crop-livestock system, in which food production (mainly maize and beans) is principally for home consumption. Altitude is 1000-1500 meters and average rainfall is 1200-2000 mm. Farm size is small (averaging 2-8 hectares), but larger than in the neighboring high density, higher altitude highlands. Herd size is around 3-10 cattle, generally herded. About 60% of the land is in annual crops, 20% in fallow, and 20% in pastures. There is very little use of fertilizers or soil conservation measures, although soils are generally low to moderate fertility. Population density averages around 230 per square kilometer, and there is substantial outmigration of men for wage employment. The system comprises much of Siaya and South Nyanza districts, and parts of Busia and Kisumu (Minae and Akeampong, 1988).

The initial D&D study identified a large number of agroforestry options for the food crops land-use system. Due to the project's limited research resources, the decision was taken to focus on interventions in annual cropland for critical problems of low crop yields, and insufficient fodder for intensive livestock production. Only the interventions for soil fertility improvement will be discussed here.

One can identify a number of alternative agroforestry (and non-agroforestry) practices to address the soil fertility problem. Although most of these interventions have some potential in the food crops system, alley-cropping was considered to offer the highest potential for green manure production and intensification of crop production on smallholdings with low fertility soils. The system could be managed either with continuous cutting back of the hedges each season, or as a rotational system in which hedges

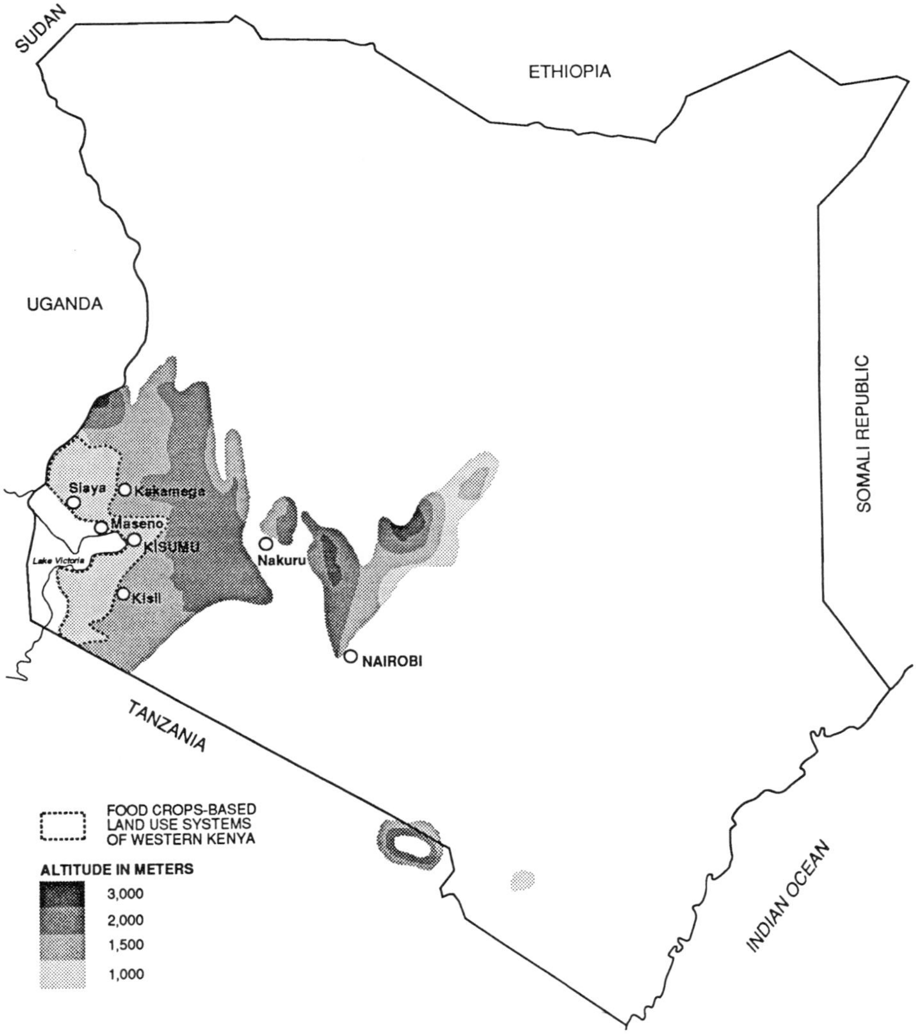

Fig. 2. Elevation map for selected regions in western Kenya.

TABLE 1

Criteria for selecting an alley-cropping intervention to improve crop production.

BIO-PHYSICAL CONDITIONS

- Sub-humid, humid climates
- Soils with low organic matter, nitrogen, low moisture-holding capacity
- Sensitive to erosion
- Crops in alleys not susceptible to hedge pests

ORGANIZATION OF THE PRODUCTION SYSTEM AND LANDSCAPE

- Plots with high land-use intensity
- Wide range of crops, subsistence or commercial production
- Crop-based or mixed crop/livestock systems
- Long-term cultivation rights and rights to product
- Ability of producer to protect plot from over-browsing
- Suitable for many niches: near farm plots under continuous cultivation, short fallow crop plots; some perennial crops
- Plot size not extremely small, unless ample evidence of significant crop yield increases and reduced risk

SPECIFIC FUNCTIONS REQUIRED

- Increase food or cash crop production through: green manure, mulch to suppress weeds/conserve moisture, soil conservation
- Supplement fodder and small-diameter fuel wood production
- Larger-diameter fuel wood or poles where rotational system possible

SOCIO-ECONOMIC CONDITIONS

- Available labor at times of cutting back and mulch application; may reduce other labor requirements
- Limited capital (other than for good pruning tools)
- Initial investment in nursery production or seedling purchase, where direct seeding not possible
- Limited access to inorganic/organic fertilizers
- Support infrastructure for high quality multipurpose, tree and shrub seed, inoculum to improve nitrogen-fixation, and/or farmer training in alley-cropping techniques

were allowed to grow into a fallow cover at regular intervals. Table 1 summarizes some of the main criteria used in selecting alley-cropping. Land-use intensities were considered too high in the food crops systems for simple improved fallow to address fertility needs well. Composting and mulching were not already accepted practices, so mulch plots were not indicated. Grass strips are not widely used in cropland, but where they are farmers appeared to be more interested in using additional biomass for fodder production. Mixed intercropping with MPTs for soil fertility would provide insufficient leafy biomass to maintain soil fertility under intensified use.

3.3 Developing alley-cropping designs for the food crops system

"Alley-cropping" is used here to describe a field plot in which crops are produced in the "alleys" between hedgerows, generally of nitrogen-fixing leguminous trees or shrubs, which are cut back regularly to provide green manure and reduce competition with crops. Alley-cropping has only been well-researched under a narrow range of bio-physical conditions and farming systems (Kang and Wilson, 1987). Alley-cropping species screening has been undertaken in humid and sub-humid highlands of Kenya, though little in the food crops system area, and there have been a few management trials with selected species (Minae and Akeampong, 1988). Alley-cropping has been adopted on a trial basis by some farmers in several parts of the zone. Alley-cropping for the food crops system could be considered a Type 2 technology, requiring some adaptive research prior to large-scale dissemination with farmers.

However, the mandate of the AFRENA project was not technology dissemination, rather it was the identification of new multipurpose trees and shrub species, varieties, and management regimes which would significantly enhance the productivity of alley-cropping designs in the land-use system. The first AFRENA trials were laid down in 1987 to address key information gaps for alley-cropping design in the food crops system.

Table 2 summarizes the main design variables for an alley-cropping technology. The target functions of alley-cropping (soil fertility improvement and secondary use for fodder) were set by the D&D and are reflected in assessment criteria used in the trials. Sites were not carefully specified, beyond constraints of altitude and rainfall regime. The principal AFRENA research site is in Maseno, Kisumu district, whose climatic and altitude characteristics are fairly representative, its soil conditions less so. As the project expands it is expected that satellite research sites on different soil types and reflecting other site differences will be established.

The first research priority was identification of more and higher-performing MPT species and varieties for alley-cropping. The main species currently used by farmers is Leucaena leucocephala, but Calliandra calothyrsus, Cassia siamea, and Sesbania sesban have also been tried. Provenances of these species have been fairly restricted and, in general,

TABLE 2

Design information for alley-cropping technologies

SITES
- Types of land (fallow, sloping, degraded)
- Rainfall, soils, altitude, wind velocity

FUNCTIONS
- Options: Increase soil fertility, restore fertility, improve soil structure, conserve moisture, control weeds, produce fodder, produce poles, produce fuel wood, control soil erosion

COMPONENTS
- Nitrogen-fixing trees which can be cut back frequently
- Associated crops/varieties

ARRANGEMENTS
- Number of rows of MPTs in hedge; between-row/in-row spacing
- Crop spacing and arrangement; fallow cycle, crop rotation

MANAGEMENT
- Time and method of tree establishment
- Density of planting, thinning, hedge formation
- Age, height, method, time, frequency of cutting back
- Management of mulch; management as a fallow
- Control of tree and crop weed, pest, disease, flowering
- Changes in crop management, fertilization
- Age and method of replanting

PERFORMANCE
- Changes in crop yield (in good and bad seasons)
- Changes in soil conditions and rates of erosion
- Yields of fuel wood, poles, mulch, fodder
- Net economic returns relative to alternatives; risk

INPUTS
- Seeds, seedlings, cuttings; labor; cash inputs needed
- Management skills required and training needs
- Market infrastructure required: land, tenure requirements

control of seed quality has been poor. Thus, several MPT species/provenance selection trials were established by AFRENA for a large number of entries, assessing growth rates, vigour, pest and disease problems, nodulation, soil nutrient status on the plot, and other characteristics.

Three experiments were also laid down to evaluate alternative arrangements and management practices for the familiar Leucaena leucocephala/maize alley-cropping system: a trial on biomass production and utilization, evaluating different levels of fertilizer/mulch combinations; effect of cutting height on tree and crop growth; and the effect on biomass production of different densities of trees in the hedge. Implications about spacing and tree population can be drawn from biomass and crop yield results in the trials (AFRENA, 1989). The collaborating CRSP (Collaborative Research Special Projects) small ruminants project, also based in Maseno, is carrying out research on the effect on biomass production and soil nutrient status of different pruning regimes on four MPT species, and evaluation of species for fodder quality and use. The response curves generated by these experiments should provide useful scientific information for extension and adaptive research projects in the zone to draw upon in the design of locally adapted, alley-cropping technologies.

However, this must be done carefully. Most of the decisions regarding treatment levels, non-treatment factors, crop management practices, and control plots conditions for these first factorial trials were made on the basis of consultation with scientists, or Ministry of Agriculture technical recommendations. They represent stylized, rather than actual farmer practice, as they are meant to have relevance across a fairly wide variety of specific farmer practices. They are intended to help explain patterns of tree and crop responses ("why" things happen), and will not always provide an accurate picture of expected responses for other locations with somewhat different characteristics ("what" happens).

The AFRENA project will allocate more limited resources in the future to the design of location-specific technologies. Only a few designs will be developed and field-tested, mainly to demonstrate the application of research results, to identify key technical problems which require more research, and to provide a focus for interdisciplinary input and training in technology design (ICRAF, 1988). Responsibility for designing alley-cropping technologies adapted to other local conditions, for a wider range of crops, and field-testing with farmers of these modified designs, will have to be undertaken by extension and location-specific adaptive research projects in the food crops system. Results from the AFRENA experiments need to be interpreted carefully and/or independently validated under location- and farmer practice-specific conditions before being promoted in new agroforestry designs.

3.4 Validating alley-cropping technologies

AFRENA scientists plan to be involved in on-farm field-testing of selected alley-cropping designs, although not until year three of the project. Adaptive research by extension and local research projects will probably emphasize testing of specific sets of recommendations--technologies--for local performance, rather than a wide range of factors and levels.

The AFRENA project will address research on inputs and institutional arrangements required for large-scale farmer adoption of proposed alley-cropping technologies only in a limited way. Locally available multipurpose tree and shrub species and provenances are among those being tested, but even with these, the existing networks for multiplication and distribution of high quality seed are inadequate. Development of seed orchards, and careful documentation of seed sources used in experimentation, are being initiated by the AFRENA project, but research to plan large-scale dissemination will have to be undertaken by others. Similarly, AFRENA may be able to support--but not lead--development and testing of technical training materials and programmes on alley-cropping for extensionists and farmers.

3.5 Areas for extension involvement

In achieving its objective of developing improved alley-cropping systems for the food crops land-use system, the AFRENA-Kenya project could clearly benefit from closer linkages with development and extension programmes, even in its primary role as provider of general scientific information. Some key areas for extension linkage include:

(a) Generating information about farmers' land-use practices to be used in selecting relevant experimental treatments, treatment levels, non-treatment factors, and control plots. This is to ensure that the experimental results are designed so as to provide extrapolable results for technology design.

(b) Generate bio-economic information about the alley-cropping designs already being tried by farmers, to set practical goals for productivity increases through research, to suggest priority design factors for research and the practical range of factor levels, and to identify key problems which could be addressed through research.

(c) Organize exploratory alley-cropping trials with farmers, on-farm component and management trials and studies, and replicated on-farm testing of alley-cropping designs proposed by researchers. This could reduce the administrative and organizational cost of such research, and could improve the quality of the research through interaction with extensionists who are knowledgeable about farmers' systems and skilled in establishing rapport with them. By tapping the knowledge base and observational skills of farmers and extensionists who are constantly on the ground, researchers can identify problems for more focused on-station experimentation and better alley-cropping design.

4 D&D IN AGROFORESTRY EXTENSION: THE ICRAF-CARE ON-FARM RESEARCH PROJECT

4.1 D&D in extension and local adaptive research projects

The D&D approach to project planning has been utilized in many agroforestry extension and community-based adaptive research projects, primarily in the stages of initial land-use system diagnosis and identification of agroforestry potentials and priorities for implementation in the project. This early extension D&D process commonly differs somewhat from that of research projects. Extension projects generally wish to work with Type 1 or 2 agroforestry technologies which can be ready within a short time period for extension with farmers. The selection of species and interventions may be limited by existing infrastructure constraints in, for example, seed distribution. Extension projects often find that the D&D discovery process can be successfully integrated with the community organization and early planning phases of the project, by emphasizing the farmers' role in identifying problems and agroforestry potentials (Buck, Chapter 5).

The "iterative" phases of the D&D process are especially desirable in the extension context, because of the lack of validated agroforestry technologies adapted to the widely varying local conditions under which extension projects operate. Where formal research projects are established, with relevant types of experiments, their outputs can be extremely useful in improving agroforestry design for extension. But it is unlikely that such data can be transformed directly into locally reliable extension recommendations without prior adaptive trials. Serious agroforestry extension projects--those which define their objectives in terms of "rural development", rather than "seedlings produced"--will generally require a permanent "extension research and development" component (Raintree and Hoskins, 1988), to monitor farmers' experience with introduced components and technologies, and to test improved agroforestry designs suggested both by farmers' experience and experimental research. A recent ICRAF review of 88 agroforestry extension projects discovered that 76% carried out systematic technology evaluation of some kind (Mueller and Scherr, in press; Scherr and Mueller, 1989). Table 3 presents a model for the process of technology

TABLE 3

Stages of technology development in an agroforestry extension project. From Scherr 1988c, p. 15.

---

1. Identification of agroforestry needs and potentials

2. Design of an agroforestry technology, tailored to local needs and conditions - a "best bet" for testing, based on:

   - Current farmer practices,
   - Farmer preferences and judgement,
   - Scientific literature, and
   - Experience in similar areas.

3. Initial trials of the "best bet" technologies on:

   - Farm plots,
   - Demonstration plots, and/or
   - Research plots.

4. Evaluation of "best bets," and development of a range of initial, site-testing "extension recommendations", by:

   - Farmers,
   - Extension staff, and
   - Technical experts.

5. Adoption and adaptation of technology by larger numbers of farmers, with monitoring and evaluation of:

   - Who adopts, and why.
   - Who doesn't adopt, and why.
   - What they adopt and how the designs perform.
   - What they modify, and how the modified designs perform.

6. Evaluation of impact of adoption of new technology, and modification of designs to modify impact on:

   - Supply and distribution of products desired from agroforestry system,
   - Resource use by land users (labor, land, cash use patterns),
   - Environmental change and sustainability of improvements, and
   - Community resources and organization.

---

development in an agroforestry extension project. This is the essence of the "re-diagnosis and re-design" concept.

4.2 The CARE-Kenya agroforestry extension project

CARE International, an international NGO, has been active for almost two decades in tree-planting projects for rural development in the tropics. In 1984, CARE International in Kenya established the Agroforestry Extension Project (AEP) in western Kenya, CARE's first project oriented toward introducing agroforestry land-use practices on smallholder farms. A full discussion of the development of this project may be found in Buck (Chapter 5).

AEP planning was influenced by ICRAF's "diagnosis and design" approach, although the specific D&D procedures were modified to meet the needs of a participatory community development project. The project identified tree species being used locally by farmers and by the forest department, and also identified promising species from other parts of Kenya. D&D was carried out with community women's groups to identify priority needs for trees, appropriate niches and configurations, and to select tree species for group nursery production.

The technical range of agroforestry interventions promoted by the project was unusually broad. The AEP operates in four agro-ecological zones in Siaya and South Nyanza districts, two in the humid/sub-humid area of the food crops system, two in drier zones where there is greater reliance on livestock and fishing. The project has promoted around 40 different multipurpose tree and shrub species, in many different agroforestry practices. None underwent formal testing and adaptation by the project prior to large-scale dissemination. Some Type 1 tree-planting technologies already found locally were indeed ready for large-scale dissemination: live fencing, home compound planting, fruit tree orchards, and boundary planting with familiar tree species.

Many Type 2 agroforestry practices were introduced, with minimum design recommendations: tree borders in cropland, fruit trees mixed in cropland, high-density mini-woodlots, trees for soil erosion control in cropland, trees for soil fertility in *Robusta* coffee, and multipurpose tree and shrub fodder banks. Farmers were, for example, given "best bet" suggestions on tree spacing, but encouraged to adjust this spacing to their own preferences. Thus, the responsibility for adaptation to local needs and conditions was initially given to participating farmers, with the project taking responsibility for monitoring their experience and communicating information about successful design innovations.

Alley-cropping for soil fertility in the dry zones was really a Type 3 technology, and has (in currently promoted designs) been unpopular. A number of tree species were new introductions to the zone, and were "screened" informally for site suitability in the early years of the project by success in nursery production, on-farm survival, and farmer evaluation.

Still other agroforestry technologies were developed by farmer participants in the project without extension promotion: mixed intercropping of timber trees in cropland, mixed intercropping of trees for staking of passion fruit, multistrata home gardens, and managed trees in pastures (Vonk, 1986; Scherr, 1988e).

When initiated, the AEP included an intensive monitoring and evaluation (M&E) programme to evaluate early farmer experience with agroforestry technologies in around 60 pilot nursery groups, and to develop more robust technical recommendations. There were regular monthly or biweekly staff meetings to discuss extension and technical problems, and to exchange information about agroforestry experiences. Field notebooks were kept in which extension staff were to note, among other things, interesting information about agroforestry technology design. In 1985, a collaborative project was initiated between the AEP and KEFRI, to set up research trials on alley-cropping, new multipurpose tree and shrub species for borders and woodlots, and a few other technologies, on six research plots in different agro-ecological zones of Siaya district (Nyamai, 1988). Small-scale, on-farm studies were undertaken in collaboration with students of the Agricultural University of Wageningen, the Netherlands. Nursery record books were kept by groups and schools. Three random sample surveys were carried out between 1985 and 1988, measuring survival rates of seedlings taken from the nurseries, with information about species, site, ecozone, etc.

For the first few seasons of the project, there were "design follow-up surveys", in which farmers' actual selection of species and configurations for planting were compared with those identified in the initial household or community D&D surveys. A number of formal surveys were also carried out with participating and non-participating community groups and households, and teachers at primary schools, although these were not fully analyzed (Vonk, 1986; Davis-Case, 1988).

The original "pilot, technology-generating" nature of the AEP was, however, overturned within little more than a year of its inception. The enthusiasm of local farmers convinced project managers that the AEP should move quickly into large-scale dissemination, and should expand operations to South Nyanza district. By early 1988, over 600 AEP-assisted nursery groups and schools had produced an estimated 9 million seedlings (with 50-70 % average survival). The intensive M&E system, set up for use with a small number of farms, became unmanageable when applied to several thousand farmers. KEFRI research plot trials continued, but were not closely linked to extension activities.

The case of soil fertility-improving interventions is illustrative. Alley-cropping was promoted by AEP mainly to address problems of low and declining soil fertility in cropland, under conditions of limited access to other organic and inorganic fertilizers. Secondary functions were identified as fuel wood production, fodder production, or as a soil erosion barrier on sloping land.

Initially, only a few designs were suggested to farmers. Because of widespread farmer acceptance of indigenous Markhamia lutea as a fertility-improving timber tree for mixed intercropping in cropland, and of Sesbania sesban as a tree fallow, these were promoted in more systematic or linear planting arrangements, which were assumed--though not demonstrated--to be more efficient. Continuous alley-cropping systems with new multipurpose tree and shrub species Leucaena leucocephala and Calliandra calothyrsus were also introduced. On-station validation trials were undertaken with KEFRI for four specific alley-cropping designs (the above species, designed with similar crop, spacing and management). "Design surveys" were used to monitor early farmer design adaptations. For example, many farmers independently opted for rotational alley-cropping systems, particularly to rehabilitate degraded croplands.

None of these farmer-adapted designs, however, has been objectively evaluated by the AEP for relative performance in meeting farmer objectives. In agroforestry, farmer acceptance or rejection of a particular technology design cannot be judged by the continued presence of trees in the field. Farmer evaluation requires many years and any new ideas farmers have about multipurpose tree and shrub components and configurations can only be tested in additional plots. Thus, although the AEP and its farmer collaborators have succeeded in generating a wide number of innovations, there is little consensus about which innovations can be recommended with confidence to other farmers in the zone.

4.3 The ICRAF-CARE on-farm research project

As the AEP expanded, it received increasing national and international attention. In 1987, the project was selected as one of six case studies in the Agroforestry Monitoring and Evaluation Methodology Project (AFMEMP) of CARE International, Food and Agriculture Organization of the United Nations (FAO), and the Ford Foundation (Buck et al., 1988). An intensive effort was made to review M&E procedures and results of the project (Davis-Case 1988; Buck and Vonk, 1989). At the same time, project staff were turning their attention to issues of technical and institutional sustainability after the departure of CARE. One key objective was to establish and validate agroforestry technology recommendations, suited to farmers' needs and conditions in the different ecozones of Siaya and South Nyanza districts, which could be integrated into the recommendations of other rural extension agencies. The AFMEMP review revealed the need for an integrated, low-cost system for monitoring and feedback, in which both extension staff and farmers were active participants.

ICRAF was simultaneously involved in identifying potential tools for technology M&E in agroforestry extension or adaptive research projects. These may be grouped by the purpose for which information is collected:

(a) for project planning and initial design of technologies;

(b) for evaluating and improving technology performance and suitability for farmers; and

(c) for evaluating the impact of specific technologies--or the project as a whole--on households and communities (Scherr, 1988a).

In 1987, the AEP established a joint project with ICRAF to develop a number of suitable and customized information tools.

"Re-diagnosis and re-design" is embodied in the second group of information tools. These are listed in Table 4, which indicates tools used by CARE AEP. "Formal" survey tools or trials are those which use formal sampling or experimental procedures, and can thus be analyzed statistically with known confidence levels.

4.4 Evaluating on-farm alley-cropping designs

Alley-cropping, predominantly with Leucaena leucocephala, has been one of the most popular interventions of the AEP. Over 50% of all Siaya district farmers involved with CARE for at least a year have adopted some variant of alley-cropping on a trial basis. A number of M&E tools are being applied by CARE, with ICRAF support, to evaluate farmer/ project innovations in alley-cropping designs. These include:

- -extension staff review workshops on technology recommendations;
- -an informal technology design/adoption survey;
- -a formal technology design survey; and
- -an informal on-farm yield assessment study.

TABLE 4

Information tools for evaluating and adapting agroforestry designs in extension projects. Adapted from Scherr (1988c), Table 3. (+) Indicates past or ongoing AEP activities. (*) Indicates being tested in ICRAF-CARE Project

---

QUALITATIVE INFORMATION, LOW MATERIAL COST, LOW LABOUR COST, LITTLE SPECIALIZED EXPERTISE NEEDED:

- Farmer comparison plots
- Farmer technology evaluation workshops
- Extension staff technology evaluation workshops *
- Extension staff field notebooks +

QUALITATIVE INFORMATION, LOW MATERIAL COST, HIGH LABOUR COST, SUPPLEMENTARY SPECIALIZED EXPERTISE NEEDED:

- Informal farmer technology evaluation surveys
- Demonstration plot technology trials
- Informal technology design/adoption surveys *
- Informal technology design/adaptation follow-up studies +

QUANTITATIVE INFORMATION, SOME MATERIAL COSTS, HIGH LABOUR COSTS, SUPPLEMENTARY SPECIALIZED EXPERTISE NEEDED:

- Informal on-farm technology trials/studies *
- Survival surveys +
- Formal farmer technology evaluation surveys
- Formal technology design surveys *
- Formal technology design/adaptation follow-up studies

QUANTITATIVE, SIGNIFICANT MATERIAL COSTS, SPECIALIZED EXPERTISE REQUIRED

- Research plot experiments
- Research plot technology trials +
- Formal on-farm technology trials * (planned)
- On-farm experiments

4.5 Review workshops on technology recommendations

A series of short workshops were held with extension staff in 1987 and 1988 to review the progress of the project since 1985 in developing technical recommendations. A systematic format was established for documenting species selection, arrangements, and management information for six major interventions, for different site conditions and farmer objectives (Scherr, 1988b).

The workshops confirmed advances in adaptation of alley-cropping design since 1985 in terms of species selection and arrangements for different ecozones. Serious gaps in knowledge were identified, however, in recommendations for management of alley-cropping systems--both as a continuous system, and as a rotational system which many farmers preferred--and performance expectations. While farmers were reported to be fairly enthusiastic about alley-cropping in the medium and high rainfall zones, there was much less interest or satisfaction with the system in the dry zones. The workshop identified a training need of extension staff; to be better informed about how alley-cropping systems work to improve soil fertility. ICRAF arranged a training session on this topic. Regular workshops of this type, held on a seasonal or annual basis, can be used to update information, identify information gaps, provide an organized forum for discussion of technical problems and potentials, and stimulate staff to make more critical observations in the field.

4.6 Pilot survey of farmer-practiced agroforestry systems

Although information on farmer technology design is extremely useful to evaluate and re-design extension recommendations, and to evaluate project impact, it can be difficult to collect systematically. This is true particularly in land-use systems where tree-growing was common prior to the project, and where such a wide range of species and technologies are being promoted in different ecozones. The AEP tree survival surveys provided some information on the species and survival rates of alley-cropping plots. The "design follow-up" forms were used consistently only in two seasons, and collected information on only a few key variables.

In late 1987, a small team of ICRAF scientists and CARE field staff carried out an informal survey of 18 farms in Siaya district, in order to work out a methodology for collecting design information on existing agroforestry plots (i.e., sub-plots of the farm where trees were planted or managed). These 18 farmers were agroforestry enthusiasts, and a total of 139 agroforestry plots were surveyed, including those established since working with CARE as well as pre-existing plots. Data were collected on date of establishment, species, configurations, and a few key management practices for all types of agroforestry technologies.

The information from this pilot survey suggested that soil fertility-improving interventions were important, but that alternative rotational alley-cropping systems and planted tree fallows were as important as continuous systems. The process of systematic observation encouraged by such a survey is valuable in itself.

Data in this study were not extrapolated to the entire group of CARE-assisted farmers, because the sample chosen was not representative. But even without formal sampling, more random farmer selection can provide a representative picture of patterns of technology design in the project area. Such surveys can also offer a structured forum for discussions with farmers about problems and advantages with particular designs, to identify fruitful areas for farmer assistance or in-depth evaluation.

4.7 Formal survey of farmer alley-cropping and border plot designs

The workshops revealed a lack of consensus by extension staff about how farmers were currently managing their trees--the starting point for improving recommendations. ICRAF helped AEP to prepare a formal farmer survey to obtain this information. In May 1988, the survey was implemented in Siaya district by AEP extension staff, for a random sample stratified by ecozone of farmers who had received extension assistance for at least a year. All alley-cropping and tree borders in cropland on 126 farms were surveyed (including ones poorly established). Detailed information was collected on farmer objectives, sites, species, tree spacing, plot size, tree establishment, management practices, and qualitative indicators of outputs and inputs (Scherr and Oduol, 1988).

The findings have implications for extension, technology design, and research. The survey confirmed a high rate of trial adoption of alley-cropping between 50 and 60 % of participating farmers. But half of the plots had tree populations below 2000 per hectare, almost certainly too low to have a significant effect on soil fertility. Nearly a third of all farmers with alley-cropping plots had not yet cut back their trees. A specific set of recommendations for rotational alley-cropping seems to be needed. Alternative recommendations are also needed for the one third of farmers whose primary objective in alley-cropping is production of fuel wood, poles, or fodder, rather than soil fertility improvement. The height at which farmers cut back their hedges ranged from 5 to 100 centimeters. Research on the effect of cutting height on subsequent tree and crop yields would be valuable. The survey also served a useful training function for extension staff in evaluating alley-cropping plots.

4.8 On-farm yield assessments

It is difficult to obtain reliable information on agroforestry plot performance through farmer survey methods, or by extrapolating yield measurements taken on research stations to farm conditions. It is preferable to physically measure inputs and outputs in farmers' fields, either after farmer establishment or--preferably--as simple trials compared with traditional practice.

ICRAF is developing methods for yield measurements in alley-cropping from such on-farm studies. A training session was organized in 1988 with the AEP extension and KEFRI research staff, to demonstrate and evaluate methods of yield assessment for tree products and crops in existing alley-cropping plots (Wolf, 1988; Jama, 1988). On a preliminary basis, crop yield data were taken from ten randomly selected farmers' Leucaena alley-cropping plots (Oduol, 1988). An expanded set of yield assessment studies, as well as formal on-farm trials of modified alley-cropping designs, are planned for 1989. Such quantitative data on performance are a pre-condition for official extension agencies to promote alley-cropping.

5 RESEARCH-EXTENSION LINKAGES IN AGROFORESTRY D&D

In the old "transfer of technology" paradigm of agricultural technology generation, ideas are generated by research scientists, tested on-station and later in on-farm trials, then given to extensionists for dissemination to farmers. As is convincingly argued by Raintree and Hoskins (1988), this model is particularly inappropriate for agroforestry, where rural communities are a major locus of existing knowledge and innovation.

There is also an important issue of efficient resource allocation. Formal sector research resources are and will continue to be inadequate to generate and field-test more than a limited number of agroforestry technologies, and to refine even fewer. Agroforestry extension projects will have to develop a capacity for agroforestry technology testing and adaptation, and effective interaction with researchers. The alternatives are to sharply curtail activities where validated recommendations are not available, or to risk disseminating information to land users which may prove to be of little benefit to them, or may even have negative environmental or economic consequences.

On the other hand, most of the information on and experience with agroforestry practices is currently concentrated in indigenous farming systems and in projects of agroforestry extension. It is desirable for technology-generating agroforestry research programmes to have an effective field presence, with a reliable network of farmer contacts, and opportunities for on-farm research on technology design and adaptation. Yet development of such independent on-farm research networks can be quite costly. Researchers could retain the old paradigm to focus only on basic/applied experimental work, leaving technology design and adaptive research to extension or other

TABLE 6

Alley-cropping development in the food crops lands-use systems of western Kenya. Summarized from Scherr 1988e. (*) Indicates planned, but not yet underway.

MULTIPURPOSE TREE AND SHRUB SELECTION

RESEARCH

- multipurpose tree and shrub species/provenance selection trials (AFRENA, CRSP)

EXTENSION

- Tree survival surveys by ecozone/soils/planting method (CARE/KEFRI)
- Survey of farmer use of existing MPTs for soil fertility (CARE)
- Evaluation of MPT performance, pest/disease problems, farmer preferences through field notebooks (CARE)

COMPONENT ARRANGEMENT AND MANAGEMENT

RESEARCH

- Effect of mulch from alley-cropping with *Leucaena leucocephala* on crop yields at different levels of fertilizer application (AFRENA)
- Effect of cutting height on biomass with *L. leucocephala* (AFRENA)
- Effect on biomass production and crop yields of differing densities of *L. leucocephala* at different distance of crop from hedge (AFRENA)
- Effect on biomass and soils of pruning regimes on 4 MPT species (CRSP)
- Evaluation of fodder quality/use for selected MPT species (CRSP)

EXTENSION

- Survey of farmer alley-cropping designs and preferences (ICRAF/CARE)
- Observation of farmer spacing and management preferences (CARE)
- Management trials for selected species (CARE/KEFRI, ICRAF/CARE) *

TECHNOLOGY EVALUATION AND TESTING

RESEARCH

- On-station alley-cropping validation trials (CARE/KEFRI)
- On-station alley-cropping demonstration trial (AFRENA)
- On-farm trials of alley-cropping for fodder (CRSP)
- On-farm alley-cropping trials (AFRENA, CARE/KEFRI, ICRAF/CARE) *

EXTENSION

- Evaluation by farmers of adapted alley-cropping designs (CARE)
- Yield assessment in on-farm alley-cropping plots (ICRAF/CARE)
- Economic assessment of selected alley-cropping designs (ICRAF/CARE)
- Identification of key research issues for alley-cropping design (CARE)

institutions, but this could lead to a serious divergence between field needs and research priorities. A viable alternative for national agroforestry researchers is to establish direct research linkages with extension programmes to provide this field base, and to support extension-based adaptive research.

The evidence from ongoing agroforestry research and extension projects in western Kenya suggests that more rapid and cost-effective agroforestry technology development could occur through better research-extension linkages. Table 6 highlights the potential complementarity between ongoing activities to improve alley-cropping technologies for the food crops system, through MPT selection, evaluation of management options, and testing of specific technologies.

But this model demands active sharing of information and responsibilities. Researchers would take responsibility for managing experimental trials and formal technology-testing for official validation, with support from extension staff in setting experimental priorities and evaluating results. Extensionists would take responsibility for monitoring and evaluation of farmer adoption/adaptation, simple component/management trials, and project-specific technology testing, with support from researchers in survey and trial design and evaluation. Different kinds of institutional arrangements could be considered, e.g., assignment of researchers to extension projects, research units within extension programmes, inter-institutional planning committees (Scherr, 1988d). In all cases, greater expertise is required of both researchers and extensionists in skills and knowledge base for technology "design."

6 CONCLUSION

Commitment to a D&D approach in agroforestry technology development has a number of implications for the planning and review process in agroforestry extension and research projects:

(a) Project objectives and strategies must be set according to farmer needs, existing agroforestry practices, and land-use system-constrained agroforestry potentials; little project planning can take place prior to initial diagnostic and design studies.

(b) Project design will be determined in large part by the level of technical information available about proposed technologies. A project dealing mainly with Type 1 interventions may follow a conventional technology transfer model, while a project dealing mainly with Type 2 or 3 interventions will require a substantial research and/or monitoring and evaluation component, and staff trained in technology evaluation.

(c) The project funding and implementation cycle needs to explicitly recognize activities related to technology design, as distinct from either formal experimental research or formal extension, and to designate responsibility for this design role.

(d) Institutional arrangements must be considered in the planning process to address identified needs for research-extension collaboration.

(e) Staffing in terms of numbers, level, and training will be determined in part by expected involvement in on-farm technology evaluation.

(f) Projects need to be structured to permit and even encourage re-evaluation and re-design of project goals and activities, in response to new technical information obtained through field experience.

In conclusion, the agroforestry diagnosis and design approach provides an exciting framework for promoting development through the intensified use of multipurpose trees and shrubs in the rural landscape. D&D explicitly emphasizes the importance of generating and disseminating information which will make agroforestry more effective in meeting land-users' needs. Its success requires institutionalization throughout the planning and implementation process.

7 ACKNOWLEDGMENTS

The author thanks the CARE Agroforestry Extension Project staff and the AFRENA/Kenya team, whose collaboration made this research possible. She is particularly grateful to Louise Buck, John Raintree and Alan Dappen, who commented thoughtfully on earlier drafts of the chapter.

8 REFERENCES

AFRENA, 1989. Experiment descriptions for the AFRENA Eastern Africa zonal research site in Maseno. AFRENA, Maseno, Kenya.

Buck, L., In press. Agroforestry Extension Training Sourcebook. CARE International, Nairobi.

Buck, L., Davis-Case, D., Mueller, E. and Ngugi, A., 1988. A synthesis of experience from the Agroforestry Monitoring and Evaluation Methodology Program. Paper presented at the AFMEMP Regional Workshop, Kisumu, Kenya, 25 pp.

Buck, L. and Vonk, R., 1989. Planning agroforestry extension projects: A case study of CARE International in Kenya's Agroforestry Extension Project. Presented at the International Symposium for Planning and Implementation of Agroforestry, Washington State University, April, Pullman Washington.

Davis-Case, D., 1988. A Case Study of CARE Agroforestry Extension Project. CARE/FAO: Nairobi, Draft, 70 pp.

Hoekstra, D., 1988. Summary of zonal agroforestry potential and research across land use systems in the highlands of Eastern and Central Africa. (AFRENA Report No. 15). ICRAF, Nairobi, 30 pp.

ICRAF, 1988. ICRAF's strategy for prototype agroforestry systems design and Testing. ICRAF, Nairobi, July, draft, 8 pp.

Jama, B., 1988. Crop yield assessments in on-farm alley-cropping schemes. Paper presented to ICRAF/CARE training session in Siaya and South Nyanza Districts, Kenya, June, Nairobi, 13 pp.

Kang, B.T. and Wilson, G.F., 1987. The development of alley-cropping as a promising agroforestry technology. In: H.A. Steppler and P.K.R. Nair

(Editors), Agroforestry: A Decade of Development. ICRAF, Nairobi, pp. 227-43.
Minae, S. and Akyeampong, E. (Editors), 1988. Agroforestry potentials for the land use systems in the bimodal highlands of Eastern Africa, Kenya. (AFRENA Report No.3). Kenya Government/ICRAF, Nairobi, March, 153 pp.
Mueller, E. and Scherr, S.J., In press. Technology Monitoring and Evaluation in Agroforestry Projects: An Annotated Bibliography. ICRAF, Nairobi.
Nyamai, D., 1988. Tentative report on CARE-KEFRI agroforestry research in Siaya District, KEFRI, Muguga, Kenya, March, 14 pp.
Oduol, P.A., 1988. Preliminary attempts to assess crop yields in alley-cropping systems. CARE-Kenya Project, Siaya District. (ICRAF-CARE Project Report No. 14). ICRAF, Nairobi, November, 7 pp.
Raintree, J.B., 1987. The state of the art of agroforestry diagnosis and design. Agroforestry Systems, 9: 219-251.
Raintree, J.B., 1989. Agroforestry diagnosis and design. Presented at the International Symposium for Planning and Implementation of Agroforestry, Washington State University, April, Pullman, Washington.
Raintree, J.B. and Hoskins, M., 1988. Appropriate R&D support for forestry extension. Paper presented to the FAO Expert Consultation on Organization of Forestry Extension, March 7-11, Bangkok, 28 pp.
Scherr, S.J., 1988a. Pilot survey of adopted agroforestry practices in the CARE Agroforestry Extension Project. (ICRAF-CARE Project Report No. 6). ICRAF, Nairobi, January, 34 pp.
Scherr, S.J. (Editor), 1988b. Current agroforestry practices and extension recommendations of the CARE Agroforestry Project. (ICRAF-CARE Project Report No. 8). ICRAF, Nairobi, April, 58 pp.
Scherr, S.J., 1988c. Factors to consider in evaluating technology performance in agroforestry projects. Paper presented at the AFMEMP Regional Workshop, Kisumu, Kenya, May 15-20, 17 pp.
Scherr, S.J., 1988d. Evaluating institutional capacity for agroforestry research. Paper for the Rutgers University ISNAR Agricultural Technology Management Workshop, Rutgers, New Jersey, July 6-8, Draft, 48 pp.
Scherr, S.J., 1988e. Designing agroforestry systems for rural development: Implications for the research agenda. Paper presented to the International Potato Center/Rockefeller Foundation Workshop on Farmers and Food Systems, CIP: Lima, Peru, September 26-30, 53 pp.
Scherr, S.J., 1989. Choosing priorities for agroforestry research. In: D. Groenfeldt and J. Moock (Editors), Social Science Perspectives in Managing Agricultural Technology. International Irrigation Management Institute and Rockefeller Foundation, New York, 34 pp.
Scherr, S.J. and Mueller, E., 1989. Technology monitoring and evaluation in agroforestry projects: A review. ICRAF, Nairobi, Draft.
Scherr, S.J. and Oduol, P.A., 1988. Farmer adoption of alley-cropping and border planting in the CARE Agroforestry Extension Project. Parts 1-3. ICRAF-CARE Project. Reports No. 15-17. ICRAF, Nairobi, December, 35 pp.
Scherr, S.J. and Pinney, A., 1989. A strategy for agroforestry technology-generating research. (ICRAF Working Paper). ICRAF, Nairobi. Draft.
Torres, F., 1987. The ICRAF approach to international cooperation. Agroforestry Systems, 5: 395-417.
Vonk, R. , 1986. Report on a methodology and technology generating exercise: the Agroforestry Extension Project 1982-1986. CARE International in Kenya, Kisumu, Kenya, November, 88 pp.
Wolf, G. , 1988. Multi-product yield assessment from hedgerow intercropping: Practical guidelines for field extension workers. Paper presented to ICRAF-CARE Project training session in Siaya and South Nyanza districts, Kenya. June 27-30. ICRAF, Nairobi, 28 pp.

Chapter 7

UNITED STATES PEACE CORPS: AGROFORESTRY PLANNING METHODS IN THE ISLANDS OF THE PACIFIC

K.J. MURRAY
Guen-Murray Associates, 3216 Winter Street, Philadelphia, PA, 19104.

ABSTRACT

In many parts of the world, adverse environmental impacts of human activity are being addressed through the application of agroforestry practices in which both food crops and trees are grown on the same land for a variety of end uses. The United States Peace Corps, a volunteer organization funded by the United States government, has agroforestry volunteers stationed around the world assisting national and local governments implement agroforestry projects. Peace Corps agroforestry programs may have many different goals including nutrition, economic development, afforestation, fire protection, watershed management, erosion control, and fuel wood production. Trees provide humans with numerous benefits, particularly in the control of soil erosion and production of important commodities -- food and fodder, fuel and timber, and medicine. Humans have taken advantage of these benefits and as a consequence, have always impacted, whether positively or negatively, the world's forests.

INTRODUCTION

Scientific inquiry and general public concern have proven that some degree of harmonic interaction between the environment and humans is of critical importance to the short and long-term survival of civilization. Without some degree of acknowledgment and respect for this interaction, humankind may find itself in a position where resources provided by the natural environment will disappear. Forests are a critical link in this relationship between people and nature.

Humans probably altered vegetation patterns on a large scale with the discovery of fire. With the advent of farming, fires were used to clear fields for livestock grazing or cultivation. American Indians used fires to clear forests, thus extending the range of the buffalo. As early as 3000 B.C., forests were exploited for the production of pottery, metals, and glass products. Wood was used to build ships, palatial temples, and palaces. Burgeoning populations, requiring enormous amounts of food and fuel, expanded into forested lands, cut the trees, and planted crops. Prior to the use of coal and petroleum, industrial needs decimated vast stands of forests.

In several industrial countries this trend has reversed. Where at one time the eastern seaboard of the United States was almost totally devoid of trees, today satellite photography shows that forests, or trees cover more area then any other land use. Of course, citizens of the eastern United States are not directly responsible for the reemergence of the forests. Very few specimens of the original forest remain. Were it not for discoveries brought on by research connected with world wars, the accompanying general affluence of the population, and recently enacted environmental regulations; these east coast lowlands, hills, and mountains may yet resemble an unnatural moonscape. Although the trend of forest destruction has been reversed in some parts of the world, other less visible forces, such as acid rain, continue to threaten forested areas.

In most cases, developing countries have not, or do not, have the luxury of allowing forests to regenerate as they have in developed countries. Forest clearing in Africa and Southeast Asia remains a serious problem. Large stands of previously unreachable tropical forests are being cleared for agriculture, cut for firewood, and harvested to supply timber to a voracious foreign market. The destruction of the forests that occurred in developed countries a century ago is occurring in developing countries today.

Human's historic and continued use of forest by-products has reduced the world's original forest area to at least one-third to one-half its original size (World Resources, 1986). At the current rate of destruction, at least 225 million hectares of the world's remaining tropical forests will be deforested and 10 to 20 percent of all animal and plant life will be lost by the year 2000. This wholesale loss of forested lands will have significant effects. Forests provide humanity with a cornucopia of benefits. Forests are living museums and laboratories that have yielded only a tiny fraction of their treasures to scientific study. A single hectare of Amazon rain forest has been known to contain up to 230 species. Many useful chemical products essential for making medicines and pharmaceuticals originate in tropical plants.

To combat this trend, scientists, planners, and government officials are seeking a balance between the protection of forested lands and their continued exploitation. This balanced approach is crucial to the success of rural development projects since it is unrealistic to assume all forested areas can be segregated and protected from the needs of people.

Agroforestry has become an accepted means for scientists, planners, and government officials to provide a balance between the needs of society and protection of the environment. The United States Peace Corps, a volunteer organization funded by the American government, is providing Pacific countries with the assistance necessary to implement agroforestry programmes. Peace Corps has worked since 1961 at the grassroots level in countries around the world in programme areas such as agriculture, public health, education, cooperatives, deaf education, and agroforestry. Peace Corps volunteers are assigned to a particular agency at the request of the host country. Ranging in age from 21 to 80, volunteers typically serve two year assignments.

Peace Corps development philosophy stresses that successful development projects must take into account the needs and desires of community members, the physical and environmental factors affecting the locality, and the availability of materials, resources, and skills necessary to carry out each step of the project. With this understanding, volunteers are trained to be prepared to face a range of different situations and challenges, and to adapt to changing circumstances. Before beginning their assignments, the volunteers are trained in cross-cultural, technical, and language skills appropriate to their assignment. This training assists them in living and working closely with the people of their host countries, and approaching development problems with new ideas that make use of locally available resources which are appropriate to local cultures.

The Peace Corps agroforestry programme is a model for other Peace Corps programmes as portrayed in the goals and objectives established in a training session manual, prepared by agroforestry volunteers and administrators working in Asia and the islands of the Pacific (U.S. Peace Corps, 1984). The following is a summary of the goals and objectives established by the programme participants.

1.1 Goals

(a) To develop a strong and confident working relationship among Peace Corps volunteers and their respective host country national counterparts.

(b) To identify and improve needed skill areas in agroforestry, site survey methodologies, and development strategies.

(c) To broaden the agroforester's knowledge of a variety of potential agroforestry techniques.

(d) To recognize the role of women in agroforestry and integrate women into community analysis and project planning for a holistic approach to human resource utilization and development.

(e) To encourage the sharing of information on specific forestry issues amongst co-workers.

1.2 Objectives

(a) Encourage the formation of groups.

(b) Gain a full understanding of the environmental, economic, and social implications of agroforestry solutions.

(c) Understand traditional agroforestry practices in Asia and the Pacific.

(d) Analyze the environmental, economic, and social aspects involved in implementing an agroforestry project on a specific site. Study the interrelationship between these factors prior to designing and implementing a programme.

(e) Utilize established extension methodologies.

The implementation of agroforestry systems by volunteers in the Pacific differ from place to place. The systems depend upon the current state of natural conditions, land-use patterns, community needs, and government policies and regulations. One Pacific island country may have young or maturing forests that are the symbol of a recovery from past mismanagement or from natural disasters and require little attention. In another country, mature forests may be under severe pressure and they need immediate protection. On other islands, agroforestry projects may be needed to reclaim denuded slopes, turning them into viable food and wood-producing projects.

2 BACKGROUND: PACIFIC ISLANDS

Those islands of the Pacific with active Peace Corps programmes include the Philippines, Solomon Islands, Papua New Guinea, Marshall Islands, Cook Islands, Tonga, Western Samoa, Fiji, and Micronesia. These islands encompass a wide variety of environmental, economic, and social conditions (topography, climatic patterns, soils, vegetation, land uses, and cultures) which influence agroforestry programme design and implementation. There are substantial differences in topography and terrain among the islands, ranging from rugged, high islands to limestone terraces and volcanic slopes. Some have flat coastal areas and steep inland mountains. Others, such as the Marshall Island group, are made up of low islands and coral atolls.

The islands encompass four major climatic types: equatorial, tropical, mountain, and monsoon. Generally, these types are characterized by seasonal patterns of wet and dry periods. Most of the island climates feature high humidity and rainfall. In the Solomons, rainfall can be as high as 8,000 millimetres per year. The island's high rainfall is due primarily to the great intensity of tropical rains rather than a higher number of days or hours of rainfall throughout the year. Rainfall of up to 15 centimetres per day is not unusual in Micronesia. The intensity of rainfall generally varies with altitude, with lower-altitude regions receiving rainfall of higher intensity.

Rainfall is perhaps the most important climatic factor affecting agriculture and forestry on the islands. It is a limiting factor even with high total annual precipitation because rainfall is unevenly distributed throughout the year. The high intensity of rains often exceed the receptive or absorptive capacity of the soils causing considerable surface runoff. In areas with high rainfall and steep slopes, the soil is especially prone to erosion.

Within varied topography of the islands, temperature is affected primarily by altitude. In Papua, New Guinea, where elevation ranges from sea level to 4,500 meters, the temperature can fluctuate from 27°C to 10°C. In general, temperature decreases with increasing altitude, although such factors as wind patterns, proximity to oceans, and ocean currents, affect temperature patterns of specific locations.

Temperature significantly affects the species that may be successfully cultivated in the islands. Cool temperature crops such as certain varieties of coffee, for example, do not grow as well in warmer, lower-altitude areas as in the cooler elevated regions. The growth of certain species may be stunted if the temperature range is not ideal. Some species are quite sensitive even to moderate temperature fluctuations.

Humidity is usually high in equatorial and monsoon climates, but falls to low levels during seasonal dry periods. There may be marked fluctuations in humidity levels from day to night. In the rain forest regions, humidity may vary from a low of 70 percent during dry days to near 100 percent at night.

The soils of the Pacific islands are of two main origins: volcanic, formed from the discharge of volcanoes; and alluvial loams formed from the build up of sediments from streams, rivers, lakes, and oceans. The characteristics and quality of soils vary depending upon climatic conditions, type of parent materials, topography, use, management practices, and age. Rain forests which cover a portion of many of the islands give a misleading impression that the soils are highly fertile. In fact, many of the inland upland soils are shallow, heavily leached, and relatively infertile. Yet, in the broad lowland and upland valleys of the Philippines, the alluvial loams or soils of volcanic origin can be very productive. The greatest soil fertility problem is nutrient leaching.

There are essentially three zones of natural plant life: mangrove forests at the shoreline, strand vegetation in the lowlands, and rain-forests in the highlands. Cultivated plants include coconut, palm, hardwoods, fruit trees, bananas, vegetables, sugar, cocoa, rice, and spices.

Land-use practices found throughout the islands are indicative of the culture. Land use not only differs between countries, but also within countries. Upland peoples in the Philippines have practiced slash and burn and integrated cropping systems for centuries. In comparison, lowland farmers typically practice a mono-culture system. On the smaller islands such as the Truks (Micronesia), traditional farmers planted the lower slopes with breadfruit and coconut trees. The upper slopes were left intact.

Population and economic pressures are changing these traditional methods of land use. In the Philippines, as lowland farmers move into the uplands in search of available land, upland farmers are forced to intensively cultivate fewer plots. In the Truks, the demand for jobs and exportable timber results in the exploitation of the steep upper slopes. The result in both cases is increased soil erosion and reduction in soil fertility.

2.1 Causes of deforestation

The causes of deforestation and the depletion of forest resources in the islands of the Pacific are not unlike those of other countries. In recent years, increasing pressure has been placed on forest lands for economic, social, as well as, political reasons.

The high population growth rate in the Pacific islands (3-5%), as well as migration among and within island regions, has contributed to increasing pressure on land and other natural resources. As population increases, the amount of forest land usually declines, because demand for land for food production and other uses intensifies. Population increases also reduce the size of individual farms or plots. Particularly for subsistence farmers, cultivation intensifies as plot size is reduced. This usually results in soil depletion and related problems such as declining crop yields. Additionally, population pressure forces many farmers onto land unsuitable for cultivation.

In some areas, disputes over land ownership have resulted in the misuse of land or poor land management practices. Without clear title to the land, people tend to be less concerned with proper management. The construction of roads and expansion of transportation networks into forested areas is also leading to destruction of forest lands. The new roads open marginal lands to lowland migrants, who often lack the skills or incentives to properly manage the land.

These factors have intensified the practice of shifting cultivation which has become common practice amongst upland subsistence farmers. In the islands of the Pacific, shifting cultivation usually occurs on steep, sloping lands; tilling with draft animals, machinery, or by hand takes place on the flat lowlands. In shifting cultivation, farmers typically slash and burn surrounding vegetation, and intensively crop a given area of land for a few seasons until the productivity of the land and crop yields decline. The farmers then leave the exhausted land fallow (uncultivated) and move on to farm elsewhere.

The combination of shifting cultivation and related practices often leads to deterioration of soil conditions and loss of protective ground cover. These practices can be particularly damaging to the land, causing rapid depletion of soil nutrients and loss in productivity, erosion, landslides, uncontrolled fires, and flooding. In the past, this method of farming was principally practiced by upland people. The method, although destructive, did not result in long-term degradation. The cleared land was quickly revegetated by adjacent forests. Previously forested land did not always return to the original state under the stewardship of the uplanders, but adequate colonization took place to retard the loss of soil and nutrients, and restore wildlife habitat.

This cycle of reforestation requires approximately 25 years. In order for soil fertility to be restored, the land should lie fallow for at least five to ten years. But because of increasing pressures for land by a growing population, in many areas of the Pacific, this "resting" time has been drastically reduced or even eliminated entirely.

Overgrazing of livestock has also led to loss of forest land and soil productivity. Animals tend to favor some forage species, leaving behind hardy weeds. Unmanaged herds tend to compact soil, thus reducing water infiltration, exacerbating runoff, and precluding the establishment of desirable plants.

Monoculture, or single crop systems, also contribute to the destruction of the forest and the depletion of soil nutrients. Many of the lowland farmers migrating into the upland areas have traditionally planted a maximum of one or two crops. This practice was brought by the farmers into the uplands with deleterious effects on the environment. Continuous corn or upland rice cultivation on steep slopes promotes erosion and reduces soil fertility.

In many regions, there has been insufficient education, promotion, coordination, and enforcement of proper forest and watershed management practices. Natural mangrove forests on many islands have been destroyed by overcutting and clearing by loggers and farmers. The revegetation of these lands is often not enforced, or is difficult to enforce due to the demand for tillable land. Farmers with poor land management skills receive limited training due to underfunded government programmes. In many cases, government agencies in charge of forestry and agriculture fail to coordinate related programmes, thus undermining long-term government objectives. Even where properly conceived programmes do exist, enforcement and public support for them may be limited.

Economic pressures greatly contribute to the depletion of the forests and the destruction of the land. Over the past 35 years the production of fuel wood world-wide has more than tripled (World Resources, 1986). Nearly 1.5 billion people in 63 countries, or about 60 percent of the people who depend on fuel wood as their principle source of energy for cooking and heating, are cutting wood faster than it can grow back. At present consumption rates, the estimated fuel wood deficit will double by the year 2000 (World Resources, 1986). This demand for timber has pushed loggers deeper into previously inaccessible places. Capital poor, small-island economies rely on this export market to pay foreign debts.

2.2 Effects of deforestation

The forests of the Pacific islands are home to some 200 million people, many of whom rely on the moist evergreen forests for the necessities of life. The forests provide them with meat, fruit, nuts, and honey; with fibers and wood for building; with fuel wood for cooking, lighting and heating; and with fodder. Deforestation is of particular concern to both upland and lowland inhabitants of the Pacific islands where forested watershed areas located upstream from agricultural lands are vital to agricultural productivity. These upland water catchments are the principal sources of water for all downstream uses, and their destruction can cause serious harm to the environment and to people who make their living from the land.

Coastal dwellers of the Pacific islands relying on fresh and salt water fish products are affected as well. Soils eroding off bare slopes into streams are eventually discharged into estuaries and mangrove swamps that serve as breeding grounds for fish. Increased sedimentation can eliminate breeding grounds, as well as destroy aquatic plant material that is an integral part of the food chain. This destruction can ultimately ruin the local fishing industry.

With the intensive cultivation of forested land currently underway in the Pacific, the natural tree cover and soil fertility may never return. Instead, shrub and grass species are becoming the dominant type of vegetation. Once established, these species are difficult to eradicate. This has already occurred on many islands with the establishment of Imperata grasses.

2.3 Environmental benefits of agroforestry in the Pacific

Under some conditions, agroforestry programmes implemented in the Pacific, using proven agroforestry techniques, have effectively provided a number of benefits.

(i) Site stabilization. Tree and plant species selected for use in agroforestry projects have deeper, more extensive root systems than other species, and function to stabilize soil. This protective characteristic is enhanced when trees are planted on sloping lands.

Litter, consisting of leaves, branches, and bark, serve as ground cover, protecting the soil surface from the pounding of heavy rains and consequent splash-erosion. The organic matter, together with the root system which penetrates into deeper subsoil, increases infiltration and percolation. This is particularly important during times of heavy rainfall.

By increasing soil stability, trees and plant cover may also reduce siltation of streams and other water supplies, including irrigation channels. This protects lowland as well as upland systems.

Extremes in water flows, very low levels or flooding, may also be regulated with properly designed agroforestry systems. Well planned placement of vegetation also serves as an effective windbreak, reducing wind erosion and damage to crops.

(ii) Soil fertility. Certain species of trees and ground cover, by stabilizing topsoil and preventing erosion, can minimize depletion of soil and soil nutrients. In addition, plants with long taproots can help recover nutrients lost to the subsoil layers through leaching and infiltration. These nutrients can then be "recycled" back to the surface as organic matter, or litter that decomposes and re-releases vital nutrients. Trees and their roots improve soil structure and thus the movement of nutrients and water by producing increased amounts of stable particles and avoiding hardpans.

Legume plants or trees have the ability to "fix," or convert, atmospheric nitrogen into usable form for plants, thus contributing to soil fertility. Legumes serve an important role in providing the essential nutrient, nitrogen, to the trees and interplanted crops. Nitrogen is otherwise depleted rapidly and can be replaced only with application of large amounts of organic material or expensive chemical fertilizers. Some legumes, especially those with long taproots, function better than others in improving soil fertility.

(iii) Micro-climate regulation. Both the tree and plant foliage, which serve as a windbreak and canopy, and the mulch created by tree litter, provide shade and thermoregulation by reducing direct solar radiation upon the soil surface and moderating water flow, thus regulating soil moisture. By reducing sunlight, the effects of high temperatures -- such as hardening or "baking" of the soil surface, which causes impenetrability and loss of surface water -- are reduced. By maintaining the temperature balance at the soil surface, the micro-climatic conditions for decomposition of organic matter and release of soil nutrients are improved. The forest canopy also buffers drastic fluctuations between day and night temperatures. A full or partial canopy will also ensure the viability and productivity of light-intolerant plant species. Because coffee trees are an understory plant, under certain conditions productivity is increased when grown under other tree species. Leafy vegetables are also sensitive to strong sunlight. Proper shading can extend their growing season.

3 PEACE CORPS AGROFORESTRY PLANNING METHODS

3.1 Levels of Planning

Peace Corps agroforestry project planning takes place at several different interdependent levels (Table 1). These four levels of planning must overlap and work together if the goals and objectives of the parties involved are to be successfully implemented. Policies and programmes such as nationwide tree-planting programmes will only become a reality if they are understood and established with strong regional and local support, and if local communities have some measure of control over the specifics of project management and implementation.

Although the Peace Corps works with national and regional governments to outline and implement goals and objectives, the role of the agroforester working at the community level is to assist the community in identifying immediate needs and in setting up a schedule and mechanism for accomplishing the various tasks necessary to achieve these needs. The volunteer develops

TABLE 1
Peace Corps levels of planning.

Level of Planning	Examples
National Policies/Programme	Set up an integrated reforestation programme
Specific National Goals	Plant 40,000 trees in 2 years
	Launch community education campaign
Regional Level	Select project areas
	Establish community nurseries
	Train workers
Projects (Local Level)	Operational and management plans Organize project management team Plan a forestry programme Set up tasks Obtain financial and other support

three project plans (Table 2) and outlines tasks within each plan in the implementation of community needs.

In establishing a realistic programme to accomplish these tasks, the volunteer will construct a multi-faceted plan, a critical path diagram that sets out the expected time-frame for completion of the various tasks and depicts the interrelationships of the various tasks, and shows sufficient room for expansion or changes to the plan as the project progresses, conditions change, and experience is gained.

3.2 Developing the agroforestry plan

(i) Community needs, goals, and objectives. The success of a forestry project depends in large measure upon the level of commitment and involvement of the community or the individual farmer. If the participants feel they have defined the project(s) most crucial to their needs and stand to gain from it, then they will be more inclined to provide full moral and physical

TABLE 2

Peace Corps project plans.

Long-Term Plan

Identify community needs.

Define specific environmental or economic problems characteristic of the area.

Identify and choose appropriate solutions.

Medium-Term Plan

Outline specific task-oriented objectives.

Organize a project team. Provide each member with defined roles and responsibilities.

Complete a feasibility study that focuses on technical and financial resources, market availability, and social and environmental opportunities and constraints.

Design a project monitoring and evaluation system.

Short-Term Plan

Set a schedule for completion of specific tasks.

Begin implementation.

Provide project management; record keeping, maintenance and work schedules.

---

support. Throughout the project development process there should be mutual respect and cooperation between the community and the forester. With this as a guideline, the agroforestry volunteer is instructed to always make an effort to encourage and welcome participation in each step of the project development process. The commonly held objective amongst agroforesters is not to "do everything" but to assist the participants in planning and carrying out their own projects, using local resources and know-how in harmony with local practices and customs. The failure of volunteer agroforesters to understand participants' goals and available resources, and to apply appropriate capital and energy-intensive technologies has led to the failure of many projects.

(ii) Identification of community structure. The volunteer is encouraged to understand the organization of the community during the initial planning phase. Areas of social importance include:

(a) Formal community organization, leaders.

(b) Informal social structure.

(c) Sex roles, defined work and leadership roles.

(d) Important traditions and beliefs.

(e) Spirit of cooperation versus individualism.

(f) Social class distinctions and differences.

(g) Level of education, skills, and knowledge.

(h) Degree of openness to innovation.

Areas of political importance include:

(a) Factors affecting the project development.

(b) Formal political structure of the area.

(iii) Identification of community needs. The first step in the identification of needs is through an "inventory" of information resources. It is essential to identify key community leaders, other development workers (current or past) within the area or in neighboring regions, government officials, and others, and develop a good working relationship with them. These contacts can prove invaluable in expediting the development of the project by helping the agroforester gain access to materials, information resources, transportation, permits, a labor pool and other support essential to the project's success.

When establishing the network of contacts and base of support for community projects and programmes, the volunteer strives to be aware of local customs. Peace Corps stresses the importance of extending the accepted courtesies and following standard protocol in all written and verbal communications. Proper respect is extended to village elders, religious leaders, or government officials, and appropriate dress and decorum is maintained. In all cases, the volunteer is encouraged to demonstrate the highest level of professionalism.

Informal surveys are used as a tool for the agroforester to become familiar with various community members. Volunteers have found it helpful to draw up a list of questions or topics for discussion in advance. These surveys can be used to obtain general information or gauge the opinions of people on various subjects.

Peace Corps agroforesters also rely on observation for a great deal of information. A volunteer's living condition within the community provides an excellent vantage point for observing needs, customs, and practices. Planting practices of farmers, for example, can be learned through observation, and such things as the informal social structure can also be observed. For example, the agroforester may offer to join the farmers when they are out working in the field. This allows the agroforester the opportunity to learn how the farmers work and what they think about their work, to observe their methods, and to gain their trust. Many farmers enjoy this companionship.

(iv) Project support. The Peace Corps agroforester makes a special effort to gain the trust and confidence of this network of contacts in order to generate interest in the proposed project and to enlist approval, support, and assistance. One method of gaining support from the various groups is through the formation of a task force or advisory group. Depending upon the particular circumstances, this group may focus on a narrow task, such as providing technical information or services to the project, or may bring together individuals or groups whose coordinated, cooperative support is needed for the success of the project, and who normally would not meet together. In some cases, the volunteer will help the group work out disagreements through negotiation or other means. Generally, the role of the Peace Corps agroforester is that of technical support.

Under ideal conditions, the types of people in the task force are not limited. If social taboos restrict certain individuals or groups from participating, then it may be necessary to talk with them on an individual basis. In many societies, women are vital forces in maintaining the family and community: they contribute significantly to agricultural and other productive activities and serve important roles as economic decision-makers. Often, however, community development and extension projects and policies fail to adequately consider these important contributions and do not consult women or solicit their support. Peace Corps agroforesters are encouraged to avoid this situation and encourage the participation of women.

(v) Socio-economic surveys. One of the first tasks faced by the volunteer agroforester is the determination of the specific needs of the community in which a project is to be undertaken. This is true even in cases where there is a specific goal, such as establishment of a fruit tree nursery, which has been set forth by government policy or plan. In order to identify community needs, the agroforester must establish an adequate base of information about the culture, local customs and practices, economic conditions, past and present agriculture and forestry programmes, and social and political factors that affect the community.

Basic information is gathered through formal interviews or informal discussions or surveys with key people. Questions concerning sensitive issues are approached carefully in order not to offend the interviewee. Biased questions are avoided.

Undertaking a socio-economic survey is one method used by volunteers to gauge project support, and identify community needs and available resources. The surveys are not the type where interviewers go from door to door asking questions. Although that type of survey can be helpful, typically an agroforester does not have the time or command of the language to undertake such an endeavor. The type of survey used by volunteers is more informal. Casual conversations, visual observations, and where possible, research, are the typical methods of fact finding. Useful information is shown in Table 3.

TABLE 3

Suggested survey subjects.

Land: total area available, distribution and tenure patterns, soil quality, topography; current land uses.

Labour: total labour supply, availability throughout the year (need for farm labour varies significantly with planting and harvesting and seasonal cycles), non-farm or competing labour needs and other sources of income, level of skills training and education.

Capital: total available, ability and willingness of farmers to invest in forestry project, availability of government or other sources of financing.

Technical Information: in-country technical and private support organizations, foreign support organizations, universities or trade schools, libraries.

Material Support: seed sources, research stations, extension workers, host country and foreign donor organizations.

Existing Systems of Production: through interviews and observation, the farmers' current system of production may be analyzed and understood. An effort should be made to determine the success or failure of earlier attempts to introduce forestry projects or new production techniques to the area.

(vi) Environmental/economic opportunities and constraints. Although the identification of community structure, needs, and available resources is the first step in outlining a project, identifying realistic environmental and economic opportunities and constraints are no less important. Gathering data on the environmental and economic conditions assists the Peace Corps agroforester to understand needs as well as realistic opportunities and constraints (Table 4).

In some areas of the Pacific this type of information is not common knowledge, or readily available, thus requiring research on the part of the forester. Regional and national government agencies are often a source of reliable information. Public and private universities have departmental libraries with extensive collections of research papers and books on topics ranging from geologic conditions to native cultures. Private relief organizations such as CARE are also a good source of information. In some cases they can provide material published by other international organizations. Volunteers will often use foreign governments with offices in the country. These often have agricultural liaisons or aid offices that are willing to provide information. Another source of information often called upon are the offices of foreign companies. With a little imagination, Peace Corps volunteers have been known to locate numerous sources of information in the most remote of locations.

(vii) Available resources. Although a survey of available resources may overlap the socio-economic survey to a certain extent, there are

TABLE 4

Data requirements for assessing opportunities and constraints.

Physical and Environmental

- Topography, climate, weather patterns, soil, water, wind; in short, factors influencing species selection and plant growth

Economic

- Economic conditions
- Land ownership, plot size, income levels, availability of investment capital, equipment
- Current and potential products and markets. A survey of demand for forest products
- Access to markets and means of transport
- Past, current, or planned economic development activities in the area
- Policies affecting level and type of production, agricultural prices
- Existence of cooperatives for marketing or production
- Availability of capital through loans or other credit schemes

distinct differences. Whereas the latter provides a more general assessment of the physical and social environment, the former is more project specific.

Separating the two has proven to provide a clear understanding of not just the opportunities and constraints of a project, but also of the tools available to implement the project.

3.3 Project Selection and Evaluation

Following the initial steps in the project planning process, situational analysis and community needs assessment, volunteers are encouraged to continue with project planning and implementation. This is generally achieved by outlining criteria for project selection and evaluation, setting up a work plan, and designing a system for monitoring and evaluation.

Since no single set of criteria is adequate for judging the feasibility and chances for success of a proposed project, project agroforesters are trained to consider multiple factors. The project must meet acceptable social and cultural criteria, conform to economic realities, and work within the particular environmental constraints (Table 5).

Volunteers are also encouraged to recognize indirect costs and benefits. Indirect costs may include the negative short-term effects of increased water runoff when trees are first planted, or aggressive competition by introduced tree species. Positive effects, or benefits, include improvements in soil fertility brought about by nitrogen-fixing trees.

For proper economic evaluation, each of the inputs, outputs, and indirect costs and benefits are identified and valued. For some, such as the environmental costs of lost soil, evaluation is quite complicated. Economic evaluation is done at the micro-, or individual project, level or the macro- or political, level. At the micro-level, only factors directly affecting the individual farmers or community are considered. At the macro-level, a wider range of factors are used. The economic evaluation is typically completed before the project has begun, during the planning phase, or as a final evaluation after the project is completed. The techniques used in all cases are similar.

3.4 Setting Up a Work Plan

The work plan has been found to be a useful way for Peace Corps agroforesters to determine proper timing as well as to outline the various needs --technical assistance, financial backing, labour, and time-- that must be met to make the project operational.

TABLE 5

Social, environmental, and economic criteria.

Environmental criteria.

- Sustainable while meeting community needs
- Enhancement of existing crop production strategies
- Conservation of the forest ecosystem and its multiple benefits
- Avoidance of competition between species
- Use of tree species and management practices appropriate to local physical and climatic conditions
- Planting and harvesting in a non-disruptive manner.

Social Criteria

- Optimal use of local human and material resources
- Community needs and community support and participation
- Understanding of self-reliant capabilities.

Economic Criteria

- Maximization of benefits at least cost
- Maximization of returns on investment
- Achievement of specified production goals
- Achievement of economic yields greater than those of the current system
- New markets for products

(i) Timing considerations. Timing considerations are crucial to the success of agroforestry projects. For this reason, volunteers are encouraged to become thoroughly familiar with the agricultural crop cycle, seasonal weather patterns, and availability of rainfall throughout the year. The schedule of tasks to be completed for the agroforestry project are, insofar as possible, timed with weather conditions in mind and complement, rather than interfere with, the other duties and responsibilities of farmers and workers. The more time-consuming, labour-intensive tasks such as land clearing or site preparation, if possible, are scheduled at times during which labour is relatively plentiful and other agricultural tasks are few.

(ii) Task-oriented planning. Task-oriented planning is a useful approach to setting up a work plan. Projects are divided into a sequence of steps, or tasks, from beginning to end, with descriptions of the activities required to complete each of the steps. The development of the work plan takes place with the full participation of the community. A table or chart is used to list the project goals as well as the various tasks required to attain them. As the tasks are laid out, the resource, material, labour, and time requirements are determined for each step of the project. Each of the tasks listed can be broken down further into subtasks, with the requisite needs and requirements listed. Possible problem areas and contingency plans should be noted as well.

3.5 Critical path in project planning

A forestry project, regardless of scale, is composed of numerous different activities which must be carried out in proper sequence and with careful timing. Even in a small project there are complex problems involving the scheduling and monitoring of activities that must be calculated so that the whole work programme proceeds smoothly within the allotted time and budget constraints.

Identifying and outlining a critical path is one technique used by volunteers to help control and schedule complex operations. It is concerned with finding the optimum way to manage a complete system, such as all the activities involved with afforestation of a particular area. The technique involves a graphic representation of the events needed to complete a task or activity in the logical sequence in which they take place. Circles represent the events and are connected to arrows, which represent the activities themselves.

The system flows from left to right; that is, the activities are drawn in sequential order from left to right. Overlapping or concurrent activities can be also represented. After the logical sequence is graphically represented, a time scale is attached. The network diagram can then be analyzed in order to determine, for example, the proper starting time for the project, or to set interim target dates for certain activities, such as planting or harvesting.

The analysis assists in determining the crucial activities that must be completed in order for the entire operation to be finished within the targeted time frame. The network can be continually updated and modified as the project progresses, and serves as a valuable reference. It helps the Peace Corps volunteer to adhere to deadlines and evaluate performance by serving as a basis or comparison between targeted and actual accomplishments.

3.6 Project monitoring and evaluation

Another crucial aspect of project management is periodic evaluation and feedback. Information is collected and recorded on a regular basis so that it can be analyzed for evaluation purposes. If the project is divided into steps, or sequential tasks, then evaluation takes place at various stages of task completion. For long-term projects, periodic evaluation sessions are scheduled throughout the year.

The information collected for evaluation provides a basis upon which the agroforesters can react to changing conditions, and thus alter the project in order to meet goals and objectives. Obtaining feedback at frequent intervals alerts agroforesters to problems while they still may be remedied. It also allows modifications in time schedules or techniques as needed. The information is also used in presentations to funding organizations, important officials, community members, or others who have decision-making power. Most funding agencies require regular reports and often have their own reporting systems established.

In order to determine whether the project is proceeding satisfactorily and whether it is meeting its stated objectives, volunteers set up a system of evaluation based on measurable terms. The "outputs" of a project are either qualitative or quantitative. Quantitative measures, such as the number of trees planted or volume of wood or fruit collected, are relatively easy to record and evaluate systematically. Qualitative measures, such as degree of community participation or satisfaction of farmers, are more difficult to accurately describe and evaluate, although they are of great importance.

3.7 Identifying sources of funding

Among the difficulties the agroforester encounters is the shortage of capital or financial support needed to undertake and carry out a project. This is especially difficult in subsistence-farming areas. Because agroforestry projects are perceived as long term and may provide little or no immediate returns to farmers or financiers, start-up capital is not always forthcoming from local sources. Furthermore, it is difficult to obtain loans from lending institutions when communities or farmer have a bad credit rating due to the failure of past programmes. If such is the case, then it is difficult to convince banks to lend capital. Agroforestry projects that provide a relatively rapid return on investment in short-rotation crops while trees reach harvestable size provide a more attractive adjunct to forestry work. A sound financial and work plan can convince banks to lend capital.

To obtain outside support, the agroforester will take an "inventory" of potential sources such as government agencies, private investors, charitable organizations, research institutions, or international development agencies. This will be done by researching local or regional library and university facilities, and then contacting the regional or national offices of these institutions. The volunteer can also seek out other community workers to inquire where likely funding sources might exist. Alternatively, the volunteer can investigate local lending institutions and farmer cooperatives, or investigate other similar projects that have been funded or financed. In addition, the volunteer can coordinate with other Peace Corps volunteers working in cooperative banks.

In developing project proposals, the Peace Corps agroforester stresses not only the environmental and economic benefits of the project, but how the community will be involved. This last element is most important. In the case of Peace Corps volunteers who may only work on a project for two years, funding agencies want to be assured the community investment is strong enough to continue a project regardless of the presence of a volunteer.

3.8 Project team selection and labour requirements

As discussed earlier, the Peace Corps agroforesters attempts to develop a good working relationship with community residents and determine the nature of both the formal and informal organizational hierarchies within the community. Once this task is completed, the volunteer begins to select the planning team.

When Peace Corps volunteers are calculating time requirements for the various project-related tasks, the time necessary for training, field visits and meetings is included. Labour requirements for tasks such as land preparation, watering, and planting are estimated based on experience gained during similar projects, if possible. Contingency plans are drawn up in case there is a shortage of workers or the project falls behind schedule.

3.9 Budgeting and record keeping

A project budget, or cost analysis, helps the volunteer estimate the amount of materials and money as well as labour required to complete the project. For accurate project related expenses, the budget will also include items that cannot be given a monetary value, such as volunteer services. "In-kind" items or donations are also included.

Record keeping is considered another crucial part of project management. Early in the planning process, a comprehensive system is set up to monitor finances and document all aspects of project activities. The ultimate goal of development projects is to develop a self-reliant capability within the community. Therefore, it is important that information be recorded in a form than can be used as a reference even after the agroforester leaves the area. Training and reference material, as well as documentation of planting times, spacing, site requirements, and problems such as insect infestations, should be recorded for carry-over, together with their effects, solutions, and any changes in plans. This may be the only permanent record available to other community workers or for project workers in other areas.

An inventory of equipment and other supplies is also kept so that the progress of the project is not hindered by shortages of needed materials. If machinery is used, then maintenance and repair records are kept. A checklist of routine maintenance procedures is considered a good way to keep equipment working properly or to spot problems before they become too serious.

4 REFERENCES

Carter, V. G. and Dale, T., 1981. Topsoil and Civilization, Norman, Oklahoma: University of Oklahoma Press.

Department of International Economic and Social Affairs, 1985. World Population Prospects: Estimates and Projections as Assessed in 1982. Population Studies No. 86, United Nations, New York, 1985.

Plumwood, V. and Routley, R., 1982. World Rain forest Destruction-The Social Factors, The Ecologist, 12 (1).

Sanchez, P.A., 1976. Properties and Management of Soils in the Tropics. New York: John Wiley and Sons.

United States Forest Service, 1986. An Assessment of the Forest and Range Land Situation in the U.S., U.S. Department of Agriculture, Washington, D.C.

United States Peace Corps, 1984. Information Collection and Exchange, Agroforestry In-Service Training, United States Peace Corps Information Collection and Exchange Office of Programme Training and Support, May.

World Resources Institute and the International Institute for Environment and Development, World Resources, 1986. Basic Books, New York.

Chapter 8

AGROFORESTRY FOR DEVELOPMENT IN KENYA: AN OVERVIEW

A. GETAHUN
International Council for Research on Agroforestry, P.O. Box 30677, Nairobi (Kenya)

ABSTRACT

Agroforestry development in Kenya is relatively new but its growth and development during the last decade has been unprecedented. Such accelerated development is largely attributable to concerted efforts between and amongst the Government of Kenya (GOK), donors, and the non-governmental organization community. These joint efforts were fortunately matched by the country's strong tradition and infrastructure for tree planting and the personal commitment of His Excellency, Daniel arap Moi, President of Kenya, in leading the nation in active tree-planting tradition. Agroforestry research and development in Kenya can be seen to have gone through several stages.

Plant introduction into Kenya has been active over the years and as a result, there are over 100 exotic tree and shrub species being grown. More than 75% of the tree and shrub species annually planted are exotic and such imbalances are currently being addressed by the government through programmes promoting indigenous tree and shrub species.

Fifteen major GOK policy interventions that have had a direct and positive effect on agroforestry development were enacted between 1971 and 1987. While the necessary policy instruments are in place and government efforts to date are commendable, government-based agroforestry programmes and activities are nonetheless weak and require donors support and the active participation of non-governmental organizations.

Equally important in agroforestry development are the Kenyan farmers, especially the Kenyan women and youths who have become active in tree planting and are now adopting agroforestry practices. Continued and growing mass-awareness of farmers regarding agroforestry as the most appropriate tropical land-use system in the face of a growing population pressure will facilitate GOK, NGO, and donor efforts.

Agroforestry developments in Kenya, especially during the last ten years, are impressive, and the future is bright. What is urgently needed now is to take stock of these efforts and lessons learned, and to clearly establish relationships and responsibilities between four groups: the Kenyan government, non-governmental organizations, donors, and farmers. By doing so, the development paths of agroforestry and its wide use for sustained development will be charted and made solid.

1 INTRODUCTION

1.1 The agroforestry foundation in Kenya

Agroforestry research and development in Kenya is relatively new but its growth and development during the last decade (1977-1987) has been unprecedented. Such accelerated development is largely attributable to concerted efforts between the Government of Kenya (GOK), the donor, and the non-governmental organization (NGO) community. Agroforestry and related activities are carried out by several government ministries and research organizations on one hand, and by development agencies and a large number of national and international NGOs on the other hand. These collaborative efforts were fortunately matched by a good national political climate, including the highest office in the republic.

Agroforestry research and development since 1977, and particularly after 1980, have mushroomed both in government and NGOs with financial and technical support from several bilateral and multilateral donor agencies. According to a survey by the Ministry of Energy and Regional Development (MOERD), by 1985 there were 13 major national organizations and 63 others active in agroforestry/social forestry and general tree planting activities in Kenya and this number has been on the increase (Energy Development International, 1985). But despite the proliferation of organizations and activities in agroforestry, it is generally felt that these efforts are still insufficient and suffer from a lack of institutionalization. There is also an inherent danger of duplication of efforts and even misdirection.

Kenya's initial interest in agroforestry as a potential technology and farm practice was based on biomass energy supply short-falls and the observed land and environmental degradation resulting from deforestation to meet increasing demands for wood, feed, and food for a rapidly growing population. Agroforestry as a multiple land-use system was thought to enable the production of agricultural crops, tree products, and livestock from the same unit of land, thereby replacing less efficient systems of farming such as bush fallow/shifting cultivation and expansive pastoral systems. As a result of this interest, in 1981 the MOERD, in close collaboration with the Ministry of Agriculture (MOA), the Ministry of Environment and Natural Resources (MENR), the Ministry of Livestock Development (MOLD) and selected NGOs, such as Kenya Energy Non-Governmental Organizations (KENGO), began a national effort to implement a coordinated programme of agroforestry research, training, and extension.

The establishment of the International Council for Research in Agroforestry (ICRAF) and its location in Kenya has had direct and indirect positive impacts on agroforestry developments in Kenya. The establishment and operation of its 40 hectare agroforestry field station at Machakos and ICRAF's expanded mandate since 1985, which led to the establishment of its collaborative programmes division (COLLPRO) to work with and assist national efforts in agroforestry, are strongly linked to the growth of agroforestry in Kenya.

The major GOK policy interventions that have a direct and positive effect on agroforestry development in Kenya include:

(a) The forest policy for Kenya developed in Sessional Paper 1, Economic Management of Renewed Growth, in 1986, directed MENR to service farm forestry activities.

(b) The Rural Afforestation Extension Scheme (RAES) (1971) was established to promote tree planting by farmers and tree planting, generally in both public and private lands outside the gazetted forest reserves, and later (1974-78) was required to emphasize watershed protection.

(c) The Chief's Authority Act (Chapter 128, Laws of Kenya) was passed to protect tree cutting and the establishment of chief's nurseries and the Presidential decree of 1980 which required every chief to establish a tree nursery in his location and promoted District Development Committees (DDC) to establish nurseries as well.

(d) Establishment of Permanent Presidential Commission for Soil Conservation and Afforestation (PPCSCA) promoted tree planting to protect the environment and which in 1984, recognized Agroforestry as a key instrument and has since then become a strong advocate for it.

(e) Parts of the Basic Land Usage Act of 1965 were revised in 1981 to allow land use in steep slopes as long as agroforestry principles were employed and this facilitated work in soil conservation and watershed management.

(f) The creation of the MOERD (1980) promoted tree planting as a renewable energy programme, especially using agroforestry to generate biomass energy.

(g) Inter-ministerial committee on agroforestry (1982) under MOERD is instituted to coordinate and promote agroforestry.

(h) Presidential Tree Fund (1985) improved financial basis for Kenya African National Union (KANU) Youth Wingers to be involved in tree planting programmes.

(i) The 8-4-4 Education (1985) required schools to teach tree planting and to maintain nurseries.

(j) Sessional Paper No.1 of 1986, Economic Management for Renewed Growth, recognized agroforestry as a sound land use and planned for 3 million hectares for the five-year planning period from 1984-1988.

(k) National Annual Tree Planting Day. In 1983 required GOK (RAES) seedling production be increased from 60 million to 200 million annually.

(l) Directorate of Personnel Management (DPM) (Treasury) in mid-1986 instructed Ministries of Energy and Regional Development, Agriculture, and Environment and Natural Resources to develop a common policy and operation for agroforestry under MOERD leadership (July, 1986) and increased GOK's manpower in agroforestry.

(m) Kenya Forestry Research Institute (KEFRI) and Kenya Agricultural Research Institute (KARI) (1986-87) become autonomous research institutes, thereby taking much of the agroforestry research responsibilities.

(n) Ministry of Planning and National Development (1986) instructed the District Development Plans of DDCs to include agroforestry in their plan and budget 1986-89 and MOERD provided the District Energy Planning Task Force.

(o) National Steering Committee for Agroforestry Research (1987) is established. The committee is chaired and coordinated by the National Council of Science and Technology (NCST). This 15-member committee includes an NGO representative (KENGO), and it establishes KEFRI as having the lead in agroforestry research.

The key GOK ministries in agroforestry and social (community) forestry are MENR, MOERD, MOA, and MOLD. The Agricultural Society of Kenya (ASK) and National Youth Service (NYS) are also active. Outside these ministries and their specialized departments, the government also encouraged and allowed research and educational institutions to develop programmes in agroforestry. Those that have now initiated agroforestry research and training and/or extension include educational institutions (The Universities of Moi and Nairobi, University Colleges of Egerton and Kenyatta) and research agencies (KARI, KEFRI, NCST/CSC).

Kenya's total area stipulated to be used for agroforestry is estimated at 3 million hectares, largely based in the medium to high-potential areas. This level of activity is needed to meet energy requirements and food security. The specific aim of this policy is to plant a minimum of ten fruit trees per family for two million small-scale farms over the planned period. Equitable and sustainable development is the objective of this national policy and the full participation of the farmers, NGOs and DDCs is the means to achieving this objective.

1.2 Agroforestry research and development stages

The developmental history of agroforestry in Kenya is indeed short and falls into two periods traditional agroforestry in Kenya and agroforestry developments before 1977, and the first decade of accelerated agroforestry development from 1977-1987.

This first decade of agroforestry research and development in Kenya has gone through four stages.

-Stage I (1979/80): Beginning of mass-awareness regarding biomass energy and environmental degradation and potential role of agroforestry.

-Stage II (1981/82): Starting of agroforestry field programmes.

-Stage III (1983/84): Expansion of agroforestry research, training and extension activities.

-Stage IV (1985/88): Kenya and AFRENA.

2 AGROFORESTRY IN KENYA AND AGROFORESTRY DEVELOPMENT BEFORE 1977

2.1 Traditional agroforestry systems in Kenya

Kenya lies on the Equator, but dry lands dominate, accounting for no less than two thirds of the total area. High potential areas account for less than 20% as shown in Fig. 1. Thus, Kenya, though it lies in the heart of the tropical world, enjoys diversified and varied climate largely due to altitude, which in turn determines rainfall.

The Kenyan highlands and their associated valleys and the humid/sub-humid coastal strip and much of the Lake Victoria basin account for the high potential lands. These areas are naturally forested and the associated agricultural practices of shifting cultivation and fallow-crop rotations could be described as the traditional agroforestry land-use systems of the region. But, with increasing population pressure, these systems have now given way to a sendentary agriculture and/or plantation (cash-crop) agriculture, including forest plantations. Outside this small but economically important region is the large arid and semi-arid area, making up much of Kenya. This zone, traditionally a pastoral region, is also undergoing change. Where deforestation is not severe, the pastoral system can be described as a silvo-pastoral system of agroforestry practices. In this zone of semi-arid areas, including the upland savannas, intensive agricultural and livestock production is increasing. Further, both arid and semi-arid areas are ecologically sensitive to intensive land use.

Within the broad definition of agroforestry systems and practices, the following types are still in use in Kenya.

(a) Windbreaks and boundary planting involve both commercial and small-holder crop farming involving many species including *Grevillea robusta*, *Eucalyptus* spp., *Casuarina equisetifolia*, *Cassia siamea*, *Acrocarpus fraxinifolius*, *Hakia saligna*, etc., largely in the highlands.

(b) Mixed intercropping and enriched fallows is a second major system in Kenya. There are two subcategories within this system: trees with crops and trees with pastures/livestock. Trees with crops systems include (1) Markhamia lutea, Sesbania sesban intercropping with maize, millet, beans and cassava in western Kenya and (2) Grevillea robusta, Markhamia lutea, Cordia africana intercropping with food crops in the central Kenya highlands. Trees with pastures/livestock include (1) Croton megalocarpus, Croton macrostachyus, Acacia abyssinica retained in pastures in the highlands, (2) Anacardium occidentale (cashew nut) and dairy association in the Kenya Coast, (3) Melia volkensii in mixed farming in dry lands, (4) Acacia xanthoploea in dairy/beef production systems in flood plains, and (5) Acacia albida, Acacia tortilis in crop and grazing lands in dry areas.

(c) A third major agroforestry system used in Kenya is home gardens and fruit orchards. These systems are found in areas of dispersed homesteads, for example in the highlands and the coastal strip. These are more complex and multi-layered near the homestead and less so in distant farms. Examples include (1) mangoes, citrus, guava, and papaya in dry lands, the Lake Victoria basin and the coastal strip and (2) avocado, plums, peaches, and apples, in the warm highlands.

2.2 Social forestry and woodlots

The reduced access to government or gazetted forests and natural bush for firewood and the motivation for cash needs, has sustained and increased the level of small farmers' planting of fruit trees and multipurpose trees. The active plant introduction, largely spearheaded by white settlers, has enabled the African farmer to have access to many new fruit and timber species as nearly all species planted are exotics or recently naturalized (Getahun et al., 1981; Getahun, 1982). The following are illustrative examples.

(i) The shamba system. This is a form of taungya, initiated during the turn of this century, has established 164,815 hectares of industrial forest plantations in the Kenyan highlands (Oduol, 1986).

(ii) Wattle plantations. Both MENR and the East African Tanning Company have established large acreages in both western and central highlands. Farmers in central highlands also plant steep slopes with wattle as a source of pole, fuel wood, or charcoal.

(iii) Plantation-crop, (tea, coffee) shade trees. This includes Grevillea robusta, Milletia dura, Acrocarpus fraxinifolius in the highlands. This system has been on the decline.

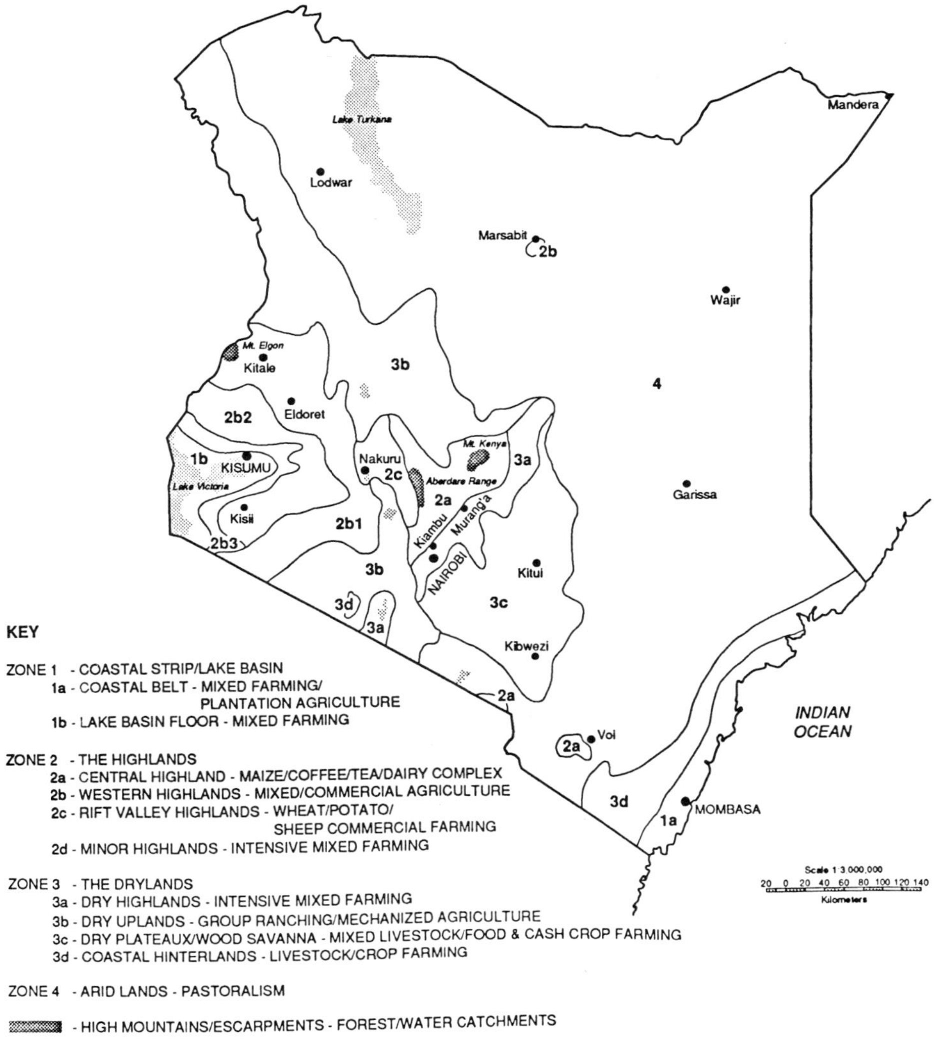

Fig. 1. Climatic regions and agricultural potential of kenya.

(iv) Woodlots. These systems are used by the railroad industry, tea estates, tobacco farms and mini-woodlots in high density highlands and include *Eucalyptus* spp., *Cassia siamea*, *Casuarina equisetifolia*, *Cupressus lusitanica*, *Acacia mearnsii*.

3 THE FIRST DECADE OF AGROFORESTRY DEVELOPMENT: 1977-1987

3.1 The role of IDRC and ICRAF's establishment

In 1977, IDRC published a report titled, "Trees, Food and People and Land Management in the Tropics", which, among other things, recommended the "setting up of an internationally financed council for research in agroforestry" to administer a comprehensive programme leading to better land use in the tropics. This resulted in the establishment of ICRAF during the same year (Cummings *et al*., 1984) and Nairobi was selected as its permanent headquarters in mid-1978.

The establishment and location of ICRAF headquarters in Kenya has served as an effective catalyst and subsequent operation of its field station in Machakos in late 1981. Its limited on-farm trials and demonstrations have enabled ICRAF to directly participate in Kenya's agroforestry development efforts and, through its more recent AFRENA zonal programme (1986), regionally in tropical Africa.

Since its establishment, ICRAF has witnessed the coming of many agroforestry programmes and the involvement of many agencies, both in the government, and in the NGO community, and to a certain extent, in the private sector. ICRAF has participated in many of these efforts. The most significant events are briefly described under the respective agencies and are summarized in Tables 1 and 2.

TABLE 1

GOK agencies/institutions and commercial firms involved in agroforestry research and development.

GOK agencies/institutions	Major Activities	Geographic focus
GOK Ministries/Departments		
Ministry of Agriculture		
Soil Conservation Agroforestry Unit	Soil Conservation	National

Ministry of Environment and Natural Resources		
Baringo/Tugen Hills	Afforestation	National
RAES	Nursery/ Afforestation	National
MENR/CARE	Agroforestry extension	Regional
Ministry of Energy and Regional Development		
KREDP		
GTZ/SEP		
KWDP		
Canada/MOERD Energy Advisory Group	Energy planning and policy	National
Ministry of Livestock Development		
Zero-grazing Programme	Highlands	National
Beekeeping Programme	Dry Areas	National
MOLD/ILCA	Coast area dairy development	Regional
SB-CRSP	Small ruminants/ agroforestry	Maseno
Presidential Commission for Soil Conservation and Afforestation	Legislative	National
NCST	Project support/ institutional development	National
Ministry of Finance (KREMU)	Resource inventory (drylands)	National
<u>Institutions/Commercial Firms</u>		
KARI	Agroforestry in land and water management	National
KEFRI	Seed services, training and agroforestry research	National
Tea Research Foundation/KTDA	Wood fuel production	Kericho, Kenya
Moi University	Nitrogen-fixing trees evaluation	Eldoret
Nairobi University	Agroforestry research	Kabete
Egerton University	Agroforestry research training and extension	Njoro
Lake Basin	Environmental conservation and tree planting	Regional
Tana/Athi River Development (TARDA)	Environmental conservation and tree planting	Regional
Coffee Research (Ruiru)	Coffee intercropping	Regional
National Horticulture Research (Thika)	Fruit trees on-farm	National
Pan-African Paper Mills	Tree planting	Zonal
E.A. Tanning Extract Co.	Wattle plantations/ charcoal	Zonal
Kenya Seed Company	Seed multiplication	Zonal

BAT	Wood fuel production/extension	Zonal
Baobab Farm Ltd.	Land rehabilitation /pole charcoal production and livestock	Coastal

3.2 ICRAF helps raise national awareness on agroforestry

In late 1980, ICRAF, in cooperation and with support from the ministries of Agriculture and Livestock Development (MOALD), Environment and Natural Resources (MENR), and the University of Nairobi, hosted the first national agroforestry seminar. The seminar brought together, for the first time, a wide variety of professionals to discuss their views and experiences on agroforestry in Kenya. A number of measures and options were recommended for future guidance in agroforestry technology development including the early establishment of a coordinating committee in agroforestry to help promote and sustain agroforestry activities among all government and non-government organizations (Buck, 1981).

The seminar confirmed that there was already a wide interest in agroforestry in Kenya and much useful data and discussion were contributed during the 10-day seminar, which included participation of experts from Sudan, Tanzania, Uganda, Burundi, and Rwanda. Kenya also benefited from the international workshops on agroforestry that ICRAF hosted during 1979 and 1980.

3.3 GOK, donors, and NGOs response

(i) Ministry of Energy and Regional Development. The Government of Kenya established the Ministry of Energy in 1980. This ministry was later (1983) merged with the Ministry of Regional Development, to become the Ministry of Energy and Regional Development, and subsequently, included the National Irrigation Board (NIB). The initial task of the new ministry was to undertake a national survey of energy needs, emphasizing renewable energy resources and to draft a national policy on energy and investment plans for this sector. The results of the survey were published in 1982, and the report recommended the use of agroforestry practices in order to increase wood fuel resources and food production as well as to arrest environmental degradation (Beijer Institute, 1982; O'Keefe and Shakow, 1981). This emphasis is reflected in the draft national energy policy. This ministry collaborated with the Ministries of Environment and Natural Resources, Agriculture and Livestock Development and selected NGOs (i.e. KENGO,

TABLE 2

Districts with agroforestry activities (1986)

DISTRICT	ECOLOGICAL ZONE	MAIN ACTIVITIES	ORGANIZATION/AGENCY
Coast Province			
Kilifi	Humid, Sub-humid lowlands	Research, Training, Extension	GOK/MOERD, MENR, MOALD, USAID
Lamu	Sub-humid lowlands	Extension	GTZ/SEP
Kwale	Humid, Sub-humid	Research, Training, Extension	GOK (MOERD, MENR, MOALD, OP
Taita/Taveta	Sub-humid Highlands	Training, Extension	GOK, DANIDA
Tana River	Aridlands	Extension, Research	FINIDA, WB, NIB
Eastern Province			
Machakos	Drylands	Research, Extension	ICRAF, KARI, KEFRI, IDRC, FAO, EEC, GTZ
Kitui	Drylands	Research, Training, Extension	GOK (MOERD, MENR, MOALD), JICA, KENGO, USAID, Green Belt
Embu	Highlands (dry)	Extension	ODA (EMI)
Meru	Highlands (dry)	Extension	ODA (EMI)
Isiolo	Drylands	Extension	ODA (EMI)
Central Province			
Nyeri	Highlands	Research, Training	GOK, USAID
Kirinyaga	Highlands	Extension	GOK
Nyandarua	Highlands	Extension	GOK
Murang'a	Highlands, Savanna	Extension	GOK
Kiambu	Highlands, Savanna	Extension	GOK, GTZ/SEP. Green Belt
Nairobi Province			
Nairobi	Savanna	Research, Training, Extension	GOK, USAID, GTZ/SEP, Green Belt
Rift Valley Province			
Nakuru	Highlands, Savanna	Research, Extension	GOK (KENGO)
Baringo	Drylands	Research, Extension	GOK, WB, FAO, DUTCH
Kajiado	Drylands	Extension	KENGO, GOK
Laikipia	Drylands	Extension	GOK, NGO, SWISS
Nandi	Drylands	Extension, Training	GOK
Turkana	Drylands	Extension	NORAD
Kericho	Highlands	Tree Planting	Tree Research Foundation
Western Province			
Kakamega	Highlands	Research, Training, Extension	GOK, BI, USAID, PEP
Busia	Highlands	Training, Extension	GOK, KENGO
Bungoma	Highlands	Extension	GOK, NGO
Eldoret	Highlands	Research	KNFU/Rotary International
Trans Nzoia	Highlands	Extension	KNFU
Nyanza Province			
Siaya	Savanna	Extension, Training	GOK, CARE, KEFRI
S. Nyanza	Highlands, Savanna	Extension, Training	GOK, CARE, ICIPE. OXFAM. Green Belt
Kisii	Highlands	Research, Extension	GOK, BI, USAID

Maendeleo ya Wanawake), and undertook the implementation and/or facilitation of the following projects where agroforestry is a major component.

(a) The Kenya Renewable Energy Development (KREDP) financed largely by United State Agency for International Development (USAID) and initiated in late 1981.

(b) Kenya Wood fuel Development Project (KWDP) largely under Dutch financing and initiated in early 1983.

(c) Gesellshaft fur Technische Zusammenarbeit/Special Energy Programme (GTZ/SEP) financed by GTZ and initiated in 1983.

(d) The Baringo Fuel and Fodder Project under Dutch government and World Bank financing, initiated in 1981.

(e) Marigat Fuel Wood and Afforestation and Extension Project supported by Food and Agriculture Organization and Australia funding and operating since 1981.

(f) The Bura Fuel wood and Agroforestry and Extension Project supported by Finish International Development Aid (FINIDA) and World Bank.

(g) District energy planning targeted towards making the districts, the district development committees (DDCs), capable of planning and meeting energy needs and where agroforestry and social forestry constitute important means of achieving the goals. This is currently being financially and technically assisted by Canadian International Development Aid (CIDA) under Ministry of Energy and Regional Development (MERD).

(ii) Ministry of Agriculture and Livestock Development. This ministry, in conjunction with the Swedish International Development Aid, established a national soil conservation programme. This programme increased fodder and browse species and the growth of fruit trees over the last decade. In addition, this ministry created an agroforestry unit in 1983 with substantial budgetary allocation in 1985/86, but the unit's field programme has remained weak except for the production and distribution of tree seedlings for farmers. Its current efforts include implementation of 3 million hectares agroforestry on 2 million small-scale farms. [Editors Note: The MOALD was divided into to two separate ministries in 1988. These are the Ministry of Agriculture (MOA) and the Ministry of Land Development (MOLD)]

(iii) Ministry of Environment and Natural Resources (Forestry Department). Since 1971, this ministry, through its forestry department, has been increasingly active in social forestry, partly because of the establishment of the RAES programme. The major objective of RAES is to create awareness of the need to plant trees for fuel wood, building materials, and soil and water conservation, and to facilitate this with the provision of technical and logistical assistance to schools, community groups, self-help groups and individual farmers. The programme reportedly operates in all 41 districts of the nation. The scheme was largely limited to community forestry activities but in 1984, RAES accepted agroforestry as within its philosophy and tradition (Permanent Presidential Commission on Soil Conservation and Afforestation, 1984). The full potentials of the nursery programme has yet to be realized.

This ministry's other programmes outside RAES include:

(a) Baringo/Tugen Hills reafforestation projects funded by the Government of France and started in May 1986.

(b) Elgeyo Marakwet Aid and Semi-Arid Lands (ASAL) Afforestation Programme, a project aimed at species trials and planting method research for ASAL in the Elgeyo Marakwet district.

(c) Embu/Meru/Isiolo (EMI) Integrated Development Project, a British government funded project with a forestry component aimed at protection of hills and catchment areas.

(d) Establishment of Nursery Training Centers and Dryland Afforestation, a project funded by the Japanese International Cooperation Agency (JICA) for training people in tree nursery technology. (This is currently under KEFRI, based at Muguga and Kitui).

(e) Fuel wood Afforestation and Extension Project, a FAO/Australian government funded forestry project. Its main activities include species trials, and establishment of demonstration facilities and tree nurseries.

(f) Kitui ASAL Programme, a forestry-based project funded by USAID and aimed at afforestation of the arid and semi-arid areas of Kitui district.

(g) Machakos Integrated Development Project, an European Economic Community (EEC) funded project with a large forestry component.

(h) Magarini Settlement Scheme Forestry Project, a forestry project financed by the Australian government and aimed at rehabilitation of the new settlement through tree planting.

(i) Taita/Taveta District Development Programme and South Nyanza District Afforestation Programme.

(j) Turkana Rural Development Project, a project supported by Norwegian Agency for International Development (NORAD), whose activities include species trials and ASAL afforestation activities in the area.

(k) Lake Kenyatta and Hindi Settlement Scheme Agroforestry Project. This project is under German financial and technical assistance.

(l) The _Euphorbia_ Project. Under Belgian assistance and included work on _Prosopis_ spp., currently inactive.

(iv) Research institutions: KARI and KEFRI. The Kenyan Agricultural Research Institute (KARI) and Kenyan Forestry Research Institute (KEFRI) were created in 1986 as autonomous research institutions of the government. Both institutes include agroforestry research and KEFRI has already established field trials at Muguga and elsewhere (Machakos, Kibwezi, Gede, and Siaya). Donors' support (i.e. World Bank, USAID, SIDA, and CIDA) is expected to be good. These institutions collaborate with ICRAF viz a viz the AFRENA programme, as well as with CARE International for Kenya in western Kenya.

(v) Institutions of higher learning: Moi University, University of Nairobi and Egerton University College. These institutions have agroforestry research and limited training and extension activities in agroforestry. Of these, Egerton University College has a more developed programme of agroforestry.

(vi) Other GOK institutions: National Youth Service, GOK prisons, and Farmer Training Centres. The National Youth Service (NYS) and GOK prisons (e.g. Mombasa and Kitui) are currently active in agroforestry species trials, seed production and zero-grazing activities. These programmes are likely to increase with more technical support from parent ministries as well as from KEFRI and KARI. The programmes are effective tools in training a large number of Kenyans from all parts of the Republic. More than 10 FTCs are currently active in agroforestry activities.

3.4 Donors and key non-governmental organizations

The list of donors and NGOs involved in agroforestry is long. Both groups effectively supplement the Government of Kenya's financial and field staff deficiency. Their involvement in agroforestry has largely come via their concern in environmental protection and biomass energy supply-and-demand management. More recently, however, both recognize agroforestry as meriting support in its own right.

The agroforestry field programmes where donors and NGOs participate with GOK are summarized in Tables 3 and 4. These programmes vary greatly in size and intensity within the respective district or agro-ecological region.

3.5 The Kenyan farmer and the Kenya National Farmers Union

Kenyan farmers, especially women and youth, hold a special place in the short history of agroforestry development. In general, Kenyan farmers, especially in the highlands, have been quick to accept and adopt new crops and practices as long as there is a good market for the product. Examples include bixa, pyrethrum, and avocado. Farmers adherence to row-planting as

TABLE 3

Major donor agencies and International NGOs involved in Agroforestry research and development in Kenya.

Agency	Donor	NGO	Major Activity	Location/Geography
ACTION-AID		x	Development/Tree Planting	Multiple locations
ANEN	x		Information/Funding	Regional
BELGIAN GOVERNMENT	x		Euphorbia Project	Baringo
CRS (Catholic Relief Service)		x	Rural development	Many Districts
			Zero-grazing, nursery	
CARE-INT. (Kenya)		x	Research, Extension,	Western Kenya
Council for Human Ecology (CHEK) Kenya		x	Beekeeping	Kibwezi
CIDA	x		Biomass Energy Advisory Service	National, S. Nyanza
DANIDA		x	Agroforestry/Afforestation	Taita/Taveta
DUTCH GOVERNMENT (KWDP)	x		Agroforestry/Biomass Energy/Fodder	Kakamega, Kisii, Muranga
ELC	x		Strengthening NGOs	National
EEC (MIDP)	x		Forestry/Agroforestry	Machakos
FINIDA		x	Wood fuel/Agroforestry	Tana River District
FAO/ADB/AUSTRALIAN GOVERNMENT	x		Fuelwood/Afforestation extension	Baringo
GTZ/SEP	x		Agroforestry/biogas/ zero-grazing	Baringo Muka Mukuu
GTZ (Forestry Group)	x		Agroforestry/FO/Tree seed	Muguga, Nairobi
ICRAF		x	Agroforestry/Research	Machakos
			Training/Information	Western Kenya
IDRC		x	Dryland AF Research/ Demonstration	Machakos
INADES-Formation		x	Training/Information	Nairobi
JICA		x	Agroforestry/Social	Muguga (Nairobi)
			Forestry Training	Kitui
NAS (Moi University)		x	N-fixing tree research	Multiple Locations
NORAD	x		Rural development/ Re-afforestation	Turkana
ODA (UK)	x		E.M.I. Project	Embu, Meru, Isiolo Districts
OXFAM (UK)		x	Agriculture/Health	Drylands (Turkana, Sambru)
Partnership For Productivity Service Fdn.		x	Training/Development	4 Provinces
Redd Barna (Norway)		x	Tree planting/Nursery	S. Nyanza, Turkana
SIDA	x		Agroforestry for Soil & Water Conservation	National
SWISS DEVELOPMENT	x		RAES	National
The Future in Our Hands		x	Afforestation	South Nyanza
UNESCO (IPAL)	x		IPAL Project - range	Marsabit
USAID-EIA	x		Energy initiatives	Africa
USAID (KREDP)	x		Agroforestry Research/ Development	21 Districts
USAID (SR/CRSP)	x		Agroforestry Research	Maseno
World Vision		x	Tree planting Training	Dry areas (country wide)
Intl. World Bank	x		Forestry Sector RAES Baringo (BPSAAP) Project	National

well as to tree planting for soil conservation has no doubt facilitated the acceptance of agroforestry as a farm practice. Indeed, new species are quickly accepted by farmers to the extent that 75-100% of the trees planted by farmers annually are exotics. Species brought in after 1980, which are already popular include: Calliandra calothyrsus, Sesbania grandiflora, Leucaena hybrids, Gliricidia sepium, Mimosa scabrella, and Flemingia congesta. This phenomenon has led the government to increase its campaign in favour of planting tree replacements with introduced rather than native species. Educational and research institutions unfortunately have excluded indigenous trees from their curricula and research agendas.

More recently, the Kenya National Farmers Union (KNFU), in collaboration with the Swedish Rotary International, has initiated assistance to farmers for tree planting in western Kenya and has completed a book, Tree Growing and Use for a Better Future. In August 1987, KNFU, in collaboration with ICRAF, hosted a two-day agroforestry workshop for the African Farmers Organization and the International Fund for Agricultural Development (IFAD). In the process, KNFU has become more aware of agroforestry and more interested in agroforestry. It is believed that KNFU's use of agroforestry to meet farmers' needs will increase.

4 A LOOK INTO THE FUTURE

There is room for both concern and optimism for the continued and potential use of agroforestry for development. The concern comes from the uncoordinated nature of agroforestry development and the lack of designated institutions such as ICRAF.

The institutionalization of agroforestry within the government, at present, is problematic. The government's attention and interest in agroforestry for development came via energy and environmental degradation issues. It is no accident that the 1984-1988 government 5-year development plan calls for the development of 3 million hectares of agroforestry farms in the country, and for the DDC to include agroforestry developments in their district development plans and development portfolios.

On the one hand, development of an agroforestry research programme (KARI, KEFRI, NCST) and agroforestry's establishment within educational curricula (Nairobi, Moi, Egerton, Kenyatta) institutions has been slow. As a result, trained humanpower development in agroforestry will continue to suffer if corrective measures are not taken. On the other hand, the initial development of agroforestry programmes in the young Ministry of Energy and Regional Development may have facilitated developments. It still has to be

TABLE 4

Major indigenous NGOs in Agroforestry Research and Development in Kenya.

NGO	Major Activities	Geographic focus
BAHAI COMMUNITY	Tree planting	Mombasa
CATHOLIC DIOCESE KISII	Soil conservation/ tree planting	Zonal
KISUM	Soil conservation/ tree planting	Zonal
SIAYA	Wood energy	Zonal
CPK Maseno Diocese	Tree planting	Zonal
Friends World Committee	tree planting/ conservation	Select districts
Green Belt Movement	tree planting/ conservation	Select districts
ISLAMIC FOUNDATION	Horticulture	Dry areas (Isiolo)
KENGO	Agroforestry/ energy conservation training/extension/information	National
Kenya Catholic Secretariat	Tree planting/ energy conservation	National
KENYA FREEDOM FROM HUNGER COUNCIL	Tree planting	Nyeri, Kiambu, Dry areas
Kenya Girl Guides Assoc.	Tree planting	National
Kenya JAYCEES	Tree planting	National
Kenya Scouts Association	Training/ ariculture/tree planting	National/ local
Limuru Boys Center	Youth training	Kiambu
Maendeleo ya Wanawake Org.	Energy conservation, tree planting	National
(KANU) Mazingira Institute	Agroforestry demonstration	District
Men of Trees	Forestry finance	Nairobi
Mennonite Central Committee	Forestry technical assistance	National
NCCK	Tree planting	National
NCWK	Tree planting	National
PFP/Kenya	Training, technical services	Western Province
Salvation Army	Tree planting	Rift Valley, Machakos
V.A.D.A.	Umbrella to 25 NGOs	National
Wildlife Club of Kenya	Tree planting	National

resolved which GOK ministry should be the national host or coordinating body for agroforestry. Many agroforestry projects are under the auspices of MENR (e.g., CARE-Kenya, DANIDA, World Bank, and FAO/UNDP) and some under the MOA and MOERD. The issue of the designated GOK ministry for agroforestry primarily affects development policies and investments. Implementation at the local and district levels is not a major problem as the policy of district focus for rural development will continue to be used.

The key and active institutions in government, however, need to be given assigned leadership and thus mandated to coordinate a specific activity and/or geographical focus. These activities include (a) tree seed services, (b) research and technology generation, (c) training and information management, and (d) rural extension and technical services.

There is a critical need to assess where agroforestry is now and to sharpen its focus. Both the Ministry of Energy and Regional Development and the Ministry of Agriculture, including KARI, have had multi-donor consultative meetings and studies to increase support to agroforestry and agricultural research in general. ICRAF has also completed a survey of agroforestry research and needs in the Kenyan and other east African highlands as part of its regional AFRENA programme and macro D&D for Kenya. SIDA/NORAD have also carried out a needs assessment of NGOs in agroforestry which has resulted in increased support to KENGO. Both the World Bank and the USAID have shown a policy emphasis on agroforestry's role in tropical environmental management and both have recently conducted field assessment studies. The ICRAF 10th anniversary "Agroforestry for Development" conference has augmented Kenya's and the world's interest in agroforestry. All these are positive developments and need to be consolidated into short-term and long-term programmes with shared responsibilities.

The current gaps in agroforestry tree seed services, formal training and agroforestry technology packaging need to be resolved. The need for an agroforestry source book for use by development experts for information on training and education, agroforestry research, development and extension in Kenya, is also urgent.

5 REFERENCES

Beets, W.C., 1985. Agroforestry in Africa farming systems E/DI for Africa Bureau. United States Agency for International Development, 62 pp.

Beijer Institute, 1982. Energy development in Kenya: problems and opportunities, Draft Discussion Document for consideration by the Steering Committee for the Kenyan Fuel wood Project, Stockholm.

Brokensha, D. and Castro, A.P., 1984. Fuel wood, agroforestry, natural resource management: the development significance of land tenure and other resource management/institutionalization systems. Institute of Development Anthropology. New York, 64 pp.

Buck, L., (Editor), 1981. Kenya National Seminar on Agroforestry. Nairobi.

Burley, J. 1982. Obstacles to tree planting in arid and semi-arid lands: Comparative case studies from India and Kenya. UNU. 52 pp.

Chavangi, N.A., 1988. Kenya forestry sub-sector review. Workshop Paper on Sociological aspects of farm forestry. Nairobi, 25 pp.
Cummings, R.W., Burley, J., Castillo, G.T. and Navarru, L.A., 1984. Report of the External Panel of ICRAF. Sept.-Dec.
Depommier, D., 1985. The role of woody vegetation in soil conservation and rehabilitation. Unpublished report ICRAF.
East, B., 1986. Social forestry extension methodologies and the introduction of agroforestry into traditional farming systems in South Nyanza District. KREDP, Nairobi, 34 pp.
Energy Development International (E/DI), 1985. A Directory of Organizations Working on Tree Planting and Wood fuel Conservation in Kenya. (Compiled by W. Mathu), 99 pp.
Energy Development International (E/DI), 1987. Agroforestry in Kenya: The Experience of KREDP. Nairobi, 65 pp.
Food and Agriculture Organization, (n.a.). Forests, Trees and People. Forestry Topics Report No. 2. FAO Forestry Department, 40 pp.
Ford Foundation, 1985. People and Forests: A report of Ford Foundation Meeting on Social Forestry and Agroforestry. Naro Moru, Kenya, 20-22 March, 1985, 43 pp.
Fortman, L. and Rocheleau, D., 1985. Women and Agroforestry: Four myths and three case studies. Agroforestry Systems 2(4): 253-272.
Fowler, A., Barrow, E., Ubara, D. and Odera, 1986. An evaluation of the CARE-International in Kenya Agroforestry and Energy Programme. Nairobi, 35 pp.
Getahun, A., Muturi, D.S. and Wamahiu, P.M., 1981. Potential fuelwood, afforestation, agroforestry species for Kenya, E/DI, MOE. Nairobi, 5 pp.
Getahun, A., 1982. Production and protective roles of Prosopis. A review. KREDP, Nairobi, 21 pp.
Getahun, A., 1983. Tree seed availability in Kenya. E/DI. KREDP, Nairobi, 17 pp.
Getahun, A., 1983. The role of agroforestry in soil and water conservation in Kenya. Proceedings of the 2nd National Workshop, Nairobi. IDS Vocational Paper # 42: 338-349.
Getahun, A. and Ndegwa, C.M., 1982. Wood fuel in Kenya: Problems and Prospects. Proceedings of the ICRAF/BAT Workshop on Agroforestry Systems for small scale farmers, Sept., ICRAF, Nairobi.
Gielen, H., 1982. Report in an agroforestry survey in three villages on Northern Machakos, Kenya. 114 pp.
Hankins, M., 1987. Renewable Energy in Kenya. Nairobi, 152 pp.
Harrison, P., 1982. The Greening of Africa: Breaking through the battle for land and food. IIED/Earthsan Study. Paladin, 388 pp.
Hoekstra, D.A. and Gelder, B. van, 1983. An Annotated Bibliography of Economic Analysis of Agroforestry Systems/Technologies. ICRAF, Nairobi, 44 pp.
Hoekstra, D.A., 1984. Agroforestry Systems for the Semi-arid areas of Machakos District, Kenya. ICRAF, Nairobi, 28 pp.
ICRAF, 1987. Profile in ICRAF in Africa. ICRAF, Nairobi, 152 pp.
Kerkhof, P., 1987. South Nyanza District Afforestation Programme. Report of the Agroforestry Department. DANIDA/Forest Department, Homa Bay, Kenya.
Kerkhof, P. and May, N., 1985. Agroforestry in kakamega. Misc., 18 pp.
Kunzel, W., 1984. Magarin Settlement area: Kenya, 56 p.
Maathai, W., 1985. The Greenbelt Movement, Nairobi.
Nair, P.K.R., 1980. Soil Management in Agroforestry. Proceedings of the Kenya National Seminar in Agroforestry, 12-22 November, 1980. 251-272.
National Task Force, 1988. Agroforestry potentials for the land-use systems in the bimodal highlands of Eastern Africa, Kenya. Jointly prepared by the Kenya Government and ICRAF and sponsored by USAID, 152 pp.
Ngugi, A.W. and Bradley, P.N., 1986. Agroforestry, soil conservation and

Wood fuel in Muranga District, Part I. KWDP. 65 pp.
Oduol, P.A., 1986. The shamba system: An indigenous system of food production from forest areas in Kenya. Agroforestry Systems 4(4): 365-373.
O'Keefe, P. and Shakow, D., 1981. Fuel wood in Kenya: The possibilities for agroforestry centers.
Owino, F., 1988. Development of social forestry in Kenya and application of agroforestry technologies. Paper presented at the Kenya National Social Forestry Seminar, Muguga, Aug. 31, 1988.
Permanent Presidential Commission on Soil Conservation and Afforestation Kenya's efforts to conserve soil, water and forests, Nairobi, Annual reports 1982, 1983, 1984, 1985.
Raintree, J.B., 1986. Agroforestry pathways: Land tenure, shifting cultivation and sustainable agriculture. Unasylva (Italia), 38(154): 32-43.
Raintree, J.B., 1983. Preliminary analysis of land and agroforestry potentials in N. Mbere Division, Embu District, Kenya. ICRAF, Nairobi, 16 pp.
Reshid, K., 1987. Agroforestry in Kenya: A bibliography with annotations. ICRAF, Nairobi, 70 pp.
Rocheleau, D.F., 1984. Land-use planning with rural farm households communities. Participating Agroforestry Research. Network on the Role of Sociologists and Anthropologists in Farming System Research. Lusaka, Zambia, 27-29 Nov., 62 pp.
Scherr, S.J., 1988. The legislation context for agroforestry development in Kenya. ICRAF, Nairobi.
Teel, D.B., 1980. A Pocket Dictionary of Trees and Seeds in Kenya. KENGO, Nairobi. 151 pp.
Thomes, D.B., 1980. Agroforestry in relation to soil and water conservation: The utilization of steep slopes In: Proceedings of the Kenya National Seminar on Agroforestry, 12-22 November, 1980, pp. 235-250.
Van Gelder, B. and Kerkhof, P., 1985. The Agroforestry Survey in Kakamega District, KWDP. Nairobi.
Waneti, F.H.M., 1981. Agroforestry practices in high potential areas of Kenya. In L. Buck (Editor) Proceedings of the Kenya National Seminar on Agroforestry pp. 287-308.
Wenner, C.G., 1980. Trees in erosion and soil conservation in Kenya, Nairobi-Ministry of Agriculture, 26 pp.
Wenner, C.G., 1981. Soil Conservation in Kenya. 7th edition. Ministry of Agriculture, 230 pp.
Williams, P. J., 1986. Much needed research (Agroforestry Research work at Mtwapa, Kenya Coast), 11 pp.
World Bank, 1978. Forestry Sector Policy Paper, World Bank, Washington D.C.

Chapter 9

PLANNING FOR AGROFORESTRY RESEARCH IN WESTERN SUDAN: APPROACHES, ISSUES AND LESSONS LEARNED--THE EVOLUTION OF A MANAGEMENT APPROACH

J. C. NOEL,
International Program Development, Washington State University, Pullman, Washington, 99164 (United States of America)

Z. A. SAAD,
Department of Forestry; Khartoum (Sudan)

W. T. BUNDERSON
Department of Natural Resource Sciences, Washington State University, Pullman, Washington, 99164 (United States of America)

ABSTRACT

Agroforestry research methodologies respond to the need to incorporate the conservation of natural resources and the sustainability of agricultural (including livestock) and/or forest enterprises as explicit objectives of agricultural and natural resource research. These evolving methodologies complement and support traditional disciplinary and commodity research and farming systems methodologies. Most agricultural research has targeted large and medium scale agricultural producers as its primary clientele. Natural resource researchers have served the marketers and consumers of these resources, with emphasis on economic returns. Some of the newer interdisciplinary approaches such as farming systems research have focused on limited resource farmers as their major beneficiaries. Agroforestry methodologies can and should be applicable to a wide range of agricultural enterprises, from large commercial to limited input agriculture.

The Western Sudan Agricultural Research Project (WSARP) has established an agricultural research network in the four western provinces of Sudan. The research programme functions in concert with its parent organization, the national Agricultural Research Corporation, to plan and implement research leading to sustainable improvements in agricultural production and productivity for the farmers and pastoralists of the regions. Within the context of a systems-oriented research approach (i.e. production systems or farming systems approaches), the WSARP developed a strategy, policies and procedures, and carried out research and planning efforts in which agroforestry figured prominently. The chapter discusses the evolution of this approach and its interface with existing organizations and institutions. Lessons learned from the development of this programme and their implications for management of agroforestry programmes and activities are discussed. Specific WSARP technical agroforestry activities and results are included in a separate chapter (see Bunderson _et al_., Chapter 10).

INTRODUCTION

A common and logical criterion by which research is evaluated is the benefits which accrue as a result of that research. Research does cost money, and someone must be prepared to either directly support it or champion its support. Therefore, research is generally designed and implemented with the intent of serving particular clientele. Different approaches to research have evolved which target different clients. Basic scientific research targets the scientific community, providing insights, understandings and tools which can then be used as building blocks for the solution of problems and other practical applications. To the extent that the targeted clientele -- the scientific community and disciplinary peers -- find the results useful, the scientist is rewarded. While some agricultural, forestry, and natural resource research has a similar basic science orientation, much has as its clientele the agricultural or forestry industries of the developed countries. With the historical direct correlation between production and economic benefits, agricultural research especially has evolved with a bias toward production, and often toward specific commodities. In the forestry and natural resource fields, much of the emphasis has been on how to discover, harvest and exploit, for economic returns, the resources of our planet.

In recent years, recognition has grown that conventional agricultural research is not meeting the needs of some important clients. As a result, alternative research methodologies are evolving which target new clients. Some of the pioneering work has been done in the developing countries. Foremost among these new approaches has been farming systems research (FSR) which has as its primary clientele the smallholder and/or limited resource producers. This method focuses research on the problems faced by the small farmer, using interdisciplinary teams of researchers to describe the farming system, diagnose key constraints to production, and conduct research, often on-farm, to address these constraints.

Early FSR approaches overlooked another important client, the government. This client heavily influenced the conditions faced at the farm level through policy, infrastructure, credit, and other factors. It also had to assume financial and managerial responsibility for the new research approach, one which was at odds with the existing (disciplinary and/or commodity oriented) systems. In many instances, small incremental changes in production resulting from the new approach were not perceived by the government or by "old guard" researchers as a significant improvement over the traditional approach. In response, FSR broadened its perspective to include extension (FSR/E) and then policy (FSR/E/P). This extended the potential impact of on-farm improvements to a larger population, and better linked on-farm research with regional and national decision-makers.

However, these approaches share a critical shortcoming. The clientele they serve need and expect short-term payback on their investments in research. Whether in Sudan or the United States, the small farmer must feed his/her family; the larger scale agricultural producer or forest industry member must derive economic benefits; the government must deliver on political promises. Failure to obtain short-term results of demonstrable benefit to these consumers of research findings can threaten their very existence.

It is increasingly evident that many of the production increases and improvements in productivity brought about through traditional agricultural research and through FSR and its derivatives cannot be sustained. In many countries, and for many commodities, production has outstripped demand. Costs of many of the inputs are escalating. Most importantly, the natural resource base -- the soils, water, forests, ranges, and the most basic biological and physical components of the environment -- are being depleted, rearranged, and polluted at an alarming rate. Thus, it is the natural resource base and the increasing number of people who realize that it must be protected and restored who are emerging as clients to be served. It is their agenda that the emerging methodologies of agroforestry seek to include in the research equation.

The Western Sudan Agricultural Research Project (WSARP), conceived in the mid-1970s and operational today, has been an ongoing experiment in the integration of each of the above research agendas. It is not an "agroforestry project", but is representative of many development efforts which can benefit from incorporation of agroforestry elements and approaches. It is a rich case study because it has grappled for over a decade with many of the very real technical, cultural, political, economic, and management difficulties inherent in implementing such a new approach. This chapter will focus on those elements that relate to management, while the companion chapter (Bunderson et al., Chapter 10) will deal with the technical programmes themselves.

2 WESTERN SUDAN AGRICULTURAL RESEARCH PROJECT: PROJECT SETTING AND CONTEXT

2.1 Overview of Sudan's environment and agriculture

Sudan straddles the Sudan-Sahelian zones of Africa. It is the largest country on the continent, with a total land area of 2.5 million square kilometers and a human population of 20 million. Topographically, Sudan is a vast plain with occasional hill and mountain ranges such as the Red Sea Hills in the northeast, Jebel Mara in the west, the Nuba Mountains in the central area, and the Imatong Mountains in the south. The country has a broad array of ecological zones which roughly follow a north-south gradient. Desert and sparse scrubland in the north give way to arid and semi-arid bush and wooded savanna in the central region, and finally to lush grasslands, tall woodlands and closed-canopy forests in the south (Harrison and Jackson, 1958). Rainfall is unimodal in distribution and follows a gradient similar to that of the vegetation, from less than 25 millimetres per annum in the far north to over 1500 millimetres per annum in the extreme south.

Agriculture contributes approximately 40% of the total gross domestic product and about 90% of the export earnings, while providing a livelihood for 80% of the population. Major crops produced include sorghum, millet, groundnuts, sesame, cotton, and gum arabic. The latter is produced largely from *Acacia senegal* trees, but also from *A. seyal* and *A. polyacantha*. Livestock represent about one-half of the gross national product but contribute a smaller percentage of total export earnings. Sudan's national herd currently numbers over 53 million head, including 18 million cattle, 19 million sheep, 13 million goats, and 3 million camels. These numbers collectively represent 30.4 million tropical livestock units, accounting for 18% of Africa's ruminant population.

2.2 Environmental setting

Western Sudan encompasses the provinces of Kordofan and Darfur, an area of about 850,000 square kilometers (35% of the country). More than 80% of the 5 million people in western Sudan depend on agriculture, livestock, and/or gum arabic for their livelihood. Most crops and virtually all livestock are produced by traditional systems, with sedentary farming dominating the former and transhumance the latter. However, crops and livestock are important in both systems. Major crops in the north include millet, sesame, sorghum, groundnuts, watermelon, and a variety of horticultural crops from small irrigated gardens. Gum arabic is collected by both crop and livestock producers. Livestock production is dominated by camels in the north and by cattle in the south. Small ruminants are common to both regions, with larger desert breeds predominating in the north and smaller breeds in the south. Western Sudan contributes to the nation's agricultural production more than 90% of the millet, 54% of the sesame, 46% of the groundnuts, 17% of the sorghum, 6% of the cotton, and 90% of the gum arabic. In Sudan's livestock sector, the west contributes 65% of the camels, 45% of the cattle, 37% of the sheep, and 32% of the goats (World Bank, 1977).

This chapter focuses on the area served by the WSARP, the central savanna zone of Kordofan at an elevation of 400 metres to 800 metres between latitudes 10°-14°N and longitudes 28°-31°E (Fig. 1). Crop and livestock activities are concentrated here due to habitation problems in the arid north, and to a lack of developed water supplies and high incidences of disease, insects, and mud in the extreme south. The climate of this zone is hot and semi-arid with a mean annual rainfall varying from 300 millimetres in the north to over 900 millimetres in the south. Rain is concentrated in a single short season which increases in reliability and length from north (July-September) to south (May-October). Considerable fluctuations in the amount and distribution of rainfall are common from year to year. High temperatures and low humidity prevail for most of the year, contributing to high rates of evapo-transpiration. Ecologically, the area can be categorized into three broad zones based on soils, vegetation, and climate. Details on these characteristics have been previously described (Harrison and Jackson, 1958; Bunderson, 1981; Bunderson et al., 1984; Hunting Technical Services, 1981; WSARP, 1982, 1985a, 1985b).

(i) The qoz sands and gardud non-cracking clays in the north and west. This region lies between 400 to 600 metres elevation and consists of wooded and bush grasslands on red stabilized sands (qoz) which alternate with non-cracking clay pediplains (gardud). Vegetation structure and composition vary considerably with the environment and history of land use.

Qoz soils in the north occur in the 300 to 600 millimetres rainfall zone and are dominated by open bushed grassland with Acacia senegal, Albizia amara, Boscia senegalensis, Combretum glutinosum, and Guieria senegalensis in the woody component. A. senegal dominates many of the drier regions of the north. Common species in the herb layer are Zornia glochidiata, Aristida, Eragrostis, Brachiaria, Chloris, Dactyloctenium, Digitaria, and Cenchrus spp.. Qoz areas in the southwest have a rainfall of 600 to 800 millimetres and are dominated by a taller, denser tree component of Combretum glutinosum, Terminalia spp., Balanites aegyptiaca, and Prosopis africana. Species of Hyparrhenia, Andropogon, Setaria, and Sporobolus are common additions in the herbaceous cover.

Gardud soils are usually interspersed with northern qoz areas, and there is a considerable overlap of plant species between the two soil types. Major woody species are: A. mellifera, A. nubica, Terminalia brownii,

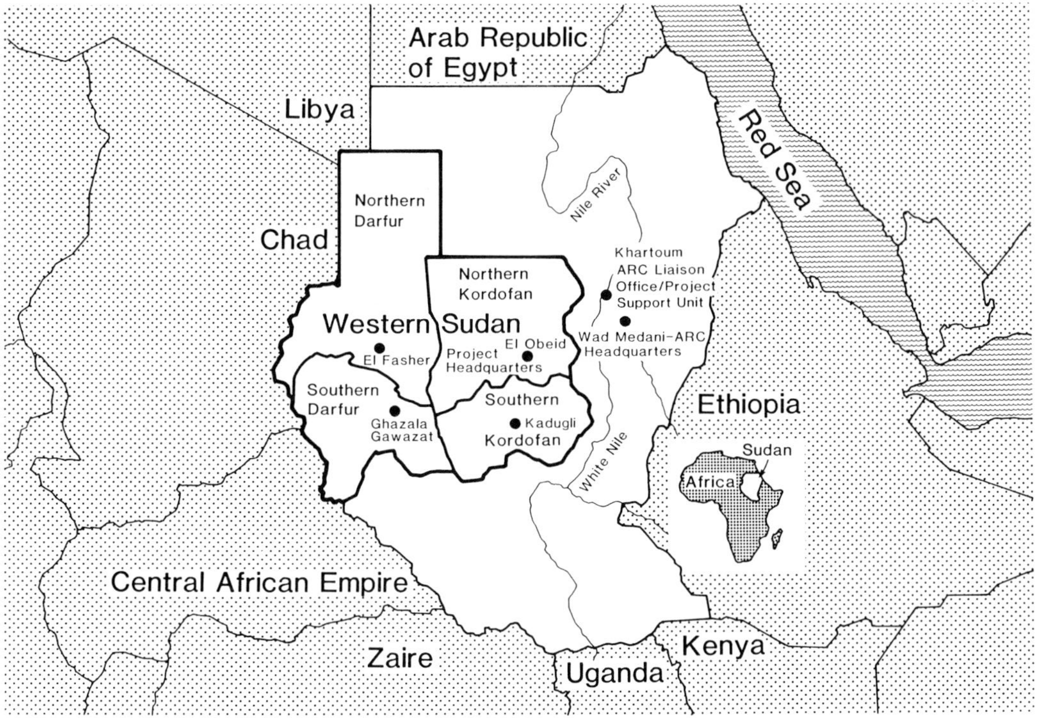

Fig. 1. Locations of WSARP research stations in Sudan.

Dalbergia melanoxylon, and Cadaba spp. Common herb species include Schoenefeldia gracilis with Aristida and Cenchrus spp.

(ii) The Nuba mountain complex in the south-central region. The Nuba Mountain area in the 600 millimetres to 800 millimetres rainfall zone consists of flat or gently undulating plains of cracking and non-cracking clays, interspersed with rocky hill ranges and their footslopes of sandy loams and clays. Elevation varies from about 500 metres in the plains to over 800 metres on hilltops. The plains account for about 60% of the land surface, with the hills and footslopes occupying the balance. Vegetation consists of a woody savanna that varies in structure and composition with changes in soils and relief.

Hill sides of rocky, sandy clays and loams consist of a woodland savanna dominated in the tree component by Combretum hartmannianum, Lannea fruiticosa, Boswellia papyrifera, and Serculia setigeria. In the herb layer, Hyparrhenia confinis, Andropogon gayanus, Loudetia togoensis, and species of Panicum, Aspilia, Aristida, and Brachiaria are common. Hill footslopes contain an open wooded savanna of Albizia amara, Combretum hartmannianum, Balanites aegyptiaca, Anogeissus leiocarpus, and Ziziphus spina-christi trees. The herb component is diverse with short-medium height species that include: Hyparrhenia confinis, Setaria pumila, Dactyloctenium aegyptium, Shoenefeldia gracilis, and species of Brachiaria, Chloris, Digitaria, Sporobolus, Aristida, Pennisetum, Spermacoce, Crotalaria, and Indigofera.

Cracking clay plains are dominated by a thorn savanna of A. seyal, A. polyacantha, Balanites aegypiaca, and Dichrostachys cinerea. The herb layer is tall with Hyparrhenia pseudo-cymbaria, Sorghum purpureo-sericeum, Brachiaria obtusiflora, Setaria incrassata, Pennisetum ramosum, Ischaemum afrum, Cymbopogon nervatus, and Rottboellia exalta. Areas of 500 millimetres to 600 millimetres of rainfall per annum, generally north of the Nuba Mountains, contain A. mellifera in place of A. seyal but are otherwise similar. Red non-cracking clays are diverse in woody species dominated by Albizia amara, Acacia gerrardii, Combretum glutinosum, C. hartmannianum, Dalbergia melanoxylon, Lannea humilis, Anogeissus leiocarpus, Sclerocarya birrea, Grewia fallax, and Dichrostachys cinerea. Major herb species include Setaria pumila, Hyparrhenia confinis, Loudetia togoensis, and species of Andropogon, Brachiaria, Chloris, Justicia, Cassia, and Commelina.

(iii) Broad clay plains in the south. This area covers an extensive flat plain south of the Nuba Mountains, with a rainfall of 750 to 900 millimetres. The dominant soils are dark cracking clays with vegetation similar to the thorn savanna plains of the Nuba Mountains. However, the area features a denser herb cover and more perennials in the genera Andropogon, Setaria, Hyparrhenia, and Echinochloa. Interspersed within the area are patches of red non-cracking soils very similar to those in the more mesic sites of the Nuba Mountains. The southern plains also contain low, broad depressions of seasonally inundated cracking clays dominated by Echinochloa pyramidalis and E. stagina. These areas are virtually devoid of trees except for Acacia nilotica, A. sieberiana, and A. seyal along their fringes.

2.3 Socio-economic and development setting

Recent trends associated largely with increasing population pressures in Sudan are dramatically altering the socio-economic, political, and ecological circumstances under which the traditional systems of production have formerly functioned successfully (Bunderson, 1985; Noel et al., 1985; McGahuey, 1986). Crop and livestock productivity are declining in response to these pressures, and the natural resource base is widely acknowledged to be deteriorating. The latter has been affected in particular by expanding cultivation, shortened fallows, widespread burning, seasonal overgrazing of fragile areas, and extensive tree cutting for cultivation, fuel wood, and building materials. These problems have been exacerbated by emerging conflicts over land use and by increasing dependence on costly external inputs.

Many previous development programmes in western Sudan shared weaknesses with those elsewhere in Africa. Their effectiveness was limited because of failure to understand the cultural, political, and ecological context of the targeted systems. Likewise, the tendency of the development community to treat symptoms in isolation from underlying causes was and is evident. Single-component technologies, based on western ideals, have been emphasized. Examples are animal vaccination and destocking campaigns, water development, breeding programmes, range management, mono-cropping, mechanization, and use of mineral fertilizers (Bunderson and Frye, 1986; Noel et al., 1985; McGahuey, 1986).

The government of Sudan continues to face severe economic and political pressures from inside and outside. As a result, economic development is stagnant, and infrastructure is limited. Within this context, research is a costly investment and such investments have not been made lightly.

3 WESTERN SUDAN AGRICULTURAL RESEARCH PROJECT: GENESIS, MANDATE, AND RATIONALE

3.1 Project genesis

Because of a recognition that some of Sudan's agricultural research and development programmes had been ineffective and/or inappropriate, reassessment of agricultural development approaches was underway at the time that the WSARP was conceived. The WSARP grew out of a number of related planning efforts. In 1974, Sudan's then Ministry of Agriculture, Food and Natural Resources (MAFNR) requested Ford Foundation assistance in assessing the status of crop and discipline research capabilities in Sudan, and in developing a plan to strengthen Sudan's agricultural research and related services. Subsequently, the International Agricultural Development Service (IADS) assisted in the further development of a master-research plan for the country, with emphasis on the rainfed sector in general and the western regions in particular (IADS, 1977). The World Bank's International Development Association (IDA) concurrently proposed a project to strengthen the Agricultural Research Corporation (ARC), a semi-autonomous parastatal organization which functions as the major agricultural research arm of the government. This project would also establish an ARC research network in the western regions (World Bank, 1977, 1979). The United States Agency for International Development (USAID) had recently reopened a programme in Sudan and had proposed a project for research and rehabilitation to decrease desert encroachment in northern Kordofan. The government of Sudan (GOS) brought all these efforts together into a single project, the WSARP, to be implemented by the ARC and funded by the GOS, IDA, and USAID (USAID, 1980).

WSARP's mandate was a blend of the agendas of each of the major contributing agencies and was "to increase the capability of the ARC to develop and test agricultural production systems that conserve and rehabilitate the natural resources and improve the standard of living of the subsistence farmers and pastoralists of Western Sudan" (USAID, 1980). Specific project outputs were to include: (a) the construction and establishment of supportive infrastructure and staffing of research stations in each of the four provinces of the west and a support unit in the capital city of Khartoum; (b) the establishment of research programmes at each station; and (c) the training of Sudanese scientific, technical, and administrative personnel. Capital construction costs and the purchase of an aircraft were financed primarily by the World Bank, with technical assistance and a portion of the construction funded by USAID. Technical assistance was provided by the Consortium for International Development (CID), with

Washington State University (WSU) as the lead university for project implementation.

3.2 Western Sudan Agricultural Research Project rationale and linkages to Agricultural Research Committee

At the time of project initiation in 1978, the ARC had a long history of involvement in agricultural research. This research was organized along disciplinary lines and focused on irrigated production of export crops, principally cotton. As a result of the earlier Ford Foundation study, ARC research programmes were beginning to undergo changes to move toward a more diverse and coordinated commodity and factor problem research specific to regional and national concerns. The ARC's historical mandate was exclusively for selected (primarily export) crops and it had not previously incorporated livestock, forestry, range, or socio-economic disciplines into its programmes.

The need to include these elements was gaining recognition at the time of WSARP's inception, and agroforestry featured strongly in the emerging plans. Integrating livestock and woody plants with crops can improve and diversify agricultural productivity on a sustained basis. Benefits to household income and welfare, as well as to the environment, result (Felker, 1978; Kang and Reynolds, 1986; McGahuey, 1986). Failure to recognize the value of livestock and trees can lead to serious problems. For example, Sudan's Gezira scheme, at one time the largest irrigated area under single management in the world, had originally eliminated all trees to prevent birds from roosting in them and causing crop damage. Livestock were also banned from the scheme to prevent crop damage. The former resulted in a fuel crisis for farmers working on the scheme. Faced with the threat of losing its labour forces, timber and fuel were brought in from afar at great expense to the scheme. The effects of the banning of livestock were more insidious. Although farm families survived without livestock, later health studies revealed nutritional deficiencies and increased child mortality due to lack of animal protein, especially milk and eggs.

The Gezira scheme was replanned to include forestry components with fast-growing tree species such as the *Eucalyptus* and *Conocarpus* spp. as well as shade trees such as mahogany (*Khaya senegalensis*) and neem (*Azadirachta indica*). Plans to include livestock as a possible component of the system were initiated, and the ARC began to train livestock researchers. In the subsequent Kenana and Rahad mechanized schemes, forestry components were included in the initial planning stages, along with fruit trees and vegetables.

However, projects which required such inter-disciplinary input were not common at that time. Forestry research had long been carried out by the Department of Forestry within MAFNR. Livestock production was isolated in a different department of the ministry, and range production was in yet another department. Socio-economic research in agriculture was limited and was concentrated at the University of Khartoum's faculty of agriculture. Various parastatal corporations and development projects also carried out research in the forestry, agriculture, and livestock sectors. Communication between and among these organizations was limited. The WSARP became a pilot project to test how these pieces could be consolidated -- under the umbrella of a permanent entity, the ARC -- to improve rainfed agriculture in western Sudan. While the project's mandate was limited to western Sudan, the ARC also looked upon this as an experiment with potential applicability country-wide.

4 EVOLUTION OF WESTERN SUDAN AGRICULTURAL RESEARCH PROJECT'S RESEARCH MANAGEMENT APPROACH

4.1 Initial Western Sudan Agricultural Research Project start-up within Agricultural Research Committee

WSARP leadership recognized that success would require changes in the way ARC normally conducted research. Given the limitations of ARC's personnel, disciplinary expertise, and resources, they also knew that collaboration with complementary organizations and projects would be necessary. However, institutional change and collaboration among bureaucratic organizations are never automatic. By nature, bureaucracies are self-protective, fully occupied, and resistant to change. Therefore, changes must be leveraged, generally through incentives and/or sanctions. With an existing pool of well-trained scientists available in Sudan and with access to a sizable pool of project resources, the WSARP decided to rely on incentives. To implement such a strategy, it was first necessary to define a system of incentives acceptable and attractive to the ARC, donors, and potential collaborators; and to determine what research programmes and methodologies the project would use to begin to develop the improved research system.

The identification of incentives was relatively straightforward. WSARP asked each of the potential cooperators, including ARC personnel and potential trainees, what would motivate them to work with WSARP and/or to consider carrying out research in new ways. Findings included: (a) additional pay, especially to compensate for living in the isolated rural settings of the West; (b) access to better facilities, equipment, vehicles, supplies, literature, and other scientific needs; (c) opportunities for training and attending scientific meetings abroad; (d) improved opportunities for publishing; (e) increased access to technicians and research support personnel; (f) increased professional recognition; (g) the opportunity to work with challenging professional colleagues; (h) good housing; (i) access to schools, transportation, and other support for their families; and (j) the opportunity for contributing to improved quality of life for the people of Sudan. Some incentives could be implemented by the project itself. Others required concurrence of the ARC, higher levels of government and/or of donors. In the case of the salary issue a presidential decree was required, sought, and eventually obtained. The nature and level of incentives remained a source of controversy throughout the life of the project, but they were and remain a powerful management tool.

4.2 Staffing and human resource development

Initially, available manpower within ARC included a variety of plant scientists, but there were no agricultural economists, foresters, social scientists, extension personnel, or range scientists, and only one animal scientist. To meet initial staffing needs, the project used the incentives identified above in a three-pronged approach: (a) key personnel were seconded from existing departments, both within and outside ARC; (b) candidates for potential M.S. and Ph.D. training abroad were identified and served in-service internships prior to beginning training programmes abroad; and (c) technical assistance personnel from the United States were hired. In order to ensure a broad future base of expertise, fields of study for the training programme included: forestry (with emphasis on gum arabic and agroforestry); soil fertility; soil/water relationships; range/livestock; range management (including forage crops); agricultural engineering (water conservation and management); agricultural economics; animal production; anthropology; rural sociology; and extension. Training programmes were designed to encourage a systems perspective, including interdisciplinary coursework. Research topics were chosen based on their potential applicability to future project research. Most research programmes either utilized data collected in western Sudan or were actually carried out by students in Sudan. The project also made arrangements to absorb additional scientists who were training abroad in horticulture, pest management, soils, and field crops (legumes and cereals) upon their return.

It should be noted that, while efforts were made to identify women for scientific posts and training, success was limited. The remoteness of the field stations and the availability of alternative career opportunities in Khartoum and/or the Gezira area were cited as major reasons. The U.S. technical assistance team included women, and local women were trained as technicians. Women farmers and pastoralists were heavily involved in the research programmes.

4.3 Research programme focus and approach

The early definition of research programmes and methodologies which included agroforestry was both by design and by chance. By initial consensus of the donors and project planners, research programmes were to utilize a multi-disciplinary systems approach and would seek both to increase production and to conserve the natural resource base. Initial design documents identified programmes which were to be established based on the major ecological areas and soil types of western Sudan. Research stations were located accordingly, with one in each of the four provinces of the west (see Fig. 1). The programmes were defined as arid livestock production at the desert fringe (El Fasher); semi-arid livestock production in the Sudanian and Sahelian zones (Ghazala Gawazat); integrated crop/livestock production on stabilized sands (Ghazala Gawazat); integrated crop/livestock production on non-cracking clays (El Obeid); and integrated crop/livestock production on cracking clays (Kadugli).

The assignment to WSARP of the director of the Department of Forestry's northern Kordofan station in El Obeid was key to the development of an agroforestry focus. Although construction of facilities was supposed to precede the initiation of research activities, WSARP management decided to initiate pilot research programmes immediately in El Obeid and at existing facilities in Kadugli, southern Kordofan. In El Obeid, the forestry department had active programmes in ecology and silviculture of *Acacia senegal*. They agreed to make their facilities and some of their staff available to WSARP pending completion of the construction of WSARP headquarters in El Obeid. WSARP used this opportunity to initiate research activities, establish a support infrastructure, and begin to incorporate systems concepts into the ongoing ARC activities in Kadugli and those of the forestry department in El Obeid. Lessons learned from these early efforts could then be applied to the entire project when the additional stations came on line.

Kadugli, southern Kordofan, the site of an existing ARC cotton and sorghum research station, was chosen as the base for the major pilot WSARP research programmes. Core scientific staff for this initial research effort included existing ARC plant breeders and agronomists, the sole ARC animal scientist, and technical assistance scientists from the United States. The latter initially included a range scientist, a social scientist, and pest management specialist. An economist, agronomist, and veterinarian/animal production specialist were later added. Forestry expertise was provided by visiting scientists from the El Obeid station.

Utilizing a series of planning meetings with regional government officials, national planning entities, ARC scientists, and related projects and development schemes, WSARP framed its research philosophy and approach and identified initial research activities. The approach selected was called a "production systems approach", in which improvements in agricultural production were to be identified, selected, tested, and introduced within the context of the total production system in which the producers operate. This approach drew heavily on the FSR methodologies which were at that time evolving. However, the WSARP approach focused to a greater degree on the natural resource base, not only as a means to define agro-ecological zones and production potential, but also within the context of multiple overlapping production systems in the area. Much of the FSR up to that time had involved systems which tended to be relatively homogeneous within a given area and with similar economic resources. The farmer's strategy was generally to increase production of a standard mix of crops (or crops and livestock) on land he/she controlled, for increased food and/or economic return. FSR started from this farmer perspective. Problem identification and potential solutions were approached primarily within the context of these rather controlled systems. The FSR terminology "on-farm research" and "within the farm gates" illustrate this focus.

In Kadugli, the initial interdisciplinary research efforts of WSARP involved range, animal, and forestry scientists, as well as the traditional mix of crop and soil scientists. This was facilitated by recruiting additional American scientists in agronomy, agricultural economics and animal health. The production systems identified did not fit nicely into concepts of "within farm gates" because livestock are highly mobile, and cropping activities themselves move over time with shifting and fallow cultivation practices. Therefore these early researchers took a broader approach as they sought to identify and describe the system(s) in physical, biological, social, and economic terms, including production strategies; to identify production constraints and opportunities; and to define research activities which would alleviate constraints and/or capitalize on opportunities. In Kadugli's "cracking-clay soils system", the researchers found that virtually all producers in the area were engaged in significant agricultural activities on multiple soil types. Utilization of the cracking clay soils themselves varied among different producer groups. Some producer groups conducted production activities in distant areas with entirely different agro-ecological characteristics. The soil, water, and natural vegetation showed evidence that conflicting use patterns of the various producers were constraining the efficient utilization of these resources. One surprising finding was that forage resources in the southern Kordofan area were actually being under-utilized on a seasonal basis.

4.4 Production strategies, systems and constraints

Based on these early studies, three major types of producer groups or production strategies were identified. Each of the groups utilized woody perennial species to some extent in their systems of production. These three production systems are briefly characterized below.

(i) The sedentary farming systems. The sedentary farming systems are dominated by crop production, but usually incorporate some livestock and/or poultry as well. The Nuba farmers of southern Kordofan are representative of this system. They raise cotton and sorghum on the dark cracking clays, groundnuts and sesame on the sandier soils nearby, and a variety of home garden crops along with small numbers of cattle, goats, pigs, and poultry. *Acacia seyal* is cleared to bring traditional and mechanized farms into production, with the residue either made into charcoal or the ashes incorporated into the soil. After a 5-7 year cycle, crop production declines and the area is abandoned for cropping, but is usually grazed by livestock. Frequent natural and human-made fires favour the reestablishment of *Acacia seyal*, a fire-climax species, completing the cycle. *Acacia seyal* is also sometimes tapped for gum. *Balanites aegyptica*, difficult to clear and valued for its fruits and *Acacia albida*, acknowledged for its shade in the dry season, lack of canopy in the rains and its ability to enhance soil fertility, also feature in the sedentary system. While WSARP was to target the traditional farmers, a subsystem of large mechanized cotton production was also present in southern Kordofan.

(ii) Transhumant pastoralist systems. Transhumant pastoralist systems have extensive (rather than intensive) livestock production as their primary objective, and producers migrate with their livestock along relatively fixed annual routes. However, these systems incorporate cropping enterprises in a systematic way into the migration pattern. The Baggara Arabs, who raise primarily cattle and small ruminants and migrate annually from southern to northern Kordofan and return, represent this group in the Kadugli area. *Acacia senegal*, which yields the highest quality of gum arabic, is a major source of income to many transhumant pastoralists as well as providing highly nutritious fodder for their livestock. *Balanites* pods and leaves are an important source of nutrition in the dry season, and *Acacia albida* is utilized as in the sedentary system.

(iii) *The true nomadic pastoral systems*. The true nomadic pastoral systems feature those migratory livestock producers who do not normally raise crops, nor do they follow a fixed orbit of seasonal migration. Fellata cattle raisers who migrate in an east-west direction and pass seasonally through southern Kordofan are an example. Camel nomads such as some of the Hawazma also follow a nomadic production system which includes the southern Kordofan region. The nomads too gather gum arabic and utilize many of the woody perennial species for their livestock. Some also transport and sell various species for building materials and for the production of charcoal. The migrational routes of both the transhumant and nomadic systems have been significantly altered by the expansion of the sedentary systems.

As the systems were identified and described, broad generic production constraints in the southern Kordofan area were identified. These included inadequate water availability, conservation and management; environmental degradation due to poor and conflicting land-use practices (e.g. expanding cultivation into marginal areas, burning, tree cutting, and reduced fallows); low soil fertility; inadequate rural infrastructure (markets, road, inputs, etc.); poor agronomic practices; livestock disease and inadequate nutrition; poorly adapted genetic stock (especially crops); crop pests and diseases; labour constraints; and conflicts between different types of producers.

However, unexploited opportunities were also apparent, including the seasonal under-utilization of forage resources in southern Kordofan. In collaboration with WSARP, Department of Forestry researchers and with related projects in El Obeid, including the International Sorghum and Millet Collaborative Research Support Project (INSORMIL) and the Western Savanna Development Corporation (WSDC), WSARP identified analogous production systems in each of the four target provinces. For example, transhumant camel producers in the arid areas of northern Kordofan utilized an agroforestry system which supplemented available *Acacia* spp. and other woody perennials with cultivated watermelons for the provision of additional moisture.

4.4 Further development of Western Sudan Agricultural Research Project research organization and management

In order to define specific research activities to address these problems and opportunities, WSARP needed to develop its research management system. With only a skeleton staff at El Obeid and no one at Ghazala Gawazat or El Fasher, it was not practical to define specific programmes for each of the identified agro-ecological areas. Producers in the Kadugli areas utilized each of the major zones already through their farming and migratory practices. The constraints identified were consistent among the different producer groups, but ranking of their relative importance and potential solutions varied substantially between groups. It was evident that sustainable increases in production in the Kadugli area would require different interventions or combinations thereof for the different production systems, as well as consideration of the relationships between them.

Based upon these considerations, the research programmes were reorganized from a strictly agro-ecological basis to a system which also considered the three main producer and strategy-based production systems (sedentary, transhumant, and nomadic). These programmes were project-wide and spanned the different research stations. Each established an interdisciplinary and interstation system and programme committee, chaired by a coordinator, with overall responsibility for that system's research planning efforts. These coordinators were analogous to the existing ARC commodity coordinators and, like the latter, each represented his programme at the national level for purposes of agricultural research planning. Each station developed a station research committee to plan and oversee programmes at the local level. Relevant system and programmes committees worked with the station committees to develop research programmes, emphasizing complementarity and minimizing duplication. An executive WSARP research committee, composed of the three programme and systems coordinators, the project director and the director general, ARC, was established to provide overall guidance and coordination. Project work plans passed from this committee to the central ARC technical and finance committees before final ARC approval. The research management organization is summarized in Fig. 2.

At the individual station level, disciplinary units were established based upon their ability to share resources such as equipment and facilities. However, research planning and many research activities were carried out in collaboration with other units, including those at other stations. Consistent with farming systems methodologies, WSARP research activities were

Fig. 2. WSARP research management organization.

biased toward applied and adaptive research. On-station research findings were moved as rapidly as possible to on-farm and in-herd/flock trials. However, somewhat in contrast to the rapid reconnaissance-type surveys favoured by FSR, a high priority was placed on the gathering of detailed information essential for understanding the natural resource base and the production preferences and strategies of the producers. Indicators of success for the research programme were broadened to include assessment of economic gains and improvements in the natural resource base in addition to (or in place of) short-term production increases.

5 DISCUSSION AND LESSONS LEARNED

WSARP'S research approach will continue to evolve, but at the present time it incorporates the basic concepts and approaches outlined above. The ongoing WSARP/ARC research efforts to serve the many stakeholders in Sudan's agricultural sector show promise, but have only begun to approach the project's ambitious goals. The organization of research programmes based on producer-oriented production systems encourages scientists to think broadly and creatively in order to identify agricultural and natural-resource problems and opportunities for improvement. However, ranking between and among the systems for purposes of research planning and budget allocation is complex.

Introduction of hedgerows and leguminous intercropping systems is expected to stabilize fragile soil while improving fertility and production of food crops in sedentary systems, thus improving the lot of the majority of western Sudan's producers. By keeping the land in production, however, transhumants and nomads may lose access to fallow forage. With export of live animals to the Middle East an important source of foreign currency to an ailing economy, transhumant producers may be more important than their relative numbers would imply. The development of systems to take advantage of underutilized range resources in southern Kordofan should assist the transhumant producers, but research has been hampered by political upheaval.

WSARP has made a conscious effort to involve not only producers, but also extension agents and actual and potential local, regional, and national policy-makers in the research planning process. This has been attempted from the early stages of the project in order to facilitate the acceptance and dissemination of research results and their inclusion into policy decisions. With few women in research or extension roles, concern remains that needs and opportunities for this important segment of the population cannot be optimally addressed. In an environment as interrelated and dynamic as that of WSARP, researchers have found that working with one set of producers or decision-makers may sometimes cause them to lose credibility with another.

Evolving donor agendas and expectations have also influenced the research programme. For example, USAID had originally planned to focus its attention in Sudan on slowing the expansion of the Libyan desert into Kordofan. New USAID political agendas in the early 1980s, coupled with their short-term funding and therefore, planning cycles, resulted in their assessing project performance on the basis of short-term production increases of sorghum and millet. Now, however, the natural resource agenda is regaining momentum, and new agroforestry projects are underway in Sudan. These build on research results from WSARP and others. The World Bank, in response to the economic crisis and cognizant of the potential for generation of valuable foreign currency by the livestock sector, began to plan interventions without full consideration for the natural resource implications of such plans. Valuable information outlined in WSARP research results and recommendations was either overlooked or ignored. Throughout the life of the project, donor attention and resource allocation have swung back and forth between the traditional sector and the larger mechanized sector. Agroforestry methodologies are appropriate to both, but, given the different levels of resources, risks faced, and production strategies of the two groups, the research activities themselves would not be identical. Thus, an improved and preferably empirical system of research ranking is needed.

In the face of such conflicting demands, it is all too easy to forget the lessons of the past. Short-term production increases have been achieved in the past in Sudan, especially in association with the introduction of mechanization and mono-cropping, in the case of sedentary farmers, or the development of boreholes or provision of vaccines for pastoralists. However, within a short period of time production has historically fallen below the pre-increase levels. These declines result from such factors as: the disruption of rotational or fallow practices, expansion of cultivation into areas incapable of supporting sustained production, inherent risks in mono-cropping, inappropriate seasonal grazing patterns, the inability of the infrastructure to absorb new production, resource degradation, and others. The result has been a steady decline of the soil, water, and vegetation resources. Recent cycles of decreased rainfall and increased population have exacerbated the problems.

Agroforestry methodologies face other operational difficulties. There is a nearly irresistible tendency for interdisciplinary teams to segregate back into their individual disciplines. This is natural and desirable in carrying out specific technical activities. An agronomist is not and should not attempt to be a forest ecologist. However, this segregation is disastrous during the stages where planning, ranking, and assessment of overall programme and impact are concerned. Likewise, the commodity orientation of the ARC has, on occasion allowed the focus to return to the strategy of maximizing production of a given commodity, rather than on systems of production in which appropriate mixes or rotations of commodities with forest and/or livestock enterprises can be better sustained. Conducting cooperative research with producers is more, rather than less, demanding than research-station programmes. In Sudan, extension personnel are scarce and lack adequate financial support, while researchers have had limited experience in working with producers. There has thus been some tendency for technical scientists to gravitate to the familiar station and laboratory setting at the expense of field activities. The socio-economic sciences are critical to the agroforestry approach as changes in producer's behaviour are required for improved systems to be adopted. Whether in Sudan or the United States such change is generally the result of economic and social benefits accrued as a result of the change. However, the marriage of biological and physical disciplines with those in the social sciences has not been effortless.

The level of financial support for WSARP's research remains a problem. Working directly with producers requires researchers to live nearby and/or travel extensively. In Sudan, investments in research stations, a project aircraft, and field vehicles were possible under the project. Meeting the recurrent operational costs from existing governmental revenues is far more difficult.

WSARP is a project, and as such is defined by time and resource boundaries. Therefore, the WSARP approach must ultimately be incorporated into ARC policies and procedures, if it is to continue. This process is ongoing, but a number of issues remain unsolved. These include policies and procedures for professional advancement within the ARC (traditionally related to individual disciplines or commodities); long-term strategies for the staffing of the relatively isolated WSARP stations, including increasing the number of ARC socio-economic and livestock personnel; improvement of communication and coordination of activities among ARC/WSARP scientists and between the ARC and other research organizations, especially those in the livestock and forestry sectors; and strategies for meeting recurrent costs.

6 SUMMARY AND CONCLUSIONS

The Western Sudan Agricultural Research Project is utilizing research planning and management approaches which incorporate and support the use of agroforestry methodologies. Critical elements include:

-a systems orientation which considers agro-ecological criteria and production strategies;

-the explicit inclusion of natural resource management and conservation as an objective of agricultural research, along with production, productivity and economic benefits;

-the use of interdisciplinary teams of researchers;

-involvement of producers and political entities in the research planning process; and

-commitment to collaboration and cooperation with other forestry, livestock, agriculture, and natural resource organizations and projects in support of mutual objectives.

Conditions in the rainfed areas of western Sudan differ from those in many other agricultural and forestry settings in the world. Limited income traditional producers dominate; access to inputs and markets is limited; complex production systems incorporating a number of different crops and livestock species prevail over single-commodity production; the environment is undergoing significant degradation; producers are scattered over a vast area; and access by an extension service is nearly nonexistent.

The operational setting for agriculture and forestry are changing world-wide. There are changes in the global marketplace, including increased competition, currency fluctuations, and changing patterns of supply and demand. Potential agro-ecological shifts loom as a result of atmospheric changes such as acid rain and depletion of the ozone layer. Farmers in the most developed countries are facing increased costs, decreasing supplies, and outright banning of agricultural inputs. The loss of top soil, depletion and pollution of groundwater, and other factors which constrain agricultural production are now problems of today, not potential problems of the future. New attention on health and nutrition is affecting agricultural practices and markets. These and other factors mandate innovative research agendas and approaches. The elements of the agroforestry research methodologies outlined above should not replace other research approaches. Rather they can be used to complement and support existing approaches, both in the developing and developed countries to support resilient, sustainable use of the earth's resources to support humankind.

7 REFERENCES

Bunderson, W.T., 1981. Annual Report of Range Research. (WSARP Publication No. 10), Khartoum, Sudan and Washington State University, Pullman, Washington, U.S.A.

Bunderson, W.T., Cook, R.H. and Fadlalla, B., 1984. Range-livestock research activities for 1982/83. (WSARP Publication No. 2), Khartoum, Sudan and Washington State University, Pullman, Washington, U.S.A.

Bunderson, W.T. and Frye, D., 1986. Adaptive strategies for successful pastoral development. In: International Rangeland Symposium. Orlando, Florida.

Bunderson, W.T., Woldetatios, T. and Gillard-Byers, T., 1985. Improved integration of legumes into the agro-pastoral system of the Nuba Mountains, Sudan, Presented at the Farming Systems Symposium, Manhattan, Kansas, October.

Cook, R.H. (Editor), 1985. WSARP 1984-85 research programme results, Volume I - Summary of Kadugli sedentary and transhumant production research programme results. Kadugli Station, Khartoum, Sudan and Washington State University, Pullman, Washington, U.S.A.

Fadlalla, B., Bunderson, W.T., Cook. R., Hashim, I. and Gillard-Byers, T., 1986. WSARP 1984-85 research programme results, Volume III - Transhumant production system research programme results. Kadugli Station.

Felker, P., 1978. State of the Art: <u>Acacia</u> <u>albida</u>, Grant No. AID/afr-C-1361. University of California, Riverside, California.

Harrison, M.N. and Jackson, J.K., 1958. Ecological classification of the vegetation of the Sudan. (Forestry Bulletin, No. 2), Ministry of Agriculture, Khartoum.

Henson, J.B. and Noel, J.C. (Editors), 1982. WSARP Workplan, Volume III. Research Plan for Kadugli and Other Stations. Khartoum, Sudan and Washington State University, Pullman, Washington, U.S.A.

Horowitz, M.M. and Salem-Murdock, M., 1986. The political economy of desertification in White Nile province, Sudan. In: Little, P.D. and Horowitz, M.M. (Editors), Lands at risk in the Third World, Westview Press, Boulder, Colorado.

Hunting Technical Services, Ltd., 1981. South Kordofan Central District Rural Development Plan. Borhamwood, U.K..

Ibrahim, F.N., 1986. Ecology and land-use changes in the semi-arid zone of the Sudan. In: Little, P.D. and Horowitz, M.M. (Editors), Lands at risk in the Third World, Westview Press, Boulder, Colorado.

Kang, B.T. and Reynolds, L., 1986. Alley farming in the humid and sub-humid tropics. In: Alley Farming Network for Africa.

McGahuey, M, 1986. Impact of forestry initiatives in the Sahel, Chemonics, Washington D.C.

Noel, J.C., Dafalla, D. and Henson, J.B., 1988. A research programme for traditional farmers and pastoralists in the semi-arid regions of western Sudan. In: Arid Lands: Today and Tomorrow. Westview Press, Boulder Colorado, pp 1189-1197.

Saad, Z.A., Gum arabic research programme for Ed Obeid. In WSARP Publication No. 15. Khartoum, Sudan and Washington State University, Pullman, Washington, U.S.A.

Sudan Agricultural Research Capabilities, 1977. Joint Team Report to the Minister of Agriculture, Food and Natural Resources, International Agricultural Development Services.

United States Agency for International Development, 1980. Western Sudan Agricultural Research Project. (Project Paper 650-0020), Washington, D.C.

United States Agency for International Development, 1980. Western Sudan Agricultural Research Project. (Project Paper Amendment 650-0020),

Washington, D.C.
WSARP, 1982. Work plan volume III: Research Plan for Kadugli and Other Stations. (WSARP Publication No. 15, Khartoum, Sudan and Pullman, Washington.
WSARP, 1985a. WSARP 84/85 Research Programme Results, Vol. II, (WSARP Publication No. 42) Khartoum, Sudan and Pullman, Washington.
WSARP, 1985b. WSARP 84/85 Research Programme Results, Vol. III, (WSARP Publication No. 43) Khartoum, Sudan and Pullman, Washington.
Woldetatios, T, Gillard-Byers, T., Bunderson, W.T., Abu Sabah, M.A. and Cook, R.H., 1986. WSARP 1984-85 research programme results, Volume II - Sedentary production system research programme results. Kadugli Station.
World Bank, 1977. Sudan Agricultural Research Project Project Staff Appraisal Report. Washington, D.C.
World Bank, 1979. Sudan Agricultural Research Project Project Staff Appraisal Report: Implementation Volume. Washington, D.C.

Chapter 10

AGROFORESTRY PRACTICES AND POTENTIALS IN WESTERN SUDAN

T.W. BUNDERSON
Department of Natural Resource Science, Washington State University, Pullman, Washington, 99164-6410 (United States of America)

A. EL WAKEEL
Western Sudan Agricultural Research Project, Agricultural Research Corporation, P.O. Box 125, Khartoum North (Sudan)

Z. A. SAAD
Western Sudan Agricultural Research Project, P.O. Box 30, Khartoum North (Sudan)

and I. HASHIM
Western Sudan Agricultural Research Project, P.O. Box 30, Khartoum North (Sudan)

ABSTRACT

Attaining agricultural sustainability remains on the forefront of Sudan's challenges. Traditional systems of crop and livestock production in the vast rain fed regions of Western Sudan have tremendous potential in helping to achieve this goal. This potential is being undermined by increasing population pressures, escalating land-use conflicts, and changing cultural, economic, and political factors. Land-use practices that long preserved the integrity of survival with non-destructive patterns of resource utilization, are fading into transient strategies of exploitation. Expanding cultivation into marginal areas is with little doubt the most destructive force affecting the natural resource base; but it is exacerbated by extensive tree cutting, increased grazing pressures, and widespread burning.

Conventional approaches to increase agricultural production in Western Sudan are possible if properly executed, but such improvements are likely to be short-lived. Most development efforts in Africa have been and continue to be preoccupied with production-increasing technologies originating from western cultures, assuming, often erroneously, that they are directly transferable. They ignore a more critical need: how to maintain sustainability under the increasing pressures and demands of rapid population growth. Fundamental to this concern is the changing nature of traditional land-use principles, especially as they relate to policy and tenure. This paper discusses WSARP's planning activities for improving agroforestry practices among traditional production systems in Western Sudan. Agroforestry is not treated as an isolated component; rather existing practices and potential interventions are presented in a system's context with research agendas imbedded in an overall action strategy. Underscoring WSARP's approach is the importance of first understanding, and second, restoring and/or improving traditional mechanisms that have ensured survival and production for centuries with little or no resource degradation. Although implementation plans for the latter remain somewhat intangible at present, until this occurs, it is both fatuous and naive to contemplate significant long-term improvements in smallholder farm or pastoral productivity.

INTRODUCTION

In the previous chapter by Noel, et al., management issues associated with the design and implementation of the Western Sudan Agricultural Research Project (WSARP) were discussed. The evolution of WSARP's approach to research was stressed in relation to traditional systems of production and the role of agroforestry. Building on that management base, we discuss WSARP's planning process for initiating and improving agroforestry practices among traditional production systems in the rain fed areas of Western Sudan. Lack of an adequate database necessitated an early emphasis on characterizing major production systems and constraints. This was based on the premise that problems must be adequately understood before attempting solutions. From these efforts, three distinct production systems were identified (a) sedentary farming communities focusing on crop production with some livestock; (b) transhumant, cattle-oriented agropastoralists migrating between well- defined wet and dry season areas; and (c) nomadic camel-dominated pastoralists in more arid environments, where movements are more frequent, migration patterns less distinct, and cultivation mostly opportunistic.

Only the first two systems are targeted here, but many of the production strategies and problems of transhumant groups are common also to camel nomads. Mention is also made of the Acacia albida farming system on the Jebel Marra massif in Darfur, even though this falls outside the region discussed in this chapter.

2 SEDENTARY FARMING SYSTEMS

Sedentary farmers account for about 60 to 70% of the population, produce 80% of the crops, and own about 20% of the livestock. There are two types of farming communities: dispersed groups mainly in the Nuba Mountain region, and clustered groups farther north and west. Agroforestry practices are integral to both communities. Woody vegetation restores soil fertility, and provides wood, fuel, quality fodder, building material, gum, fruit, medicines, fibers, dyes, and shade. Most tree uses are derived from range or fallow land, but certain species are also retained on cultivated farms.

2.1 Dispersed communities in the Nuba Mountains

These groups represent numerous tribes collectively known as the Nuba, but with similar origins, traditions and agricultural practices. Farming communities consist of dispersed sedentary neighborhoods typically located on or in close proximity to hill ranges. Crop and livestock activities generally span the entire topo-edaphic gradient from rocky, sandy loam hillsides to deep cracking clays in the plains. A brief description of the system is provided below based largely on information from HTS (1981), WSARP (1982; 1983; 1985) and Bunderson et al. (1984).

(i) Crop production. Production activities are dominated by a bush-fallow system of hand-hoe cultivation. Apart from private or government schemes, there is no mechanization or animal draft, and few if any external inputs. Farming is concentrated in the cracking clay plains, with sorghum and sesame as the principal field crops. The distribution and production of major crops raised by traditional households are shown in Table 1. Cultivation in the plains involves a bush-fallow cycle where land is cleared of native vegetation, cropped for 6 to 12 years, and abandoned for a similar period (or longer) when soil fertility or infestation by *Striga* no longer support acceptable yields. Systematic crop rotations are nonexistent. Some intercropping of sesame and/or cowpeas with sorghum is practiced, but at low intensities. Major woody species in fallow vegetation include *Acacia seyal*, *A. polyacantha*, *Balanites aegyptiaca*, and *Dichrostachys cinerea*. *B. aegyptiaca* is also retained on cultivated land because (1) it is difficult to remove, (2) it is protected by law, (3) it yields useful products (edible fruits, fuel wood, fodder, and medicines), and (4) it does not appear to compete with crops, at least at low densities.

Nuba farmers also cultivate house gardens or "jubrakas" and "near farms" which form an integral part of their production activities. These sites are located on sandy clays and loams adjacent to the household, and function for added food security and income during the "hungry" period prior to the later harvest of the far fields. Relative to the cracking clays, crop husbandry is more intensive here because cropping is more continuous, soils are less productive, and labor is more readily available due to proximity to the household. Manure collected and stored from animal enclosures or huts is spread by hand for fertilizer. On cultivated hillsides, crude terraces are formed by clearing and aligning rocks at intervals along the contour to reduce water run-off, soil erosion, and loss of manure. Crops are also protected from livestock damage by thorn fences that are erected yearly from cut branches of the coppicing tree *Ziziphus spina-christi*. This tree is commonly retained and nurtured for this purpose as well as for its edible fruits, quality fodder, and wood (valued for furniture and farm tools).

TABLE 1

Crude protein contents of common forage plants in the Nuba mountain region. Data for each species are for the late wet and dry seasons respectively; in some woody species, dry season values were higher because of new leaf growth 2 months before the onset of the rains, a common characteristic of many species in the region. NA indicates that the species was not available or growing during that season. NC indicates that forage samples were not collected. (Source: Bunderson, unpublished data).

HERBACEOUS SPECIES		WOODY SPECIES	
Grasses		Leguminous Trees/Shrubs	
Aristida hordaceae	5.49/2.93	Acacia albida	NA/18.61
Brachiaria abtusiflora	5.68/3.45	Acacia gerarrdii	NC/15.88
Brachiaria xantholeuca	15.94/4.77	Acacia nubica	NC/27.28
Dactyloctenium aegyptium	11.40/4.56	Acacia polyacantha	28.44/NC
Echinochloa pyramidalis	6.17/2.91	Acacia senegal	26.03/NC
Hyparrhenia confinis	5.21/2.71	Acacia seyal	17.00/18.65
Hyparrhenia pseudocymbaria	8.30/2.99	Albizia amara	16.87/18.79
Setaria incrassata	6.63/2.66	Bauhinia rufescens	17.32/11.91
Setaria pumila	6.22/4.30	Dichrostachys cinerea	16.26/14.47
Sorghum purpureo-sericeum	4.47/3.14	Piliostigma reticulatum	11.98/14.44
Forbs(present in wet season)		Non-Leguminous Trees/Shrubs (leaves)	
Alysicarpus glumaceus	15.37/NA	Anogeissus leiocarpus	14.24/ 7.86
Ocimum basilicum	16.30/NA	Balanites aegyptiaca	17.53/10.07
Rhynchosia minima	7.72/NA	Combretum hartmannianum	13.33/12.12
Spermacoce chaetocephala	9.94/NA	Gardenia lutea	11.12/13.61
Zornia glochidiata	17.92/NA	Ziziphus spina-christi	20.35/11.25
Crop Residues			
Sorghum leaves	4.24/3.42		
Sorghum stalks	NA/2.50		
Sorghum tillers	NA/8.64		
Sesame leaves	NA/4.18		
Cowpea leaves	21.52/16.30		
Groundnut tops	NA/9.82		

(ii) Livestock production. Livestock production among most Nuba tribes is based on a sedentary system of management dominated by goats and cattle. Sheep and swine are also raised to varying degrees according to individual preferences and the influence of Islam. Livestock are important for protein supplementation, socio-cultural functions, and investment. They are also valued for risk reduction, diversification, and farm expansion. Grazing management is characterized by sequential use of hilltops and footslopes during the growing season, and the clay plains after the main sorghum harvest. Following the harvest of farm plots, crop residues are grazed by livestock on a communal basis. Groundnut tops are usually the only residue commonly harvested and stored for supplemental feed. By the mid-dry season, crop residues are depleted, and livestock move to fallow and range vegetation. Fallow land is particularly valuable since resident village herds generally have better access to and control over these areas. For much of the dry season, animals frequently travel 20 to 30 kilometres per day in search of food since forage proximate to water is rapidly depleted.

Browse provides an important source of quality feed on a year-round basis, especially in the dry season when the quality of herbaceous vegetation drops below maintenance requirements (see Table 1). Herders commonly lop off branches of trees to feed their herds when browse availability is limited or out of reach. Species usually lopped in this manner include Anogeissus leiocarpus, Balanites aegyptiaca, Bauhinia rufescens, Combretum hartmannianum, and Ziziphus spina-christi. The latter is usually cut when erecting fences around home gardens and near farms. Trees are also cut for livestock enclosures in different sites depending on herd locations. Acacia seyal and A. polyacantha are the major species used in the plains, while in the hills, enclosures are usually made from Acacia gerarrdii, Anogeissus leiocarpus, Boswellia papyifera, Combretum hartmannianum, and Lannea species. During the rains, smoke fires from freshly cut tree branches are used to minimize biting insect problems for both herders and livestock. Woody vegetation from both fallow and rangeland also provides other useful products and services as mentioned earlier.

(iii) Other agroforestry activities. Major off-season activities include extensive fencing of gardens and near farms, household repairs, and the production of charcoal, wood, timber, and thatching from native vegetation for domestic use and sale. A limited number of producers with access to permanent water grow small, irrigated gardens of fruit trees (mainly mangoes and citrus species) and vegetables, principally for cash.

2.2 Clustered communities

Much of the information presented here is derived from WSARP (1982; 1983), Reeves and Frankenberger (1982), and Saad (1982). North and west of the Nuba Mountain region are farmers representing various ethnic groups and tribes, but with a distinct settlement and cultivation pattern. Households numbering 6 to 1000 or more tend to be clustered in villages of straw huts accessible to water points. Crop production is practiced mostly on stabilized red sands (qoz) and to a lesser extent, red non-cracking clays (gardud). Both soil types are low in fertility. Qoz sands have poor water and nutrient-holding capacity, and are susceptible to wind and water erosion. The ease of working goz sands permits more extensive cultivation, and encourages expansion into rangeland. Gardud soils have better potential for retaining water and nutrients, but their compacted surface makes cultivation difficult, and severely impedes water infiltration, resulting in sheet runoff.

(i) Crop production. Cultivation involves shifting or bush-fallow practices, sometimes with Acacia senegal trees planted in fallow land for gum arabic production. Animal traction and mechanization are rare. Villages are surrounded by cultivated areas dominated by millet, sesame, sorghum, and groundnuts in mixed or pure stands. Millet is the major field and food crop. Minor crops grown include cowpeas, okra, watermelon, and kerkadee or roselle (Hibiscus sabdarifa). These are frequently interplanted with sesame and sorghum. Vegetables and fruit trees (mainly citrus and mango) are also produced under irrigation in numerous small gardens along seasonal streams, or permanent water, but the availability of such sites is limited. Crop produce is commonly sold for cash or traded in kind.

Landholdings commonly average 10 to 15 hectares per household. More than half of this is cultivated at any given time, indicating insufficient fallowing. Fallowing is the principal means of restoring soil fertility, and is usually practiced after 3 to 5 cropping seasons. Many farmers on sandy soils plant Acacia senegal in their fallow land for gum arabic production, an important source of household income. Farm integration of A. senegal is the most extensive form of planted agroforestry in the country. In addition to dominating the world's market of gum arabic, A. senegal in Western Sudan serves a major role in soil conservation, restoring fertility, and providing fuel wood and fodder (Table 2).

(ii) Livestock production. Livestock play an important role in the system, but are usually raised on a small-scale. Herds consist mainly of goats, sheep, chickens, and a few cattle. Donkeys are the main beast of burden, although horses are also used by some farmers. Most households own a donkey and several goats. The availability of year-round drinking water and of dry-season fodder are major concerns in raising livestock. Crop residues and commercial sorghum are important fodder sources for work animals. Other livestock subsists largely on range vegetation beyond the village's perimeter of cultivated land. Among farmers raising gum arabic, small ruminants are permitted to graze in the fields during the years of gum production; they may also browse the gum trees to a limited extent. In addition to gum arabic production, many households supplement their income through various off-farm

TABLE 2

A typical planting sequence involving A. senegal in the goz sands of the north and west Sudan (annual mean rainfall 250-500 millimetres).

Years	Activity
1-4	Field crops of millet, sesame, sorghum, and groundnuts are planted.
5-7	Acacia senegal seedlings are planted and established.
8-10	The harvest of gum arabic is begun and continued.
11-13	Gum production declines; A. senegal trees are then cut and used for building or charcoal production.
12-15	Crop cycle is repeated with planting of field crops as above, but A. senegal trees are not removed.
16-18	Cropping activities stop, and coppicing of A. senegal is encouraged.
19-23	Gum harvest begins again, after which trees are removed for wood and fuel, and the cycle repeats beginning with year 1 above.

activities. Tree-related uses include producing charcoal and bricks which are then sold in village markets or urban centers.

(iii) Acacia albida farming systems. The lower slopes and highlands of Jebel Marra mountain in Darfur Province are the most common areas in Sudan where Acacia albida are integrated with local farming practices. Although used in other parts of western Sudan, A. albida is much less abundant than on Jebel Marra. The Fur are the major group in the area. They cultivate semi-permanent terraced fields of millet and other subsistence crops under multi-purpose trees dominated by Acacia albida, but including species such as Cordia abyssinica and Ziziphus spina-christi (Wickens, 1969; Miehe, 1986). Trees have been retained for their beneficial effects on food production, as well as for wood fuel and fencing material. The system has supported a densely settled population for centuries, but recent out-migration has caused a gradual return to shifting cultivation (Miehe, 1986). Present subsistence farming in the region is more exploitive in nature resulting in decreased tree cover, increased danger of soil erosion, and a decline in soil fertility. Given the steepness of the terrain and the areas higher rainfall, a plan of action is urgently needed with agroforestry as a central thrust.

3 TRANSHUMANT SYSTEMS

Transhumant groups form about 30% of the region's rural population, and produce over 70% of the livestock. They are predominantly semi-nomadic pastoralists, although the cultivation of major cereal and oil seed crops serves a limited support function for household subsistence. The majority of transhumants in Kordofan belong to the Baggara Hawazma and Mesiriya tribes who are Islamic and Arabic-speaking. Basic characteristics of pastoral transhumance are presented below based on information from Wilson (1976a; 1976b; 1976c), WSARP (1982; 1984; 1985b), Bunderson et al. (1984), Horowitz and Salem-Murdock (1986), and Ibrahim (1986).

Transhumance among the Baggara involves seasonal migrations between specific northern and southern ranges in response to changing forage conditions and the incidence of disease, pests, and mud. This is a necessary adaptation to a harsh environment, where rainfall and forage conditions vary greatly on a temporal and spatial basis; the inherent mobility of transhumance also permits higher livestock densities than could be supported under sedentary management. Resource use by tribes or lineage groups is determined largely by territorial control over specific water points and/or seasonal pastures and migration routes. During migrations, ties with sedentary relatives are re-established and strengthened to promote greater security and complementarity in their activities. Much of the cultivation of crops is undertaken by sedentary relatives or hired labor, more rarely by household members.

Livestock management among Baggara transhumants involves strategies geared for reducing risk in a harsh, dynamic environment, rather than maximizing production. Milk forms the basis of their livelihood (as in other African pastoral systems) since an economy dependent upon meat could support only a small fraction of the existing pastoral population (Jahnke, 1982). Unreliable livestock and grain markets also limit dependence on meat production. Cattle are the dominant stock raised and the major producers of milk, but sheep and goats are integral parts of the system. Multi-species herding increases the diversity of animal products and uses, and permits more efficient use of available ecological niches and human resources. Herds are generally kept as large as the labor costs of herding and watering allow. This is instrumental in maintaining a minimum herd size for milk production (and hence household survival) in the dry season or other times of stress, and for herd rebuilding after drought or epidemic disease. Mixed herds also limit losses if one species is struck by disease or drought, and small ruminants frequently enable producers to recover from heavy cattle losses. This is related to the greater resilience of sheep and goats to drought and poor feed, as well as to their early physiological maturity, rapid reproduction, low unit price, small size, and fewer management costs. Thus, despite the apparent focus on cattle, small ruminants help ensure a constant source of protein and income to households, and decrease the risk of drastic reductions in herd size and productivity.

4 DIAGNOSIS OF RESOURCE AND PRODUCTION CONSTRAINTS

The previous chapter by Noel et al. summarized broad generic constraints affecting crop and livestock production in Western Sudan. Improved inputs and services, exposure to new technology, and formal lines of credit are generally unavailable to traditional smallholders. Major obstacles include poor infrastructure and extension services, policy regulations on imported goods, low household capital, and lack of information on new or improved technologies to cope with rapidly changing socio-economic, political, and ecological conditions. Our discussion here is limited to constraints relating to potential agroforestry opportunities, i.e., those affecting agricultural sustainability. In this regard, a sound assessment of the primary resource base was considered fundamental. Research by WSARP on the resource ecology of the area was addressed in the context of existing land- use practices in order to ascertain the impacts, if any, on their current and potential productivity. Constraints are not separated by system due to many commonalties, even though the interventions used to solve them may differ. Considerable attention focuses on communal land-use strategies and their deterioration in recent times, since these have significant implications for current and future crop and livestock production.

4.1 Land tenure

Uncultivated land is used freely and communally. There are no traditions concerning western concepts of range management or pasture improvement, or of land enclosure for fodder, fuel, or timber production. Nevertheless, pastoralists have used common ranges for centuries without degradation. In the past, frequent drought, disease epidemics, inter-group conflicts, and restricted water supplies helped to limit herd sizes, usually well below the carrying capacity. Historical strategies of herding and resource use (most still in operation today) represented a de facto, but sound range management system (Horowitz and Salem-Murdock, 1986) based on (a) mobility, similar to modern concepts of short-duration grazing, (b) controlled access over traditional water sources and/or seasonal ranges, and (c) limitations on herd size by the labor demands of herding itself, particularly watering. Movements were determined by several factors, including the sensitivity of milk yields to changing feed supplies, the quality and abundance of fodder and water, and the incidence of disease, pests, and mud. These strategies optimized animal health, nutrition, and productivity under conditions that are highly variable and unpredictable.

In contrast to pastoralists, farmers enjoy a more formal but still rudimentary system of land tenure. Although the government holds residual rights to all unsurveyed and unregistered land, farming is primarily usufruct in nature. Rights to cropland can be claimed by individuals, or inherited, as long as the crop-fallow system of cultivation is maintained. For regions with limited resources, use rights can be rented, purchased, borrowed, or obtained as gifts. In areas previously uncultivated, usufructs rights can be acquired simply through land clearing. Thus, although cultivation in some areas may be a marginal activity with detrimental impacts, land claims of cultivators are generally honored over the objections of pastoralists.

Recent events are leading to increased human population pressures which are drastically altering the context under which traditional systems functioned successfully for ages. As pointed out by Ibrahim (1986), the impacts of changing land-use patterns fall into two basic categories intensification of traditional land-use due to increasing populations, and concentrated human settlements around secure water supplies and other services; and changes in methods of land-use, and the rural economy in general.

Rising sedentary populations are reducing the mobility of pastoralists, and expanding cultivation is encroaching on traditional ranges. Sandy areas to the north and west are particularly vulnerable to degradation and wind erosion due to their ease of cultivation. This is aggravated by shortened fallows for maintaining or restoring soil fertility. Water developments by the government increase access to under-utilized areas, reduce labor costs, and improve human health; but they are open to the general public which precludes regulatory access to the water and adjacent land. This leads to reduced mobility which encourages sedentarization, overstocking, and over-cultivation. Without effective regulatory processes, improved veterinary and agronomic inputs only exacerbate the problem, despite their obligatory function to society and agricultural development. Ultimately, human and animal carrying capacities decline and social conflicts over land-use increase. Attention to mechanized farming over the past few decades has also had severe impacts. Large tracts of land, taken from traditional producers, have been cleared of trees, and recommended rotation and fallow practices have rarely been followed in the short-term interest of profit- maximization. These exploitative practices quickly render land unproductive, leading to their early abandonment and reduced capacity for restoration.

In summary, the former dominance of territories by tribes or lineage groups for specific purposes has greatly diminished, resulting in conflicting land-use pressures from competing entities. As such, traditional producers, particularly pastoralists, face problems of land-use management which resist swift and simple solutions. A stage of forced transition is clearly evident where rational patterns of land-use are gradually being replaced with opportunistic forms of exploitation, largely for basic survival interests, more rarely for economic gain. Development efforts to improve crop and livestock production must carefully consider the changing context of resource use and its ramifications for sustained productivity.

4.2 Status of the natural resource base

Quantitative information on the natural resources of the region shows a general trend of deterioration. In particular, the tree cover is declining in density and diversity, and its skewed age-structure shows that replacement is not keeping pace with tree cutting, heavy browsing pressures, and natural mortality (Bunderson et al., in preparation).

Destructive forces involve a combination of factors. The most severe is the expansion of crop production in marginal areas using practices incompatible with resource maintenance (Bunderson, 1985; Ibrahim, 1986). Resource degradation is particularly acute in the northern zones which are fragile to prevailing cropping activities, although resilient to traditional grazing practices. Land preparation involves complete tree felling and removal of grasses and forbs, followed by hand-hoe cultivation. Several weedings during the growing season ensure suppression of natural vegetation. After harvesting, crop residues are removed for fodder or other uses, leaving the land bare to the elements of the long dry-season. Within a few years, soils are eroded and depleted of nutrients and all plant cover is destroyed, a feat that would take decades (at the very least) for livestock to accomplish. Under these conditions, opportunities for plant recovery and seedling establishment are greatly impaired. This is especially true of cultivated sites which are more subject to escalating pressures from sedentary human and livestock populations for fuel and forage due to proximity to water. This scenario of degradation is exacerbated by the wet season grazing pressures of migratory livestock populations; excessive lopping of trees by herders for supplemental animal fodder, indiscriminate tree-cutting for building and charcoal/brick making, and intentional burning. Important tree species that are declining are *Acacia senegal*, *Combretum glutinosum*, *Albizia amara*, *Cadaba* spp., and *Terminalia brownii*. The results are causing problems of soil erosion, high run-off, declining fertility, and acute shortages in wood, fuel, and fodder.

Although the situation farther south is less severe, similar patterns are emerging in the more densely populated areas due to increasing demands on land for crops, wood, fuel, and fodder. These areas are usually found on coarser-textured soils on or around hill ranges which are better suited to human habitation than the clay plains due to fewer insect, disease, and mud problems in the wet season. Coarser soils also support greater tree diversity, with many species favored for fuel, construction, and other wood products. Increasing demands on trees by village households are severely reducing the tree cover in these areas (Bunderson *et al*., in preparation). Species which are disappearing include *Acacia albida*, *A. polyacantha*, *A. gerrardii*, *Albizia amara*, *Anogeissus leiocarpus*, *Balanites aegyptiaca*, *Dalbergia melanoxylon*, *Prosopis africana*, *Sclerocarya birrea*, *Terminalia laxiflora*, and *Lannea* spp.

4.3 Problems of soil fertility and erosion

Soils over much of Western Sudan are deficient in organic matter, nitrogen, and phosphorus. In the face of improper land-use, they are also susceptible to erosion, especially from water on steep hillsides, and from wind on sandy soils in the arid north. Bush-fallow practices that once permitted sustained crop production are being threatened by extended periods of cereal cropping, shortened fallows, and few if any inputs. Development of productive fallow vegetation is further impaired by excessive tree cutting, frequent burning, and intensive grazing and browsing. Cropping systems that depend on bush-fallowing to restore fertility require a proper balance of fallow to cultivated land, usually at a ratio of 3-4:1. This represents an important constraint for the following reasons:

(a) households must maintain and have available relatively large areas of land to support continued cultivation;

(b) labor costs of clearing fallow land are high, and frequently lead to delayed planting with adverse consequences on crop yields; and

(c) forced reductions in the fallow period, without soil fertility inputs, result in declining crop yields.

This situation is likely to deteriorate as increasing human populations place greater demands on arable land, and options for use of imported inputs become more limited and expensive.

4.4 Livestock nutrition deficiencies

The productivity and nutritional status of sedentary and transhumant livestock in Western Sudan have been studied by several researchers including Wilson (1976a; 1976b; 1976c), Fadlalla, et al. (1985), Cook and Fadlalla (1985), Bunderson, et al. (1984), and Hashim (1988b; unpublished report). Traditional sedentary and transhumant herds generally reflect slow rates of growth and maturation, extended parturition intervals, and relatively high mortalities. Inadequate dry-season nutrition is the most critical factor influencing ruminant production, with lactating and late pregnant females most affected. Both sheep and cattle exhibit pronounced deficits in protein, energy, and phosphorus during the last 3 to 4 months of the dry season, when weight losses of 20 to 25% are common. Forage supplies are further affected by limited water supplies, widespread burning, tree- cutting, land-clearing for cultivation, and shortened fallows.

5 RESEARCH PROGRAM

WSARP has been implementing a comprehensive research program aimed at increasing or maintaining production among smallholder farmers and pastoralists. Potential strategies to achieve this goal include intensification, diversification, expansion, or improved integration among specific components, provided there is little misuse of the resource base. The key to successful improvements is to ensure that they are compatible with the needs, interests, and resource capabilities of the target producers. Crop research has included on-station and on-farm testing of improved crop varieties, cultural practices, fertilizer inputs, pest and disease control, animal draft, intercropping of improved food and forage legumes, and mulching. Livestock research has focused on the production, quality, and benefit-cost ratios of various feed supplements to improve the dry-season nutrition of cattle and sheep. This has been coupled with research on strategic measures for disease control due to its interactive effects on animal health and nutrition.

Promising agroforestry interventions are discussed below as having complementary, multiple benefits, as opposed to isolated, single-purpose functions. A preliminary list of tree species with potential for various agroforestry uses in Western Sudan is provided in Table 3.

5.1 Improved fallows

Improved integration of leguminous trees and shrubs among traditional systems in Western Sudan offers potential for improving crop sustainability, as well as yielding a range of useful tree products, notably fuel wood, building material, and high quality fodder.

In terms of crops, the major goal is to stabilize or improve the agronomic resource base. Alley cropping offers poor prospects because of its complex labor-intensive management, and its dependence on high yields of nutrient- rich leaves. The former detracts from its adoptability, while the latter is unlikely in the harsh environments of Western Sudan. In regions where fallow periods are declining at the expense of crop and soil productivity, significant improvements can be made through improved fallows. This can be done by deliberately planting one or more valuable perennial species to shorten the "unproductive" fallow period without sacrificing the ability to restore productivity (Vergara and Nair, 1985; Weinstock, 1985). Selective tree planting simply ensures more rapid development and dominance over slower-growing and less desirable species in the natural fallow. The result could effectively neutralize current impacts of shortened fallows by ameliorating the influence of external factors, such as proximity to viable seed sources, and patterns of rainfall, grazing, and fire. This modification of an already existing system entails minimal socio-cultural change with little or no disruption of subsistence food production.

In the sandy qoz soils of the north and west, improved fallows with Acacia senegal are already practiced. Historically, this has been promoted based on income generated from gum arabic production. Considerable expansion in the tree's integration could be achieved by emphasizing other benefits, notably soil restoration, fuel wood and fodder production. For the cracking clays of the Nuba Mountains, similar practices could be advocated, but with different species. The most likely candidates are Acacia seyal, A. polyacantha, and Balanites aegyptiaca because of their environmental adaptability and multiple uses or products (see Table 3).

In recent years, B. aegyptiaca has attracted attention as a potentially lucrative source of medicines, pesticides, edible oil, animal feed, nuts, soap, and fuel, as well as hard-currency export revenues. Much of these products are derived from its fruits, which yield 125 kilogrammes per tree. Perhaps the most important is diosgenin which is a major source of steroidal

TABLE 3

Some important tree species with agroforestry potential in western Sudan. Identification code WF-wood fuel; WW-woodworking; BU-light building uses/furniture; F/B-high quality fodder and browse; CM-charcoal making; SB-shelter belts; SF-soil fertility; CC-crop compatible; LF-live fencing; G(f)-gum for food industry; G(ci)-gum for cosmetic industry; RF-reforestation; IR-insect repellent; LFP-live fence posts; FR-edible fruits; SC-soil conservation; TI-timber; TA-tannin products; MA-match making; M-medicinal; TH-thatching; SH-shade.

SPECIES	USES
Acacia seyal	WF, CM, F/B, G(f), SF, BU
Acacia albida	CC, F/B, SF, SB, WF, BU
Acacia polyacantha	CC, F/B, SF, WF, G(f), CM, BU
Acacia mellifera	LF, CM, WF, F/B, SB, SF
Acacia nilotica	BU, WF, CM, TA, TI, M, SF
Acacia senegal	CC, F/B, SF, SB, WF, BU
Azadirachta indica	BU, SB, WF, TA, F/B, IR, RF, SC
Balanites aegyptiaca	BU, CC, WF, CM, F/B, FR, M, SC
Boswellia papyrifera	G(ci), MA, BU, LFP
Borassus aethiopum	BU, TI, FR, TH
Commiphora africana	LF, CM, WF, F/B, SB, SC
Cassia siamea	BU, SF, CC, SB, RF, WF, CM
Combretum hartmannianum	BU, F/B, CM, WF, SC
Leucaena leucocephala	BU, WF, CC, SF, SC, F/B
Prosopis juliflora	LF, CM, WF, F/B, SB, SF
Prosopis africana	BU, WW, WF, TA, F/B, SF
Sclerocarya birrea	BU, WW, WF, CM, F/B, SH, FR
Tamarindus indica	BU, TI, F/B, WF, FR, SH
Terminalia brownii	BU, WF, CM, F/B, SC
Ziziphus spina-christi	SB, SC, BU, CM, WF, F/B, FR, LF
Ziziphus mauritiana	SB, SC, BU, CM, WF, F/B, FR, LF

drugs such as corticosteroids, contraceptives, and sex hormones. Although many substitutes are used today, diosgenin could dominate the world market if offered at a competitive price with a constant supply. It is estimated that Sudan alone could produce 1,200 tons of diosgenin per annum, enough to satisfy half the world demand and earn an export income of $36 million.

5.2 Permanent tree intercropping

This is an extreme form of the improved fallow, and represents simultaneous combinations of annuals and perennials that transform shifting or bush-fallow cultivation into more permanent, continuous agroforestry practices. A good example is the A. albida based farming systems on Jebel Marra in Darfur, as well as other regions of Africa, including the Sahel and Southern Africa. The benefits of the tree to crops, livestock, soils, and the micro- environment have been reported fairly extensively and need not be repeated here (Felker, 1978). As pointed out earlier, other trees deliberately retained on cultivated land in Western Sudan include Acacia senegal, Balanites aegyptiaca, and Ziziphus spina-christi. However, except for gum arabic plantings, this is an opportunistic practice, i.e., trees found growing naturally are nurtured and cared for, but there is no intentional planting. In many areas, these trees are declining in density because of slow natural replacement in the face of increased pressures from human and livestock populations. Building on existing practices, this situation offers the opportunity of introducing systematic interplanting of these trees to achieve a more uniform impact with enhanced sustainability and self-sufficiency. Research should investigate planting and spacing requirements of different species with management practices and provenance testing for optimizing multiple outputs and services. Initial efforts should focus on indigenous species already in use, but other species should also be screened.

5.3 Improved home and irrigated gardens

Despite their unrealized potential, home and irrigated gardens are neglected as an important, integral part of traditional farming systems, largely because of their small size and apparent insignificance. Yet they play a vital role in food production and income generation during critical periods. Home and irrigated gardens offer diversification, cash cropping, enriched diets, risk minimization, and relatively high returns to labour. Their small size and proximity to the household permits more intensive management in terms of labor and other inputs; hence they offer a comparative advantage for successful agroforestry improvements. Suggested interventions include live fences from species of Acacia, Prosopis, and Ziziphus to better protect these plots from animal damage, as well as to provide fuel wood, fodder, green manure, and/or fruit. Multi-purpose species such as Acacia albida, Azadirachta indica and various fruit trees (e.g., citrus species) could also be integrated within or around the gardens and household compound to provide similar products. Fodder banks of suitable species could be produced for supplementing a few dairy animals to improve milk supplies for domestic consumption or possibly for sale. Alternatively, banks could be raised for green manure or mulch. These interventions may require additional on-station research; or they could proceed with farmer participation depending on the level of technology development, species choice, and the interests of the producer. In short, greater attention to house and irrigated gardens using a combination of horticultural, agronomic, and agroforestry practices could substantially improve the well-being and sustainability of farm households.

5.5 Shelter belts

Shelter belts already exist around certain towns and villages in some of the arid regions of Western Sudan. The idea is to reduce the effects of wind on soils and crops, while supplying supplementary sources of wood and fuel for domestic needs, as well as for activities such as curing tobacco. For smallholder farmers, adoption will depend on compatibility and perceived benefits. Research by WSARP is currently investigating various attributes of shelter belts (Saad, 1982):

(a) screening suitable species, and combinations thereof, including various spatial arrangements;

(b) effectiveness at reducing soil erosion, sand movement, and moisture loss;

(c) the potential for providing fodder, shelter and shade to animals, as well as fuelwood, building poles, and fence posts to farmers

(d) demarcation of land holdings;

(e) overall effects on improving the environment and crop sustainability, and

(f) behavioral responses of farmers to the concept and practice of shelter belts.

Species under study include *Acacia albida*, *A. senegal*, *A. mellifera*, *Balanites aegyptiaca*, *A. nilotica*, and *Prosopis chilensis*.

5.6 Gum arabic research

Much has already been said about promoting *Acacia senegal*, but the importance of gum arabic deserves special attention. Sudan is the world's major supplier of gum arabic, and tapping *A. senegal* trees for gum has for a long time been an important post-harvest occupation of traditional farmers in Western Sudan. Attempts to improve production have led to research on (a) ecological and silvicultural aspects of *A. senegal* (Saad, 1982), (b) yield improvements through genetic means (Saad and Day, 1988), (c) gum chemistry (Anderson, 1985), (d) nodulation characteristics (Habish, 1970; Habish and Khairi, 1970), and (e) gum production and trade (Adamson *et al*., 1974). More recent research is examining the nutrient dynamics of the *A. senegal* system and the implications for improving soils and crop yields (Day and Saad, 1987).

Despite the interest in the tree, and its economic importance to Sudan, production levels of gum have fallen in recent years due largely to a combination of drought and government price controls on farmer sales. The latter in particular has acted as a strong disincentive for farmers to grow gum arabic trees. As land degradation continues to accelerate in the northern and western regions, and foreign exchange earnings decline, mechanisms to promote the increased production and integration of A. senegal trees into the farming systems of the region are more acutely needed.

5.7 Supplemental feeding with leguminous tree pods

Under existing patterns of communal land-use, conventional methods of range management and improvement are not feasible for obvious reasons. An alternative to improve or maintain animal production is to supplement breeding stock with low-cost, nutrient-rich supplements. Feeding trials conducted by WSARP scientists using crop residues, groundnut tops, native hay, annual legumes, and sesame seed cake have met with limited success. This has been due to one or more factors, including high-cost of production, limited availability, conflicting labor demands, and/or the low quality of the feed supplement. Hashim (1988a) recently investigated another potential supplement, namely pods from indigenous leguminous trees. Promising results in terms of weight changes were obtained with non-lactating sheep whose diets were supplemented with pods from A. albida and Dichrostachys cinerea. These were fed every other day after normal grazing for a 7-week period in the dry season. Crude protein content of the pods were 16.5% and 10.9% respectively and in vitro dry matter digestibilities of 43.9% and 50.5%. Although further research is needed, leguminous tree pods show potential as a quality, low cost, abundant source of feed. They can be easily collected during the off-season from native fallow or range vegetation and fed to important stock during critical periods of stress. A potential also exists for improving their availability by planting trees in fallow or farm land.

6 SUMMARY

Planning strategies are discussed for improving the sustainability of smallholder farming through appropriate agroforestry practices. The approach emphasizes the context of the farmer, recognizing the values of traditional practices, and the ecological and socio-economic conditions which limit flexibility. Increased adoptability of interventions is suggested through incremental improvements of existing systems. This is examined with reference to several types of agroforestry interventions: improved fallows;

improved home and irrigated gardens; permanent tree interplanting; supplemental livestock feeding with leguminous pods; shelter belts; and improved gum arabic production. However, we would like to emphasize the recognition of a larger concern, though we offer little direction on how it should be addressed. Improved agricultural sustainability is unlikely to be achieved until land-use issues that embody traditional concepts and values are adopted or re-instituted in more effective forms. In this regard, land-tenure concerns should not be confused with western concepts of private ownership, particularly for pastoral dominated societies. In Sudan and indeed much of Africa, converting communal rangeland, for example, into private holdings would only impede the necessary mobility of animals and increase the probability of resource degradation. It would also greatly reduce the human and livestock carrying capacity of the area. An approach is clearly needed to restore or strengthen traditional mechanisms for rational resource-use. Among other things, this will require political initiatives that empower producers with authority to better control land-use practices based on skills and knowledge developed over centuries.

7 REFERENCES

Anderson, D.M.W., 1985. Gums and resins, and factors influencing their economic development. In: Wickens, G.E., Goodins, J.R., and Field, D.V. (Editors), Plants for Arid Lands, George Unwin.

Bunderson, W.T., Hashim, I.M., and El Wakeel, A., in preparation. Woodland density and species diversity in the Nuba Mountains, Sudan.

Bunderson, W.T., Cook, R.H., and Fadlalla, B., 1984. Range-livestock research activities for 1982/83. (WSARP Publication No. 29) Khartoum, Sudan and Pullman, Washington.

Bunderson, W.T., Woldetatios, T., and Gillard-Byers, T.E., 1985. Improved integration of legumes into the agro-pastoral system of the Nuba Mountains, Sudan. In: Farming Systems Symposium, Kansas State University, Manhattan.

Cook, R.H. and Fadlalla, B., 1985. In-herd livestock feeding trials. In: WSARP 84/85 Research Programme Results, Vol. III, (WSARP Publication No. 43) Khartoum, Sudan and Pullman, Washington.

Cook, R.H., Bunderson, W.T., Fadlalla, B., and Henson, J.B., 1983. Application of Farming Systems Research and Development to an extensive sedentary livestock production system in Southern Kordofan, Sudan. In: Farming Systems Symposium, Kansas State University, Manhattan.

El Amin, H.M., 1973. Sudan Acacias. Forestry Research Institute Bulletin No. 1.

Fadlalla, B., Cook, R.H., and Hashim, I., 1985. Studies on the nutritional status of traditional livestock. In: WSARP 84/85 Research Programme Results, Vol. III, (WSARP Publication No. 43) Khartoum, Sudan and Pullman, Washington.

Felker, P., 1978. State of the art: Acacia albida as a complementary, permanent intercrop with annual crops. Prepared under contract for U.S.A.I.D, Grant No. AID/afr-C-1361. University of California, Riverside.

Habish, H.A., 1970. Effects of certain soil conditions on nodulation of Acacia spp. Plant and Soil, 33(1):1-6.
Hashim, I.M., 1988a. Forage and tree legumes. Annual report for 1987/88. WSARP, Kadugli.
Hashim, I.M., 1988b. Live weight change in sheep supplemented with seed pods from two leguminous trees in South Kordofan, Sudan in the dry season. Paper presented at the workshop on the utilization of forages and crop by-products, PANESA and ILCA, December, Lilongwe, Malawi.
Horowitz, M.M. and Salem-Murdock, M., 1986. The political economy of desertification in White Nile province, Sudan. In: Little, P.D. and Horowitz, M.M. (Editors), Lands at risk in the Third World, Westview Press, Boulder, Colorado.
Hunting Technical Services, Ltd., 1981. South Kordofan Central Districts Rural Development Plan. Borehamwood, England.
Ibrahim, F.N., 1986. Ecology and land-use changes in the semi-arid zone of the Sudan. In: Little, P.D. and Horowitz, M.M. (Editors), Lands at risk in the Third World, Westview Press, Boulder, Colorado.
Jahnke, H.E., 1982. Livestock production systems and livestock development in tropical Africa. Kieler Wissenschaftsverlag Vauk Kiel.
Mieshe, S., 1986. Acacia albida and other multipurpose trees on the fur farmlands in Jebel Marra highlands, Western Darfur, Sudan. Agroforestry Systems, 4:89-119.
Reeves, E.B. and Frankenberger, T., 1982. Socioeconomic constraints to the production, distribution, and consumption of millet, sorghum, and cash crops in northern Kordofan, Sudan. International Sorghum and Millet Cooperative Research Support Project (INTSORMIL). University of Kentucky, Lexington.
Saad, Z.A., 1982. Gum Arabic Research Programme for El Obeid. (WSARP Publication No.15) Khartoum, Sudan and Pullman, Washington.
Saad, Z.A., 1983. Acacia senegal research program. (WSARP Publication No. 14) Khartoum, Sudan and Pullman, Washington.
Wickens, G.E., 1969. A study of Acacia albida del (Mimosoideae) Kew Bulletin, 23:181-202.
Wilson, R.T., 1976a. Studies on the livestock of Southern Darfur, Sudan: II. Production traits in cattle. Trop. Anim. Health Prod., 8:47-51.
Wilson, R.T., 1976b. Studies on the livestock of Southern Darfur, Sudan: III. Production traits in sheep. Trop. Anim. Health Prod., 8:103-114.
Wilson, R.T., 1976c. Studies on the livestock of Southern Darfur, Sudan: II. Production traits in goats. Trop. Anim. Health Prod., 8:221-232.
WSARP, 1982. Work plan volume III: Research Plan for Kadugli and Other Stations. (WSARP Publication No. 15, Khartoum, Sudan and Pullman, Washington.
WSARP, 1983. Work plan volume IV: Initial Research Proposals, Kadugli Research Station. (WSARP Publication No. 18) Khartoum, Sudan and Pullman, Washington.
WSARP, 1984. Research Programme for the Kordofan Region 1984-1985. (WSARP Publication No. 22) Khartoum, Sudan and Pullman, Washington.
WSARP, 1985a. WSARP 84/85 Research Programme Results, Vol. II, (WSARP Publication No. 42) Khartoum, Sudan and Pullman, Washington.
WSARP, 1985b. WSARP 84/85 Research Programme Results, Vol. III, (WSARP Publication No. 43) Khartoum, Sudan and Pullman, Washington.

Chapter 11

PLANNING AND IMPLEMENTING AGROFORESTRY FOR SMALL-HOLDER FARMERS IN MALAWI

A.R. SAKA
Agroforestry Commodity Team, P.O. Box 158, Chitedze Research Station, Lilongwe (Malawi).

W.T. BUNDERSON
Department of Natural Resources, Washington State University, Pullman, Washington, 99164 (United States of America).

Y. MBEKEANI
Agroforestry Commodity Team, P.O. Box 158, Chitedze Research Station, Lilongwe, Malawi.

and O.A. ITIMU
Agroforestry Commodity Team, P.O. Box 158, Chitedze Research Station, Lilongwe, Malawi.

ABSTRACT

Malawi is a land-locked country in southern Africa with an economy strongly based on agriculture. Relative to other African nations in the region, Malawi has demonstrated stability and growth in agricultural production. However, this is now being threatened by land pressures from a rapidly expanding human population, coupled with the insurgency problems in neighboring Mozambique. The latter has caused Malawi's former trade routes to the sea to be severed, so that exports and imports must now be transported at high cost through Zambia, Tanzania, Zimbabwe, and South Africa. It has also led to a great influx of Mozambique refugees, resulting in additional pressures on the nation's food supply and natural resources. These problems have dramatically altered the context of agriculture in Malawi. The country is currently faced with declining agricultural production, escalating costs of fossil fuel based inputs, deficits in foreign exchange earnings, deteriorating natural resources, and increasing shortages of fuel wood and building materials. As a nation dependent on agriculture, Malawi appreciates the urgency for developing production-increasing technologies that preserve the integrity of its natural resources. This chapter describes the process of instituting a national agroforestry programme to help improve the sustainability and self-sufficiency of small-holder agriculture in Malawi. Details of programme structure, organization, and function are discussed in relation to a strategic plan of action, the essential elements of which are:

(a) developing Malawian institutional capabilities for undertaking applied and adaptive research in agroforestry;

(b) identifying and ranking small-holder problems and research needs as they relate to agroforestry potentials;

(c) establishing effective research-extension linkages through technical support and training, joint-programme planning, and collaborative field activities;

(d) integrating natural resource disciplines with those of agriculture to reduce duplication and to maximize impacts on small-holder farms;

(e) enhancing technology adoption through initiatives that complement the existing practices and available resources of small-holder farmers; and

(f) promoting national and international support for agroforestry through greater public awareness of its impacts on agriculture and natural resources.

INTRODUCTION

Malawi is located in southern Africa between latitudes 9^0 to 18^0 S and longitudes 33^0 to 36^0 E (Fig.1). It occupies an area of 11.8 million hectares including Lake Malawi, the third largest lake in Africa. Of the land area, 5.3 million hectares are available for agriculture, with the remaining 4 million hectares represented by natural and human-made forests, game reserves, swamps, and hills.

1.1 Environmental characteristics

The climate in Malawi is tropical, warm, and semi-arid to sub-humid. Malawi has four major agro-ecological zones based on climate, altitude, and vegetation. These include the Shire Valley in the southern Rift Valley, lying below 400 metres above sea level, with a mean annual rainfall of 500-750 millimetres, and a semi-arid savanna grassland and thicket. The Lakeshore Plains are situated along Lake Malawi, lying below 800 metres above sea level, with a mean annual rainfall which varies from 750 to 1000 millimetres, depending on locality and proximity to highland areas. Natural vegetation is dominated by an _Acacia_, wooded savanna grassland and bushland thicket. The third agro-ecological zone is comprised of medium-altitude plateaus between 800 and 1600 metres above sea level, and covering about 75% of the country. The mean annual rainfall in these areas is 800 to 1200 millimetres and vegetation consists of _Acacia_-_Combretum_-_Piliostigma_ wooded savanna mixed with _miombo_ woodland dominated by species of _Brachystegia_ and _Jubernardai_. The high-altitude plateaus and localized mountains between 1600 and 3000 metres above sea level are the fourth agro-ecological region. The mean annual rainfall in these areas ranges between 1200 to 2000 millimetres with low to moderate temperatures. Vegetation consists of montane grasslands in the upper elevations, with _miombo_ woodland lower down.

Soils are variable ranging from coarse-textured sands and shallow rocky soils on hillsides, to latosols and fine-textured alluvial clays. Most soils are inherently low in fertility and exhibit substantial deficiencies in nitrogen and phosphorous.

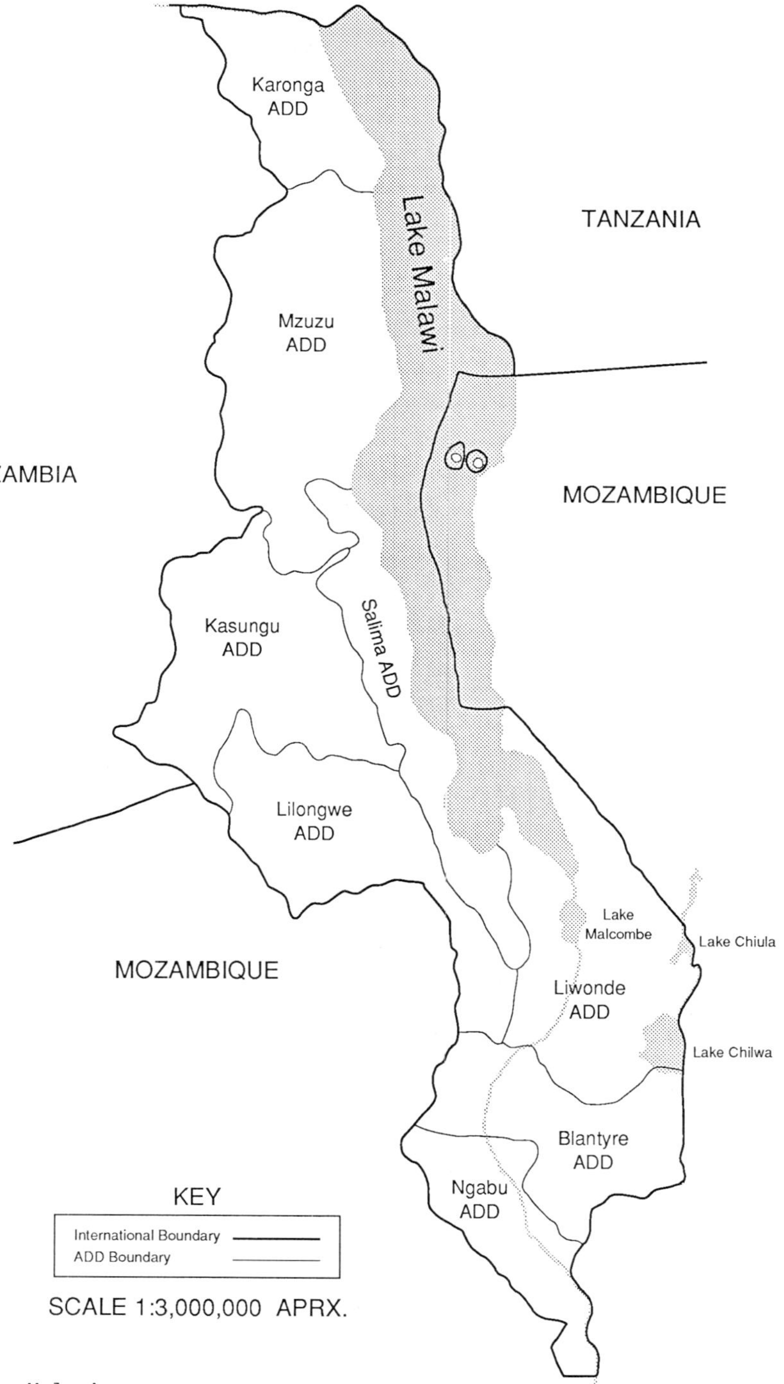
Karonga
ADD
Lake Malawi
TANZANIA
Mzuzu
ADD
ZAMBIA
MOZAMBIQUE
Kasungu
ADD
Salima ADD
Lilongwe
ADD
MOZAMBIQUE
Lake
Malcombe
Lake Chiula
Liwonde
ADD
Lake Chilwa
Blantyre
ADD
Ngabu
ADD
KEY
International Boundary
ADD Boundary
SCALE 1:3,000,000 APRX.

Fig. 1. Malawi.

1.2 Population and agricultural economy

Malawian population is now 7.8 million with an average density of 83 people per square kilometer and a growth rate exceeding 3.2% per annum, both considered high even for Africa. About 90% of the population is rural. The population, however, is unevenly distributed with a density of 29 people per square kilometre in the north, 81 people per square kilometre in the central region, and 110 people per square kilometre in the south. Malawi has one of the better economic performance records in the region, despite recent declines in real gross national product. Agriculture dominates the economy, providing about 40% to 50% of the GNP, 85% to 90% of the foreign exchange earnings, and involving 85% of the labour force. Agricultural production is based on small-holder farming on customary land and estate farming on leasehold land (largely private). The estates produce about 70% of the export trade. Small-holder farmers produce the remaining 30% of the export trade and contribute 85% to total agricultural production. Major food crops are maize, groundnuts, cassava, various pulses, and rice. Ruminant livestock in 1982 numbered about 900,000 cattle, 760,000 goats, and 110,000 sheep. Principal exports include tobacco, tea, sugar, coffee, groundnuts, cotton, macadamia nuts, and maize. Current estimates indicate that agricultural production is expanding at 3.5% per annum, and that demands for fuel wood and other tree products are increasing at similar rates (Nkaonja, 1985).

Under existing land and population pressures, Malawi is facing serious problems of increased erosion and surface runoff, soil-nutrient depletion, overgrazing, deforestation, diminishing ground water supplies, and loss of biological diversity. Natural resource degradation is being accompanied by concomitant reductions in crop and livestock productivity along with shortages in fuel wood and building material. These problems are exacerbated by the increasing dependence on imported fossil-fuel based inputs, which together with high transport costs (caused in part by the severing of trade routes to the sea through Mozambique), have further depressed the GDP and foreign exchange earnings. For small-holder farmers, other constraints to the improvement of performance include limited price incentives, lack of appropriate technologies, and poor research-extension-farmer linkages. The overall result is undermining the self-sufficiency of agriculture which is the Malawian lifeline to the future.

2 RATIONALE FOR AGROFORESTRY

Agroforestry is a collective term for land-use systems involving the deliberate and complementary integration of trees with crops and/or animals, spatially or temporally, on the same land unit (Anonymous, 1982; Chandler and Spurgeon, 1979). Trees in such systems are intimately linked to agriculture. They play a vital role in sustaining and improving crop yields by helping to maintain the soil and water base, and by reducing noxious weeds, pests, and diseases. Trees ensure environmental stability by mitigating the effects of climatic irregularities, storms, and winds. They reduce soil erosion, improve water infiltration, moderate stream flows, and decrease evapotranspiration. Trees restore and recycle soil nutrients and improve organic matter content. They provide shelter for humans and animals and yield an extraordinary array of products essential to nutrition, health, and trade. Besides wood for fuel and building materials, trees provide oils, gums, latexes, resins, tannins, waxes, dyes, fibres, and high quality animal fodder. Many foods, such as fruits, nuts, spices, edible oils, and vegetables, are derived from trees. They also provide a major but largely untapped source of modern and traditional medicines. The many services and functions of trees clearly make their proper management and integration essential to humankind and to the world as a whole.

Although agroforestry is an age-old practice, it is only now emerging as a science. Care must therefore be exercised not to view agroforestry as a solution to all problems of land-use management. However, empirical evidence suggests that agroforestry can provide a sound ecological basis for increased crop and animal productivity, more dependable economic returns, and greater diversity in social benefits on a sustained basis. Given present circumstances in Malawi, increased agricultural production must come from (a) developing alternative trade routes to the sea, while restoring traditional ones, and (b) intensifying and diversifying the productivity of resource-poor farmers who practice cultivation under marginal conditions. For developing countries, the latter is perhaps the greatest challenge yet faced by national governments and the international development community (Wolf, 1987). There is ample evidence today that sustained and improved productivity depends upon the level of soil organic matter, the type of vegetative cover, and the carrying capacity of the natural resource base (McGahuey 1986). Thus, agricultural intensification will require more, rather than less, effort devoted to maintaining the natural resource base.

Recognizing Malawian dependence on the sustainability of agriculture, the Ministry of Agriculture recently constituted agroforestry as an integral component of its research and extension efforts. The organization, function, and activities of a newly formed agroforestry commodity team within the Department of Agriculture Research are described below.

3 ORGANIZATION AND FUNCTION OF THE MINISTRY OF AGRICULTURE

The Ministry of Agriculture (MOA) is responsible for the overall policy formulation, planning, coordination, and management of small-holder agriculture. It also provides direction and guidance to parastatals and all other organizations involved in agricultural production in the country. One of the primary goals and functions of the MOA is a comprehensive development effort aimed at increasing the farm productivity and quality of life for Malawian small-holder farmers. This falls under the umbrella of a National Rural Development Programme (NRDP).

For administrative purposes, the ministry is divided into five organizational units:

(a) Department of Agriculture Research (DAR)
(b) Department of Agriculture (DOA)
(c) Department of Animal Health and Industry (DAHI)
(d) Planning Division (PD)
(d) General Administration (GA)

3.1 Department of Agricultural Research

The DAR is responsible for conducting all research targeted for the small-holder farmer in Malawi. The DAR is headed by the Chief Agricultural Research Officer (CARO) who is assisted by two deputies, one of whom is responsible for research programmes (DCARO-RP), and the other for technical services and administration (DCARO-TSA). The DAR is organized into seven commodity groups, each headed by a national research coordinator (NRC). The commodity groups are cereals; grain legumes, fibers, and oilseeds; livestock and pastures; soils, land husbandry, and agricultural engineering; horticulture; technical services; and adaptive research.

Commodity groups are composed of several commodity teams, representing a specific commodity or discipline, and headed by a commodity team leader (CTL). The CTL is an active research scientist located at a research station where that particular commodity is especially important. The CTL is technically responsible to the NRC who in turn is responsible to the CARO through the DCARO-RP or DCARO-TSA. Applied research is conducted at various research stations located in different climatic zones. All agricultural research stations are directed by a senior agricultural scientist who is responsible to the PS through the DCARO-RP, the CARO, and finally the controller of agricultural services for the National Rural Development Programme (CAS-NRDP).

3.2 Department of Agriculture

The Department of Agriculture is basically the extension arm of the ministry, and largely falls under the National Rural Development Programme. Its chain of command and structure are similar to that of the DAR. The DOA is headed by a chief agricultural officer (CAO) supported by a deputy agricultural officer (DCAO). The DOA has several components: (a) land husbandry; (b) extension and training; (c) crops, inputs, and marketing; (d) agricultural communications; (e) food and nutrition; (f) women programmes, and (g) credit. Each component is headed by an assistant chief agricultural officer who reports to the CAO. The national agricultural extension system is decentralized into 30 rural development projects (RDPs) each subdivided into several extension planning areas (EPAs). The planning, administration, and management of the extension programme at the field level is carried out by eight Agricultural Development Divisions (ADDs), each headed by a programme manager who reports directly to the CAS-NRDP.

4 INTEGRATION OF AGROFORESTRY RESEARCH WITHIN THE DAR

4.1 History of agroforestry initiatives in Malawi

Aside from existing farmer practices, attempts at managed agroforestry in Malawi began around 1982, largely under the Forestry department, but also by Bunda College, the Tobacco Research Authority of Malawi (MTRA) and land husbandry in the ADDs under the DOA. The Forestry Department has focused on afforestation for pole and fuel wood production in different silvi-cultural zones of the country, with screening tests on a range of exotic and indigenous species (Nkaonja, 1985). The Forestry department is also examining the potential of interplanting certain tree species with maize and other crops (Forestry Research Institute of Malawi, 1987). The major emphasis was on *Eucalyptus* species, though results thus far indicate that the genus is not compatible with agricultural crops. MTRA also investigated the planting of *Eucalyptus* with tobacco to provide fuel wood for curing (Tsonga, 1983). Research at Bunda College of Agriculture is investigating various woody species such as *Gmelina arborea*, *Leuceana leucocephala*, *Cassia siamea*, *Cajanus cajan* for compatibility with crops (Edje, 1982; Chiyenda and Materechera, 1987). The land husbandry section of the DOA has initiated work on alley cropping and contour hedgerows on steep hillsides in some ADDs. These are unreplicated trials with a demonstration-focus. *Leucaena leucocephala* has been emphasized to improve soil fertility, reduce fertilizer costs, and decrease soil erosion. In Lilongwe and Kasungu ADDs, contour buffer strips of trees, grass, and hedgerows are also being tried for soil conservation on steep slopes.

4.2 Establishment of a national agroforestry commodity team

Although the interest and progress made with agroforestry in Malawi are commendable, coordination is seriously needed among the various bodies engaged in the activity. There is also a notable lack of quantitative research in the design and evaluation of appropriate agroforestry technologies. Because of its integrative nature, a need was articulated for designating agroforestry under the Ministry of Agriculture. One result of this was the formation of a national task force in 1984 to facilitate agroforestry research and development among and between the Ministry of Agriculture, the Ministry of Forestry and Natural Resources, and the University of Malawi. Under the reorganization of the DAR, agroforestry research was constituted as a new discipline within the land husbandry, soils, and agricultural engineering commodity group. This led to the establishment of an agroforestry commodity team in October 1986. Technical assistance and limited funding support have been provided by USAID under contract to the Consortium for International Development, with Oregon State University as the lead institution assisted by Washington and Colorado State Universities. The DAR under the Ministry of Agriculture is otherwise responsible for all staff recruitment and operational support. The team is composed of a Ph.D. in soil physics (Team Leader), two B.S. holders in general agriculture (both currently on advanced degree training in the USA), one Ph.D. in natural resource ecology (US Technical Assistant), and three field assistants trained at the level equivalent to a trade school for agriculture. The commodity team is based at Chitedze Research Station, 16 kilometres from the capital of Lilongwe. The team identifies and develops appropriate agroforestry technologies for small-holder farmers, coordinates agroforestry efforts countrywide, integrates agroforestry research with other commodity research entities, and provides technical support to the ADDs for more effective extension activities in agroforestry.

5 AGROFORESTRY COMMODITY TEAM: OVERALL GOALS AND ACTION PLAN

Given the agricultural-related problems facing Malawi, restoring degraded land to productive use, while ensuring that no more is put out of production, has become crucial. Priority has been given to integrate tools of natural resource management to improve and diversify agricultural production in arable regions, while reducing resource degradation in marginal areas. This must be accompanied by adaptive farming practices to local needs and ecological conditions, controlling erosion and deforestation, and regenerating the vegetative cover. The primary goal is to institutionalize an agroforestry programme for long-term sustainability of agriculture in Malawi.

5.1 Objectives

The basic objective of the agroforestry programme is to identify new or improved agroforestry technologies which address priority farmer constraints and which complement existing agricultural practices. Many agricultural development programmes in tropical Africa have attempted to replace components of traditional systems with modern or high-input farming methods from temperate zones, usually with destabilizing results.

Examples include the rapid decline in the productivity of tropical soils under continuous cultivation despite supplementary fertilizers (Allan, 1965; Le Mare, 1972; International Fertilizer Development Center, 1985; Spenser 1985), and the failures of mechanized farming (Phillips, 1960; Moorman and Greenland, 1980). Such programmes focus on the simple transfer of technology to solve apparently similar problems in developing countries. These efforts usually fail to distinguish between cause and effect, resulting in the symptomatic treatment of problems, and their subsequent reoccurrence, frequently in more acute forms (Bunderson and Frye, 1986).

The nature, and poor understanding, of eco-climatic and socio-economic relationships in Africa, have also imposed limitations on the direct transfer of technology, while local knowledge has often been ignored or discounted (McGahuey, 1986). As resource-poor, risk-avoiding farmers, most small land holders lack capital to support recurring purchases of inputs such as fertilizer, pesticides, or machinery. It is no coincidence that small, land-holder farmers in Africa have, over the centuries, evolved well-adapted responses to the vagaries of their environment, evidenced by their long and continued survival. Thus, it appears likely that advances in Africa's small land-holders agriculture will only be realized within a development framework defined by sound traditional and scientific knowledge, guided by the achievement and failures of the past.

The proposed action plan involves a new coordinated approach to address natural-resource management issues in agriculture. It provides strategies that are flexible; that are responsive to carefully determined needs; and, above all, that are realistic and practical.

5.2 Strategies

At present, there is considerable in-country support and interest in agroforestry, especially in the DAR, DOA, and ADDs. The Forestry Department, MTRA, and Bunda College are also enthusiastic and have been active in the field. In order to maintain this momentum, early successes are essential. For this to occur, proposed agroforestry interventions must first be biologically sound, and second, they must be economically feasible and relevant to the objectives, priorities, and resource capabilities of the small-holder farmer. This is especially important given the fact that most ADDs are now implementing agroforestry activities with little or no support. Under the current NRDP initiatives, all eight ADDs in the country have been mandated to conduct agroforestry demonstration trials through the Land Husbandry branch of the DOA. However, no agroforestry technology has been adequately tested under Malawi conditions for successful extension to small-holder farmers. The ADDS also lack trained personnel in agroforestry. This scenario is potentially dangerous, because if unproven agroforestry technologies are extended to farmers, the risk of failure is high, particularly if the management aspects are poorly understood by extension staff. The result could easily dampen or even destroy the existing enthusiasm for agroforestry in Malawi. In order to effectively address these issues, the strategies of the agroforestry commodity team are:

(a) to develop and strengthen the research capabilities of the agroforestry commodity team through improved organization, planning, relevant training programmes, and recruitment of qualified and motivated staff;

(b) to establish and strengthen linkages with national and international groups and institutions that are actively engaged in agroforestry;

(c) to raise ministry, public, and international commitment and support for agroforestry research and extension as a priority activity in Malawi;

(d) to identify priority problems and research needs, emphasizing interventions having high potential for impact on small-holder farming conditions;

(e) to evaluate the status of agroforestry research in and outside Malawi, and to help coordinate all national agroforestry activities;

(f) to identify or improve technologies through applied on-station research activities which address documented priority problems;

(g) to improve extension capabilities in agroforestry and linkages with research by assisting in training, providing technical and logistical support in designing and evaluating potential technologies, and developing collaborative activities that are jointly planned and executed at local levels; and

(h) to increase the adoptability of agroforestry by making incremental improvements in existing small-holder farming practices using technologies that are complementary, simple, practical, and visibly effective.

A number of anticipated outcomes include:

(a) provision of an effective framework for planning, implementing, and evaluating agroforestry activities in Malawi;

(b) ensured active participation and integration of agroforestry activities within the DAR, DOA, university, Ministry of Forestry and Natural Resources, and other commodity groups;

(c) strengthened institutional capabilities in Malawi to promote self-sufficiency and integration of agroforestry research and extension;

(d) increased small-holder participation and adoption of agroforestry technologies;

(e) improved food and fuel wood security for small land-holders farmers and the nation as a whole through more sustainable agricultural production practices and enhanced economic growth; and

(f) restored productive capacity of land and sustainable use of natural resources.

6 TRAINING OF AGROFORESTRY PERSONNEL

6.1 Short-term training

Several short-courses and workshops hosted by ICRAF have been attended by staff in the agroforestry commodity team over the past two years. These include: a six-month fellowship for the commodity team leader at ICRAF headquarters in Kenya, a 3 to 4 week course on ICRAF's D&D methodology, and a workshop on research designs and methods for agroforestry technologies potentially suitable for farmers in southern Africa. Additional agroforestry training courses and study tours have been attended by research personnel in the DAR as well as extension staff in the land husbandry section of the DOA. Other international training opportunities have been communicated to the DAR for both research and extension staff involved with agroforestry.

6.2 Long-term training

Master degree training programmes have been initiated for two of the team's senior staff. They are being trained in crop and tree ecophysiology in the Departments of Agronomy/Soils and Natural Resource Sciences respectively at Washington State University. Field research in Malawi is proposed for both degree programmes. In addition, the nomination of two senior extension officers working at the field level have been suggested for advanced degrees in agroforestry-related fields.

6.3 In-service training

Informal training within the agroforestry commodity team has been provided in computer operations, applications, and software packages; nursery management, tree identification, and germplasm collection and screening; agroforestry research designs and data collection and analysis; work-plan and budget planning; and preparation of scientific papers and reports.

6.4 Training workshops for ADD extension personnel

A workshop in Malawi has been proposed to train ADD extension staff in basic agroforestry concepts and technologies relevant to local conditions as well as to the research activities being conducted by the agroforestry commodity team.

7 LINKAGE DEVELOPMENT

The basic premise for establishing linkages with other groups and institutions engaged in agriculture is to optimize use of scarce resources, to avoid unnecessary duplication, and to maximize impacts on small-holder production. Within the DAR, collaborative research activities are being conducted with other research teams to evaluate the benefits of integrating trees with various commodities. The teams currently involved include maize, soils, livestock, pastures, grain legumes, horticulture, microbiology, and adaptive research. The latter group is particularly important for evaluating and modifying technologies generated at the research station for adaptation to farmer conditions. Outside the DAR, linkages are being established to help coordinate agroforestry activities with the DOA and ADDs (specifically land husbandry, women's programmes, and extension AIDS), the forestry department, Bunda College of Agriculture, the Tobacco Research Authority, and ICRAF. The agroforestry commodity team has also provided assistance in the design of agroforestry components for ongoing or proposed donor-funded projects by World Bank, FAO, EEC, GTZ, and UNDP. Linkages with these groups are vital for identifying, developing, and disseminating agroforestry technologies successfully to the targeted farming communities. In all cases, the initiation of collaborative activities was preceded by the identification, diagnosis, and ranking of problems which could be addressed by potential agroforestry technologies.

7.1 Problem focus and research priorities

Assessments of farming characteristics, problems, and needs in different regions were based largely on the abundant information already existing in the country, and involved active input from the collaborating groups. Where information was deficient, especially regarding agroforestry practices and potentials, additional surveys have been conducted or proposed. These include surveys of Malawi's plateau areas by the agroforestry task force in collaboration with ICRAF, informal surveys of the ADDs by the agroforestry commodity team, and recent diagnostic surveys of the ADDs coordinated by the adaptive research group. The latter surveys are based on guidelines developed by CIMMYT, but now include an agroforestry component designed by the agroforestry team.

Analysis of the available information reveals that in many regions of the country, small-holder farmers are faced with four fundamental problems: (a) low or declining soil fertility and crop yields, (b) lack of capital or credit to purchase mineral fertilizers and other inputs, (c) fuel wood shortages, and (d) fodder deficiencies. Hilly regions are also experiencing severe soil erosion and surface water runoff. The initial thrust of the agroforestry commodity team is geared to address these problems in collaboration with the groups mentioned above. The context of the small-holder farmer is central to this effort, with a focus on making incremental improvements in existing practices rather than introducing radical transformation. Technology attributes under scrutiny include the relative perceived advantage, compatibility with local culture, low technical complexity, researchability, and observability (adapted from Rogers and Shoemaker, 1971).

8 CURRENT RESEARCH AND EXTENSION ACTIVITIES

Based on documented priority problems and needs, activities by the agroforestry commodity team fall into two basic categories: applied research under experimental conditions with relevant commodity research teams within the DAR, Bunda College, forestry department, and MTRA; and technology testing and/or on-farm trials with adaptive research and extension personnel at the ADD level. Specific activities are summarized below.

8.1 Germplasm collection, screening, and multiplication

A number of indigenous tree species are already retained on small-holder farmland for various purposes, but are rarely established artificially. The more common include Acacia albida, A. polyacantha, Erythrina abyssinica, Kigelia africana, Markhamia obtusifolia, Pericopsis angolensis, Piliostigma thonningii, and species of Albizia, Brachystegia, Combretum, Ficus, Pterocarpus, and Ziziphus. Most are selected for their compatibility with crops in supplying fuel wood, building materials, edible fruits, and other tree products from the same land cultivated to crops. Some also enhance crop yields or provide important dry-season fodder. Although farmers may practice selective retention and thinning of trees on their farms, they have no direct control over the species, number, and density of trees occurring there naturally. In addition, the optimal spacing and density of trees are not known for most species. Deliberate tree planting on farms is generally restricted to fruit species such as mango, pawpaw, and bananas. Exotic trees are also planted for fuel wood and building poles in small woodlots or around farm households. These commonly include Cassia siamea, Eucalyptus species, Gmelina arborea, Melia azedarach, and Toona ciliata.

Given the limited number of tree species identified for use in agroforestry, the objective has been to collect, screen and multiply woody species with promise for various agroforestry functions in Malawi. The team has collected or assembled seed from over 90 species of exotic and indigenous plants. Complementary activities are being conducted where necessary to determine needs for scarifying different species. Nursery facilities have been developed at Chitedze for screening and multiplication purposes. Screening criteria on existing germplasm collections are in progress and focus on compatibility with crops and characteristics related to the desired tree use or function. Within this framework, evaluations are being made on ecological adaptation, growth, vegetative productivity and quality, regenerative ability, multi-purpose functions, and ease of establishment. The collection and screening of germplasm will be a long-term activity.

8.2 Tree-crop interface studies with Acacia albida and maize

In order to develop improvements in existing agroforestry practices, information on the interaction between trees and crops is needed, along with existing or potential management strategies. The objective here is to investigate the effects of mature Acacia albida trees on soils and maize to better quantify the benefits of this tree. On-farm studies are being conducted in two locations with extension personnel in land husbandry. This will help foster research-extension linkages, and will provide information on the functions and uses of Acacia albida from the farmer's perspective. Future studies will include other indigenous trees.

8.3 Promoting Acacia albida on small-holder farms

Although quantitative research is limited, Acacia albida is well known for its beneficial effects on cereal crop production and soils (Porteres, 1954; Charreau and Vidal, 1965; Jung, 1967; Dancette and Poulain, 1968; Wickens, 1969; McGahuey, 1986). Acacia albida is also valued for fodder and to a lesser extent for fuel wood (Boudet and Riveriere, 1967; Charreau and Nicou, 1971; Felker, 1978). In terms of adoptability among small-holders, the tree has other advantages, notably:

(a) Farmers in Malawi (and elsewhere in Africa) have long appreciated the value of the tree, and use it opportunistically wherever it is found; its more systematic integration will lead to greater improvements in crop and livestock production on a broader scale without changing cultural traditions.

(b) Unlike many agricultural technologies which require long-term research and testing, the well-documented benefits of Acacia albida permit its immediate extension into areas known to be suitable to its growth; this effort can be expanded as other suitable environments are identified.

(c) Acacia albida is indigenous to Malawi and is much more widespread than commonly believed, although its present abundance and distribution need precise definition.

(d) A density of only 25 to 40 mature trees per hectare is needed for affecting crop and soils uniformly, and for providing abundant high quality fodder during the dry season. This greatly reduces costs for establishing the tree, and minimizes land allocated to it.

(e) Defoliation at the onset of the rains deposits a fine mulch of leaf litter, enriching the soil and reducing the evapo-transpiration, while allowing adequate light penetration to crops for normal photosynthesis. Other tree species require intensive labour and management to reduce shading and to spread the leaf manure on the soil.

The Acacia albida programme attempts to address, simultaneously, four fundamental problems facing small-holder farmers in Malawi. These problems are low crop yields due to declining soil fertility and organic matter content from continuous cropping with little or no inputs; increasing costs and declining availability of imported mineral fertilizers; dry season fodder deficiencies among ruminant livestock due to the low supply and quality of available grazing resources, which are aggravated by overgrazing and/or the inability and cost of growing adequate feed; and fuel wood shortages due to increased deforestation

The integration of Acacia albida has high potential for markedly improving the productivity and sustainability of small-holder agriculture in Malawi. This potential depends on three key issues: the environmental limits of the tree's growth, the development and on-farm testing of an effective extension message, and a concerted commitment by the ministries of agriculture and forestry/natural resources to support the effort as a priority activity. These issues are being addressed by a coordinated programme led by the agroforestry commodity team and involving the following groups:

(a) the DOA, specifically land husbandry, extension aids, and women programmes (responsibilities: assistance in developing, testing, and disseminating Acacia albida to farmers at the ADD level);

(b) the DAR, namely adaptive research, soils, cereals, and livestock (responsibilities: assistance in quantifying the benefits of Acacia albida, improving practices associated with it, and promoting further research); and

(c) the forestry department (responsibilities: provenance testing, determination of the environmental limits Acacia albida in Malawi, and establishing seedlings country-wide in the department's network of nurseries).

8.4 Alley cropping

The concept and practice of alley cropping is well documented (Kang and Reynolds, 1986; IITA and ILCA, 1986). It basically attempts to adapt the age-old practice of bush-fallowing to present-day circumstances by allowing a higher intensity of land use, while retaining the functions of nutrient recycling, weed suppression, and erosion control. It can also yield high quantities of fuel wood and fodder.

The agroforestry commodity team is undertaking a series of experiments on alley cropping to determine its potential in Malawi with respect to choice of hedgerow species in different agro-ecological zones; inter and intra-hedgerow spacing; pruning management in terms of time, frequency, and height; the effectiveness of leaf prunings as a green manure and mulch for reducing inorganic fertilizers, weeds, and soil-moisture losses; the benefits of using leaf prunings for fodder versus green manure or mulch; the fuel value of the prunings; and methods to improve adoptability by reducing labour and management costs. Large on-station experiments are in progress involving relevant commodity research teams. Numerous less complex trials are also being conducted at the ADD level with adaptive research and/or extension personnel to test hedge species and practices which appear especially promising.

8.5 Contour alley cropping

This technology simply involves alley cropping on the contour of hillsides, but with the primary function of controlling soil erosion and runoff. Trials underway are examining hedgerow species and inter-row spacing on hillsides with land husbandry extension staff in the Lilongwe, Salima, and Kasungu ADDs.

8.6 Multi-storey alley cropping with maize

Under Malawi conditions, alley cropping practices with single hedgerow species are unlikely to eliminate the need for inorganic fertilizer. They are also limited in terms of diversified outputs of functions. The primary objective of this activity is to evaluate the potential of upper storey trees, properly spaced among hedgerow species, to supplement hedgerow prunings with natural or induced leaf fall to enhance soil fertility and crop productivity; to increase the diversity of tree products and uses from farmland; and to offer a more adoptable alternative for integrating valuable but perhaps slower growing multi-purpose trees with crops. Research is currently confined on-station, and involves evaluating seven tree species that are superimposed on typical alley cropping arrangements.

8.6 Maize response to green leaf manure from trees

The objective here is to compare the differential response of maize to green manure from trees vs. nitrogen fertilizer. The most economical and biologically effective practice will be determined with respect to (a) tree species, (b) leaf application levels and frequencies, (c) application methods, and (d) management issues, particularly labour costs. Changes in soil-nutrient and soil-moisture conditions will also be monitored, along with leaf assessments from hedgerow and individual plants of the tree species involved. The latter will determine tree establishment needs for an adequate response from maize on a per unit area basis.

Two trials are being implemented on-station using woody species which perform well in many parts of the country. Another involving Leucaena is being conducted in Ngabu ADD. The latter attempts to find an alternative use of Leucaena which was formerly grown by farmers in the area as a cash-fodder crop, the market for which no longer exists.

8.7 Live fencing and boundary demarcation

The aim of this activity is to investigate multi-purpose woody species with potential for live fencing and boundary demarcation around farms and home gardens. Desirable characteristics include soil conservation and provision of fruits, fodder, green manure, and fuel wood or poles.

Sisal (Agave sisalana) is already used by small-holder farmers for live fencing around households and irrigated vegetable gardens, along stream banks, and in poorly drained basin areas. Limitations are that sisal performs poorly in waterlogged soils and in the Shire valley which is too hot and dry; it has a shallow, adventitious root system which competes with crops (creating significant edge effects for small gardens); and sisal products are limited to its high quality fiber. Alternative species identified as potentially suitable for live fencing have been raised and planted in two trials, one on-station, and one in Ngabu ADD where damage to crops and tree seedlings by livestock is severe. The species include Acacia nigrescens, A. polyacantha, A. tortilis, A. albida, A. nilotica, Caesalpinia pulcherima, C. decapetala, Erythrina abyssinica, Prosopis juliflora, Pyracantha coccinea, Ziziphus mucronata, Z. mauritiana and Z. spina-christi. Increased plantings of fruit, pole and fuel species on farm boundaries and around the household are also being encouraged in cooperation with the horticulture group and the forestry department.

8.8 Contour buffer strips

The objective of this technology is to investigate how fruit trees, hedgerows, and grass strips can be integrated along the contour of cultivated hillsides to reduce soil erosion and surface runoff. Additional functions include enhanced soil fertility from hedgerow prunings and supplementary income or products from fuel wood, fodder, and tree fruits. Collaborative trials have been initiated with Lilongwe and Kasungu ADDS in hilly regions to evaluate the species combinations and spatial arrangement of the buffer strips for addressing the stated objectives.

9 SUMMARY

Although coordinated agroforestry efforts in Malawi have only just begun, a sound foundation has been developed with respect to building the country's institutional capabilities in the field. This has been achieved through an approach which focuses on (a) quality training and recruitment of staff, (b) strengthening research-extension linkages through collaborative activities, (c) promoting inter-disciplinarity within and among national institutions, (d) identifying priority farmer problems and needs, and (e) developing or improving agroforestry technologies relevant to small-holder farm conditions.

10 REFERENCES

Allan, W., 1965. The African Husbandry Man. Oliver and Boyd, Edinburg.

Anonymous, 1982. What is agroforestry? Agroforestry Systems 1: 7-12.

Boudet, G. and Riveriere, R., 1967. Emploi practique des analyses fourraeres pour l'appreciation des paturages tropicaux. Senegal, Institut d'Elevage et de Medecine Veterinaire des pays Tropicaux (mimeographed).

Bunderson, W.T., and Frye, D.O., 1986. Adaptive strategies for successful pastoral development. In: International Rangeland Development Symposium, Orlando, Florida. February 1986.

Chandler, M.T. and Spurgeon, D., 1979. International Cooperation in Agroforestry. ICRAF, Nairobi.

Charreau, C. and Nicou, R., 1971. L'amelioation du profil cultural dans les sols sableux et sabloargileux de la zone tropicale seche Ouest-Africaine et ses incidences agronomiques. Agron. Trop., 76: 565-631.

Charreau, C. and Vidal, P., 1965. Influence de l'Acacia albida Del. sur le sol, la nutrition minerale et les rendements des mils Pennisetum au Senegal. Agron. Trop., 20: 600-625.

Chiyenda, S.S. and Materechera, S.A., 1987. Agroforestry in Malawi: Experiences. Paper presented at the National Agroforestry Workshop, Mangochi, Malawi, January 26-30, 1987.

Dancette, C. and Poulain, J.F., 1968. Influence de l'Acacia albida sur les facteurs pedoclimatiques et rendements des cultures. Sols Africaines 13(13): 197-239.

Edje, O.T., 1982. Agroforestry: Preliminary results of interplanting Gmelina with beans, maize, and groundnuts. Luso: J. Sci. Tech. 3: 29-32.

Felker, P., 1978. State of the art: Acacia albida. Grant No. AID/afr-C-1361, University of California, Riverside, California.

Forestry Research Institute of Malawi, 1978. Agroforestry trials of the Rural Fuel wood and Polewood Research Project. Paper presented at the National Agroforestry Workshop, Mangochi, Malawi, January 26-30, 1987.

International Fertilizer Development Center, 1985. Fertilizer research programme for Africa. IFDC, Muscle Shoals, Alabama.

IITA and ILCA, 1986. Alley Farming Network for Tropical Africa. IITA, Ibadan, Nigeria and ILCA, Addis Ababa, Ethiopia.

Jung, G., 1967. L'influence de l'Acacia albida Del. sur la biologie des sols dior. Bambey, Senegal, Centre National de Recherches Agronomiques (Mimeographed).

Kang, B.T. and Reynolds, L., 1986. Alley farming in the humid and sub-humid tropics. In: Alley Farming Network for Tropical Africa.

Le Mare, P.H., 1972. A long-term experiment on soil fertility and cotton yield in Tanzania. Exp. agric. 8: 299-310.

McGahuey, M., 1986. Impact of forestry initiatives in the Sahel. Chemonics, Washington D.C.

Moorman, F.R. and Greenland, D.J., 1980. Major production systems related to soil properties in humid tropical Africa. In: Priorities for Alleviating Soil-Related Constraints to Food Production in the Tropics. IRRI, Los

Banos, Philippines: 54-77.
Nkaonja, R.S.W., 1985. Fuel wood and Polewood Research Project for the Rural Population of Malawi. (Forestry Research Record No. 62) Malawi Government.
Phillips, J.F.V., 1960. Agriculture and Ecology in Africa. Praeger, New York.
Porteres, R., 1954. The improvement of agricultural economy in Senegal. African Soils III(1): 13-51.
Rogers, E.M. and Shoemaker, F.F., 1971. Communication of innovations. The Free Press, New York.
Spenser, S.C., 1985. A research strategy to develop appropriate agricultural technologies for small farm development in Sub-Sahara Africa. In: H.W. Ohm and J.G. Nagy (Editors). Appropriate Technologies for Farmers in Semi-Arid West Africa, Purdue University, West Lafayette, IN.
Tsonga, E.W., 1983. An investigation into the feasibility of intercropping _Eucalyptus_ and flue-cured tobacco. Agroforestry Workshop, June 28-July 1, 1983, Lilongwe, Malawi.
Wickens, G.E., 1969. A study of _Acacia albida_ Del: (Mimosoideae). Kew Bull. 23: 181-200.
Wolf, E.C., 1987. Beyond the Greed Revolution: New Approaches for Third World Agriculture. Worldwatch Institute, Washington D.C.

Chapter 12

LANDSCAPE DEVELOPMENT SCENARIOS FOR PLANNING AND IMPLEMENTING AGROFORESTRY: A CASE STUDY IN THE SEMI-ARID LANDS OF EASTERN KENYA

J.W. JANSENS
Department of Landscape Architecture, Agricultural University Wageningen, Gen. Foulkesweg 13, 6703 BS Wageningen (The Netherlands)

ABSTRACT

Within the dryland agroforestry research project, a planning study was carried out in 1984 and 1985 for improvement of the environmental quality of a semi-arid rural landscape in eastern Kenya. The purpose of this study was to contribute to the development of a procedure for planning and designing semi-arid landscape through agroforestry techniques. The formulation and design of recommendations in the form of a planning strategy for the rehabilitation of eroded grazing lands was another goal.

Field research in the catchment basin, study area revealed that severe land-use pressure had caused considerable soil erosion, wood shortages, and reduced vegetation regrowth. Continuation of these processes would seriously reduce the carrying capacity of the area. Field trials in on-farm conditions were set up to test a combination of principles and techniques from agroforestry to soil and water conservation. A landscape development strategy was formulated and two scenarios were therefore designed: the "cooperation scenario" and the "independence scenario." They were based on two major socio-economic processes occurring in the area: continuing self-sufficiency among the majority of the first generation farming families and incorporation of the richer farmers in the regional market sector; and increasing cooperation and self-help among the young, and relatively poor second generation farmers.

Based on the "cooperation scenario", a short-term landscape development plan was worked out. By selecting this scenario, the group that was most in need of help and training would benefit most from the project. The scenarios were designed to make decision-makers more aware of possible research and land development options and the related consequences for land-use development processes. The scenarios and the plan made for the catchment area are discussed in this chapter. The procedure followed is evaluated and various recommendations are made on planning of agroforestry projects.

1 INTRODUCTION

What can a planning and design approach contribute to agroforestry interventions for the improvement of the quality of human life, in general, and for farming conditions in particular, in the semi-arid lands of eastern Kenya? In order to find an answer to this question several studies were carried out, starting in 1983, in Kenya's drylands by the Department of Landscape Architecture of the Agricultural University Wageningen, The

Netherlands, and the International Council for Research in Agroforestry (ICRAF), in Nairobi (Van den Hoek, 1983; Rocheleau and Hoek, 1984). In 1984 and 1985, these studies were followed by a landscape development plan in which landscape planning, agroforestry, and soil and water conservation were linked. This study was carried out within the dryland agroforestry research project (DAR project), a regional project in the drylands of Machakos district. The DAR project and the field work for the study were carried out under the umbrella of the Kenya Agriculture Research Institute (KARI).

1.1 Kenya's development policy

Throughout the 1980s, Kenya has been focusing more and more on integrated development at the regional level. This policy has evolved into a district focus policy that strongly emphasizes development at the district level. At this level authorities are being established with relatively free mandate for district development. In the Machakos district, this policy has resulted in the establishment of the Machakos Integrated Development Programme (MIDP). Within this district development programme, the Kakuyuni catchment area has been one of the target areas for rural development. A 1.6 million cubic metre dam, a medical field centre, and several schools have been built in this catchment area since 1982, and the main provincial road has been improved. The economic and social status of the area has gained considerably from this development and various other activities, such as the establishment of the Undugu Society farmers' training centre and the DAR project.

1.2 The dryland agroforestry research project

The DAR project started in 1983. The project was a joint initiative of the forestry branch of KARI and the national dryland farming research station (NDFRS) at Katumani near Machakos town, with scientific backstopping by ICRAF. Technical support was provided by agricultural specialists of the MIDP. The goal of the DAR project is to develop improved systems for agroforestry land management, which should be appropriate for the mixed farming systems characteristic of Kenya's semi-arid zone, with particular attention to the farming situation in Machakos district. With the proposed agroforestry systems, the project intends to improve the quality of life of the inhabitants of this dryland region (Sang, 1984).

Research has been carried out through on-station and on-farm trials. At the NDFRS-Katumani, research plots have been established for alley-cropping tests under undisturbed (non-farming) circumstances. In order to test and apply selected appropriate agroforestry techniques in the field under practical farming conditions, the Kakuyuni catchment was chosen as the research area for on-farm trials. In this field research, the project has been focusing on the establishment of a tree nursery and a vegetable gardens, and the testing and selection of appropriate tree species on farm plots through:

(a) protection of existing vegetation and growth measurement,

(b) planting of hedgerows and windbreaks along farm boundaries, and

(c) planting of square blocks of 100 seedlings on several grazing lands with various soil types.

In 1984, KARI and ICRAF decided to develop a long-term research plan for the Kakuyuni catchment area. This research plan was intended to explain the techniques to be developed and tested, as well as the location of test sites in the catchment area. Furthermore, the plan was to establish a schedule for the timing of this research within the context of the total project. A preliminary outline for further research and planning was given in a proposal formulated by ICRAF (Hoekstra et al., 1984). The case study presented in this chapter is based on that proposal (Jansens, 1986).

1.3 Goals

At the outset of the study for a landscape development plan for Kakuyuni, it was felt that an adequate methodology for landscape planning in arid and semi-arid regions was not yet fully developed. Existing planning and implementation approaches in the various fields of rural development (land evaluation, watershed management, forestry) are often oriented from a single disciplinary perspective and pay little attention to the actual levels of organization and planning. In the diagnosis and design (D&D) methodology of ICRAF (Raintree, Chapter 2) however, this shortcoming is greatly alleviated. The D&D methodology has been developed for farm-level applications and has been successfully applied at the level of a watershed area as well (Hoek, 1983; Rocheleau and Hoek, 1984). Simultaneously, through ongoing research and studies carried out within the Department of Landscape Architecture at the Agricultural University Wageningen, an approach for interdisciplinary planning and implementation of agroforestry in a framework of rural and regional development was gradually being worked out (Hoek, 1984; Tjia and Tjiook, 1985; Duchhart and Nauta, 1986; Jansens, 1986). Building on the experiences in this study an effort is made to further develop the planning approach for agroforestry interventions at the level of the watershed. The goals are:

(a) to contribute to the development of a method for planning and designing of landscapes in semi-arid regions, in particular through the application of agroforestry and soil and water conservation techniques; (this goal is related to the ongoing research at the Department of Landscape Architecture of the Agricultural University Wageningen), and

(b) to formulate and design a plan for the testing and implementation of agroforestry techniques on-farm in the Kakuyuni water catchment area in Machakos district, Kenya (this goal related to the activities of the DAR project).

The method, the planning strategy, and the scenarios are intended to make local decision-makers aware of the possible agroforestry research options and the related consequences for land-use development processes.

2 THE CASE STUDY

2.1 Method

The research and planning method can be divided into four phases (Fig. 1.), that were run through in an iterative fashion during the analysis and design process. In this way, analysis results were directly used in planning and designing and frequent feedback on data requirements and planning concepts was obtained from all phases.

The first phase consists of interpretation of aerial photographs, map studies, and ground-truth surveys combined with interviews with farmers and local government officials. Preliminary mapping and designing were also undertaken to obtain a better understanding of the state of the existing landscape and land use in Kakuyuni.

In the second phase, collected data were used for a classification of landscape units and farming systems, for a tentative land evaluation, for erosion hazard assessment, and for preliminary conclusions on trends concerning needs, opportunities, and restrictions. Detailed studies on erosion processes and soil and water conservation techniques were started in this phase of the study and field trials were designed and implemented.

In the third phase, the landscape analysis and the results of the detailed studies and field trials were used to assess the planning goals; select priority areas for intervention; select target groups; and select appropriate techniques for land conservation, spatial organization, and land management. In the final phase, assessment of priorities and target groups was used to formulate two alternative landscape development and management scenarios, as well as an action plan for short-term implementation of agroforestry field trials.

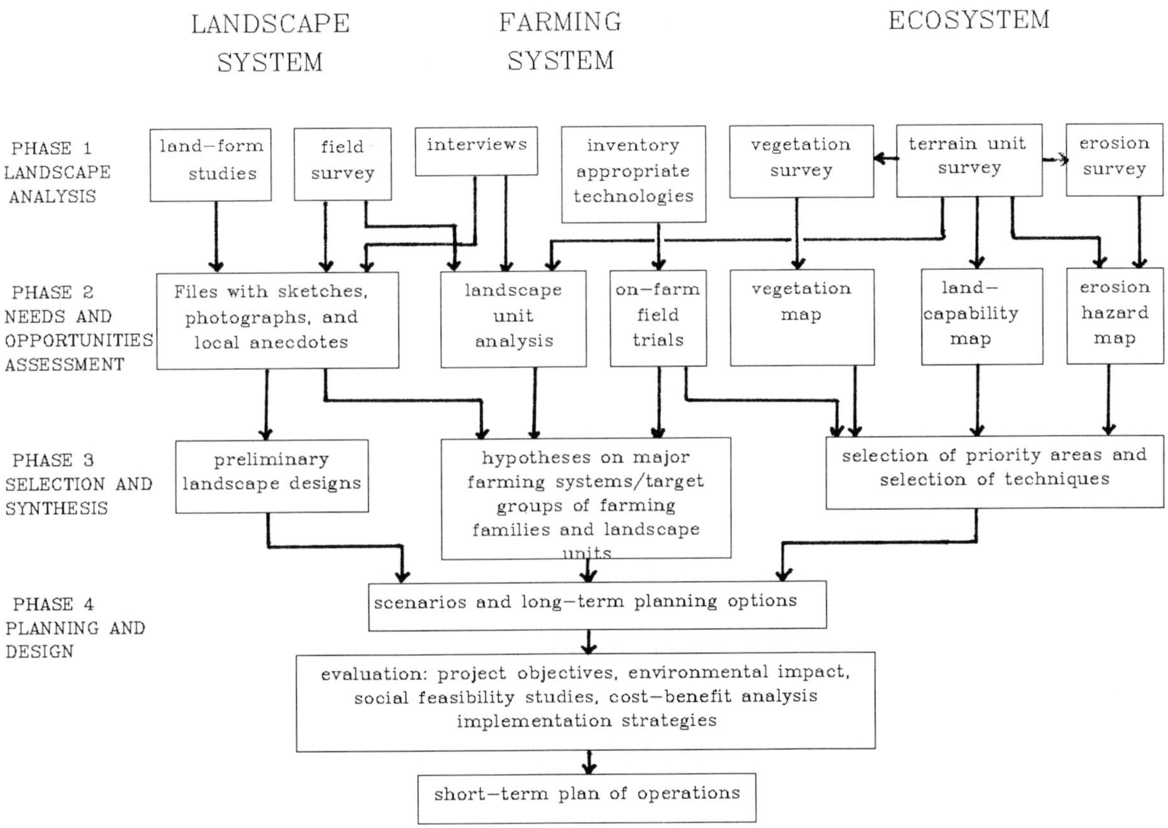

Fig. 1. Research and planning methodology.

2.2 Landscape analysis (phase 1)

(i) Physical features. Kakuyuni is located on the top of the Yatta plateau in a semi-arid climate zone, some 150 kilometres east of Kenya's capital Nairobi. The Yatta plateau is (on average) 5 kilometres wide and 250 metres higher than the surrounding rolling savannah landscape at an elevation of 1200 metres above sea level. The study area comprises a small water catchment and two plateau sections at both sides of the valley (Fig. 2). The study area is almost rectangular and is 2 kilometres wide by 6 kilometres long. Slopes vary between almost flat to 40 percent. Soil erosion by water is a major land process.

Soils in the catchment area are well drained, vary from shallow to deep and are dark red, sandy clay-loams. The valley floor consists of sandy colluvium, as a result of sedimentation of debris from the severely eroding slopes. Rocky, shallow soils and gravels are found on the higher ridges and plateau edges. The plateau is characterized by depressions filled with vertisols. There is severe soil erosion in the valley. The climate of the area is characterized by a bi-modal rainfall pattern, with averages of 305 millimetres in the October-December season and 195 millimetres in the March-May season (Hoekstra *et al*., 1984). The rainstorms are unpredictable and locally variable. Generally, the intensity of the precipitation is high. The indigenous vegetation consists of dense shrubland of the *Acacia-Commiphora* type in the valleys and on the slopes, and a moderately dense savannah vegetation with trees, herbs, and grasses of the *Acacia*-*Combretum* type on the plateaus (Schork,1980).

(ii) Settlement history. For the past 55 years, rapid population growth and decrease of productive arable land in Kenya's highlands have forced the Wakamba people to move to the unoccupied drylands. These dryland areas have much lower agricultural potential. In these areas, the Wakamba continue their system of permanent dryland agriculture combined with some livestock farming (Hoekstra *et al*., 1984). Currently, the population density in Kakuyuni is greater than the average population density of the whole agro-climatic zone. Since the 1950s, the catchment area has been gradually converted into a landscape dominated by human habitation. At present, all the land is occupied and legally adjudicated. The location of Katangi market and increased public and private services in the area have created new marketing opportunities for the farming community. This, in turn, has given the community freedom from their economic, social, and cultural isolation. Since 1984, however, the MIDP has withdrawn its control over the catchment area.

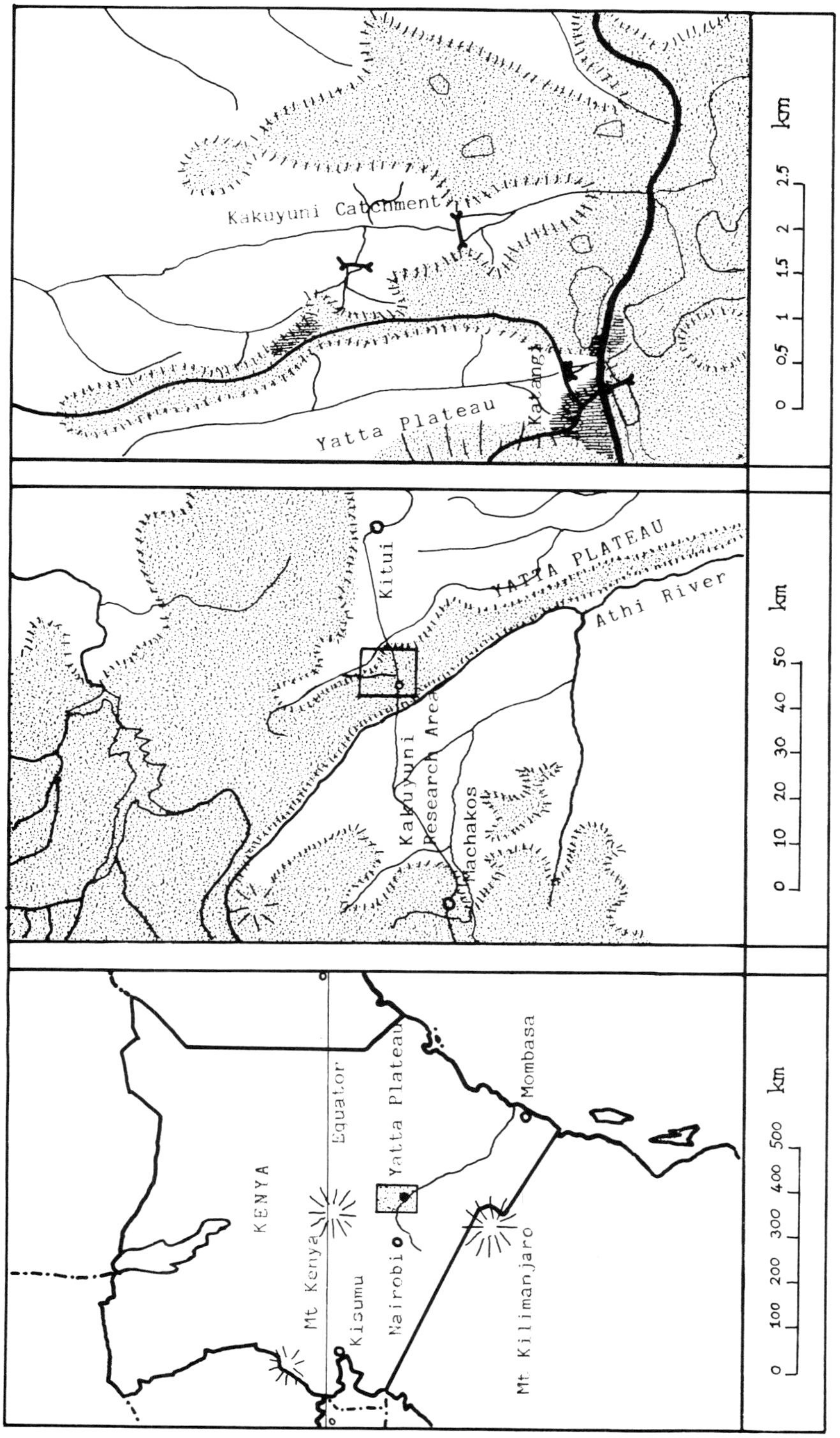

Fig. 2. Study area.

2.3 Assessment of needs, opportunities, and restrictions (phase 2)

(i) Landscape pattern and agricultural systems. The landscape pattern of the study area is dominated by a more or less north-south running river valley, which forms a central axis in the area (see Fig. 3). Farm houses are scattered throughout the area, but tend to be concentrated along roads and paths. The inheritance system and recent land adjudication have caused a considerable fragmentation of land. Consequently, many small farms are surrounded by a patchwork of small remnants of the ancient bushland (often where soils are unsuitable for agriculture and where there are steep, rocky slopes), croplands, and grazing areas. The patchwork pattern which emerges is related through landforms, soil type, the accessibility of the terrain, and also to the proximity of the farm compounds.

The main land-use type in Kakuyuni is agriculture. Apart from arable cropping and livestock farming, families are engaged in farm forestry, charcoal production, woodcarving, beekeeping, brick making, and local construction works. Croplands are found on the undulating parts of the plateaus and on the upper and lower slopes of the valley. Grazing is spread all over the area and can be found predominantly on the slopes of the valley and on the crests, ridges, and depressions of the plateau. The bushland is used for browsing by goats and the collection of fuel wood.

Interviews with the farm families indicate that farm sizes vary considerably, ranging from about 3 hectares to nearly 120 hectares. About 75 percent of the farms are small (3-30 hectares). Observed landscape features were useful in distinguishing landscape units which were representative of existing farming systems and socio-economic differentiation within the farming community. Special attention was paid to two particular groups of farming families: the first generation of farming families, which is older, more experienced, and richer is located on the plateau; and the second generation farming families, young and relatively poor, are found predominantly in the valleys. The differences between the farming systems and the related landscape pattern is described in Table 1.

(ii) Processes and trends: needs, opportunities, and restrictions. Both the landscape and society in Kakuyuni catchment are in rapid and perpetual change. Rapid population growth resulting from the settlement of new, young families and, subsequently, a high birth rate and increasing local marketing activities are obvious features. Consequently, the increase of basic human needs escalates at the same speed.

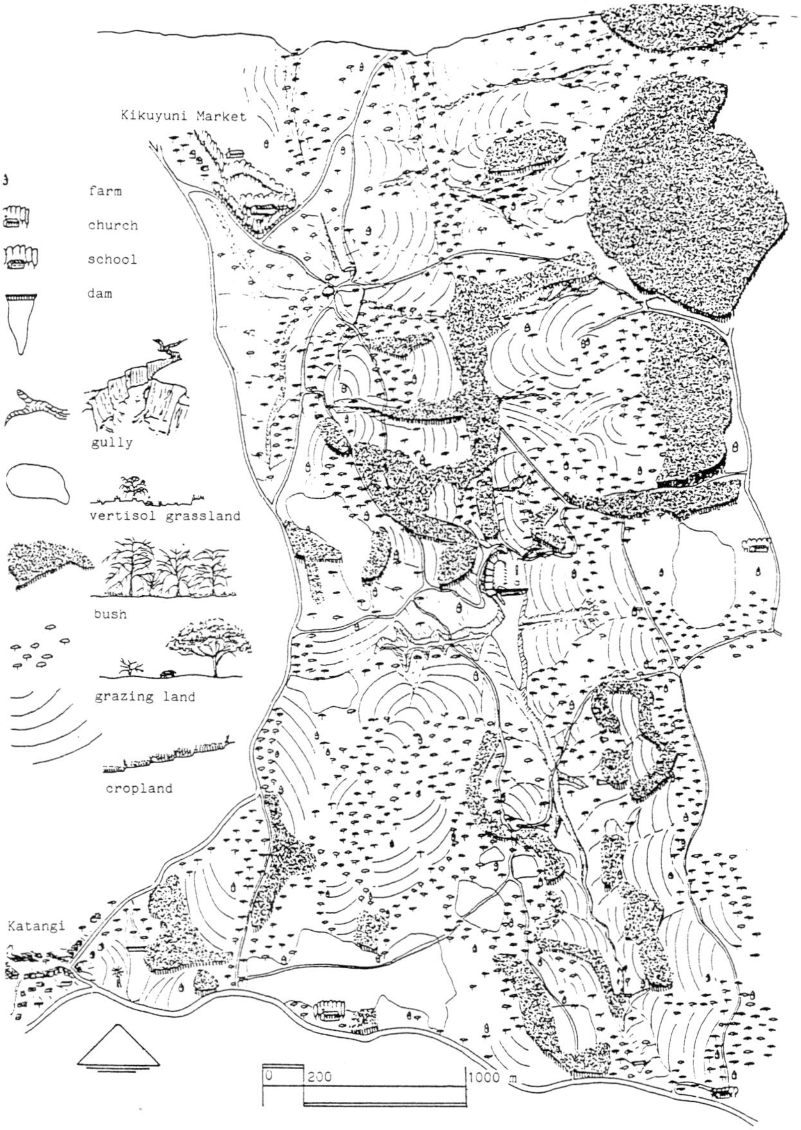

Fig. 3. Landscape pattern of the study area.

TABLE 1

Comparison between farming systems and landscape pattern.

FARMING FAMILY TYPE	FARMING SYSTEM	LANDSCAPE PATTERN
First generation farmers, older, experienced and richer	Arable cropping (maize, millets, and sorghum) as major produce; livestock farming less important; labour and capital limiting factors	Large-scale, mostly non-terraced plots and with bushland not paddocked grazing land
Second generation farmers young, inexperienced, and poorer	Livestock farming relatively important; some arable cropping for subsistence; available productive land, and capital limiting factors	Small, fenced and sometimes terraced plots small bush-land and grazing area; some multiple, integrated land use and zero-grazing

A number of environmental impacts are associated with this rapid socio-economic development. Bushland is being removed. What vegetative cover remains is dwindling because of over-grazing, trampling, wood gathering, and arable cropping. Hardy, unpalatable shrubs and herbs are invading both grazing and croplands. Soil erosion by water and wind is affecting the whole area and locally creating gullies and badlands. Interception of water by vegetation and infiltration in the topsoil is poor. Additionally, surface run-off becomes high as a result of the hardened topsoil and the sloping topography. Because of these impacts, during the rainy season standing water in the vertisol depressions restricts the accessibility of these areas.

The estimated annual soil loss in the valleys is about 125 to 150 tonnes per hectare. Much of the sediment is collected in the main water reservoir of the catchment area. This 1.6 million cubic metre reservoir, built in 1982, will be completely silted up some time between 1990 and 1995 if the present erosion and sedimentation is allowed to continue. This will be fatal for the development of agroforestry in Kakuyuni. Since the dam also provides drinking water for human consumption, this process is a major potential threat for further development of a wider region.

Notwithstanding the decline of natural production factors in the area, rapid socio-economic development encourages self-organization and cooperation among the farmers, especially among the younger generations of farm families. The older and richer farmers are more self-reliant and independent and are seldom engaged in self-help cooperation.

Based on the interviews with farm families and land-use studies, the most important needs and wishes of the inhabitants of Kakuyuni were assessed. Major issues to be solved in the development plan are the shortages of productive land, fodder and fuel resources, and the increasing deterioration of the land through soil erosion, especially on grazing lands. These resources have to be used more efficiently in order to meet local production requirements. Limited productive land areas, wood shortages, and soil erosion are prevalent in the valleys and are therefore directly related with the cooperating, poorer, and second generation farming families. This group of farm families relies considerably on its livestock. This logically related to the conditions of their land. The sloping, rocky areas in the valley are mostly unsuitable for arable cropping, but are suitable for grazing. The farmers, however, need to practice some arable subsistence cropping. Therefore, they often clear the bush on slopes that are unsuitable for cropping and highly susceptible to soil erosion. Because their land area is relatively limited, the remaining areas are severely over-grazed, especially at the end of the dry season. Moreover, daily movement of the herds to the drinking places causes considerable damage to the standing vegetation and to the soils along the animal tracks.

Other needs mentioned during the interviews were organizational and economic production factors, such as local marketing potential, credit facilities, proper legislation and adjudication of land, availability of tools, seeds and seedlings, and the general lack of cash money. An additional consideration is that infrastructure conditions in Kakuyuni are poor. Main roads are unsafe for pedestrians and animals. Deep gullies along the roads form dangerous traps for all users. Micro-climate in the catchment area is harsh. Roads, paths, and gathering places are often not shaded, and storms annually damage farm buildings. During most of the year, the availability of clean water is limited, especially in Katangi market. Finally, the larger farmers are accustomed to cultivating large areas for arable cropping and therefore always experience a lack of humanpower in the cropping season. Most of the farm families have not yet received any training in intensification of farming methods or in the application of agroforestry techniques. Consequently the development of local humanpower will be necessary for the adoptability and continuity of the proposed agroforestry interventions in Kakuyuni.

2.4 Selection and synthesis (phase 3)

(i) Goals for planning. Planning and design of a landscape development plan for Kakuyuni catchment can help find ways:

(a) course to safeguard and develop the area's resources and its carrying capacity, in order to meet the basic needs of the inhabitants,

(b) to create an environment with favorable conditions for human life in Kakuyuni,

(c) to develop appropriate techniques in the fields of agroforestry and soil and water conservation

(d) to select tree species appropriate to the environmental conditions of the area, and

(e) to encourage the inhabitants to carry out these planning objectives independently through training, self-help activities and close cooperation and good communication with project officers.

(ii) Selection and synthesis of data. Since it is not feasible to protect and improve every hectare of the whole water catchment simultaneously, it is necessary to set priorities for the planned interventions. Moreover, there is no need to work on every plot, because the selected intervention may have a positive effect on neighbouring areas, and in this way those areas will improve too.

Three approaches were used for the assessment. The first approach was based on the classification of erosion processes and the restoration capability of the land. Priority areas for intervention have been assessed by overlays which show the physical constraints of the environment of Kakuyuni, such as present erosion hazard by water and wind, and wood shortages. The initial land evaluation, where conflicts between land suitability and land use were evaluated, supported the assessment of erosion hazard areas and priority areas for intervention. Deteriorating processes were classified and ranked in emergency classes. Linked with the location of these processes, the priority areas were classified high, middle and low priority. An emergency classification was used to indicate which areas had priority in the implementation of rehabilitation activities.

Land classes were directly linked with locally approved techniques for land protection and rehabilitation. By matching the land capability map with the map of priority areas, potential rehabilitation techniques could be selected for each particular area. The resulting map shows priority areas for intervention. This map illustrates that the grazing lands and paths in the valleys and on the plateau edges have to be protected and repaired (Fig. 4, map A).

In a second approach to the selection of priority areas, the entire land-use system was studied. In this system, the central dam area and water reservoir are central for the region as a whole. Because of the present siltation of the reservoir and deterioration from cattle tracks, this system is seriously threatened. It is of vital importance for the development of the farming system and all other land-use activities, if not for the well-being of the total community, to secure this area. Furthermore, because so many paths cross this area and because the plant nursery of the DAR project is located near the dam, this central area has been given the highest priority for future action. The emphasis placed on the dam area means that the slopes of the whole water catchment area upstream of the reservoir have high priority in order to be able to preserve the future need for water in the area (see Fig. 4, map B).

In the third approach to the assessment of priorities for a planning strategy, the socio-economic farming system was examined. As indicated in the landscape analysis, two major farming systems can be distinguished in Kakuyuni. From a strategic point-of-view, cooperation with the second generation farm families, who mainly live in the valley, may offer the best opportunities for testing and implementing agroforestry techniques, since these people seem very cooperative and are relatively well organized (see Fig. 4, map C).

2.5 Planning and design (phase 4)

(i) Design principles. Apart from a process oriented planning approach through step-by-step interventions within the DAR project and the selection of priority areas and target groups, various other design and planning principles form a basis for the final plan. A study of how the landscape was perceived by local people, shade, shelter, and access to the land seemed to be important qualitative determinants in of the planning process. An image of order, efficiency, and productivity within the landscape seems to be appreciated in this community. Except for this functional image of the landscape, however, people indicated that they like the bushland and the rough terrain with its views and appreciate the variety of cropland, grazing, bush, and settlements. These issues were considered during the design of a spatial structure for agroforestry techniques for the entire area.

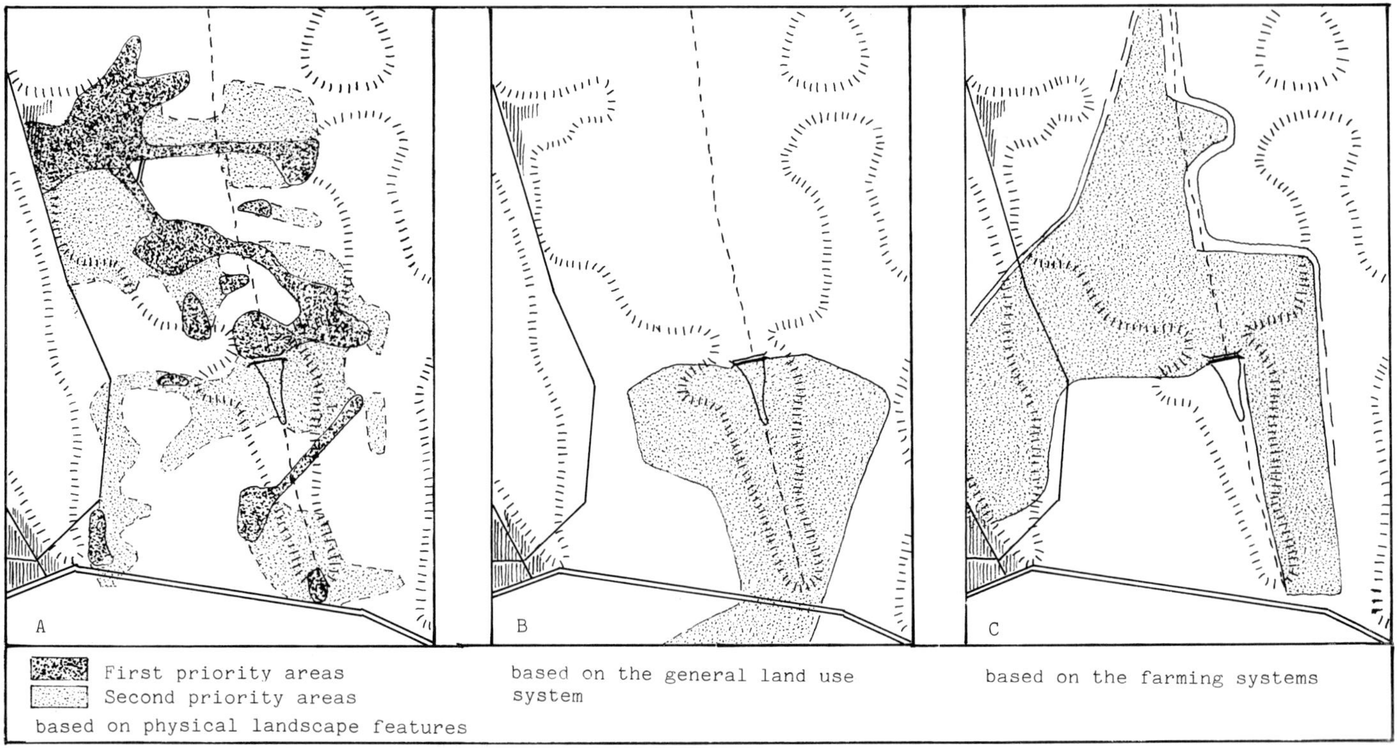

Fig. 4. Priority areas.

Practically, this framework consists of tree rows, hedges, and reafforestation of plots with a varying density of scattered trees. Linear elements like tree rows and hedges are alien forms in the traditional landscape of scattered trees and clumps of bushland. Yet, because of their efficient and functional form (erosion control, fencing, and land demarcation) and regional effect in micro-climate improvement (shading on paths and wind breaks), they can by very suitable and adaptable in the future landscape of Kakuyuni.

(ii) Evaluation of on-farm trials. On-farm trials were designed for two severely eroded grazing lands in order to assess the adoptability of the techniques with farmers. Combinations of techniques were selected with farmers that seemed most appropriate for the local land-use system. The trials were designed in such a way that various soil and water conservation techniques could be tested in combination with agroforestry techniques. As such, the trials were conceived as integrated techniques for grazing-land improvement and land rehabilitation. The trials formed a first step towards the design of new experiments of the DAR project and formed the basis for a feasibility study of the landscape development plan.

Based on these trials, the following conclusions can be drawn regarding planning and design for the whole catchment area.

(a) In most cases, not all precipitation from one rain storm can be retained safely. This is because the rains are usually very intense (one or two rain storms with more than 50 millimetres per hour occur annually). This means that proper diversion of the run-off is at least as important as water harvesting on the slopes for on-the-spot infiltration. Small micro-catchment and infiltration pits, however, can play a role in water conservation at the level of a slope or a plot (Ssekabembe, 1984).

(b) A feasibility study is always essential for success for the design of a field trial or of an implementation technique for farm improvement (e.g., run-off and soil-loss estimations; checks on the availability of seedlings, tools, labour resources, and supervisors; tests in the field on particular issues such as the location and accessibility of farm entrances, fences, and paths; and the location and form of contour lines and micro topography).

(c) Work sessions have to be organized carefully. Field work has to be mapped in advance in a legible form for field supervisors, or has to be planned out in the field before the working party arrives. Work times and phasing of the work have to be discussed with the groups in advance. Phasing of the implementation of the work is particularly important when tools and labour requirements are limited.

(d) Finally, the planning and design for grazing-land rehabilitation is very time consuming and expensive for local farmers in Kakuyuni. Self-help needs to be encouraged and people do find out that cooperation in soil and water conservation, reafforestation, and grazing land rehabilitation in more effective, more efficient, and cheaper than individual initiatives. People can also be trained by the project supervisors or by their fellow farmers in the application of new techniques. Field plots established under such conditions can then be used as demonstration areas for a wider region. For researchers, the field trials form a good opportunity to develop and evaluate a more efficient use of various techniques in a practical situation.

(iii) Two scenarios. Based on the independence and self-sufficiency of first generation farmers, and on the increasing awareness of the importance of cooperation, especially among the poorer, second generation families, two concepts for planning were developed. The concepts or scenarios describe a possible imaginative future for the area.

The discussion of these scenarios and plans is directed at a debate on advantages and disadvantages of the chosen starting points and help planners and project officers become more aware of the consequences of particular moves of the project or particular "autonomous" processes in Kakuyuni. The scenarios can thereby help decision-makers with selecting the best way to mutually adapt the farming system, the needs of the community, and the opportunities and constraints of the environment. Hence, it will be possible to establish research trials, with a functional impact on farming practices in the area.

(iv) Scenario A: "Independence scenario". According to this scenario, people in Kakuyuni will be well educated and the population density should not increase. Farming families manage to maintain and to develop their resources in a sustainable way. Cash-crop production will become a substantial part of the farming system. Livestock farming is limited and part of an integrated farming system in which animals are useful for their manure and draught power. The animals will be fed on crop residues and cut and carry fodder supplies.

A clear distinction is made in the landscape between public and privately-owned land. Public paths will lead directly to the market places and the main public water reservoirs. Public works will be done by self-help groups and governmental services.

(v) The implementation plan for scenario A. In this scenario, the farm families are approached independently by the DAR project officers, since cooperation between the farmers is not anticipated. No measures will be undertaken in which several farms will be involved at the same time. Only those private areas that are so deteriorated that severe damage can be done to neighboring places or areas of public interest, will have to be planned and repaired by a public service.

Public gathering places such as markets, school yards, tap points, and drinking places for livestock will receive major attention in the plan. Tree planting and erosion control will be the most important inputs on these places. Along the public paths, hedges, and trees will be planted for soil stabilization, fencing, shade, and shelter. This human-grown vegetation will provide a certain degree of spatial organization and may provide better orientation in the landscape.

The DAR project will focus primarily on research. Research will be dominated by development and selection of appropriate techniques for the protection of public areas and for the development of individual farming systems by the private owners. The individual farmers might be approached in a "Training and Visit Programme" and they will receive help by the DAR project only if they have asked for it. Organization and guidance of group work is limited to involvement in the seed orchard and tree nursery of Kakuyuni and to the repair of the damaged public places (Fig. 5).

(vi) Scenario B: "Cooperation scenario". In this scenario, the standard of living is to stay at 1985 levels. However, the population density is forecasted to increase significantly. The population will be dominated by many small, young families. Since the people realize that cooperation is the only way to relieve the problems of erosion and lack of supply of natural resources, they will share grazing land, wood lots, paths, and water supplies. They will also help each other through cooperation in self-help groups.

The farming system will still be mixed in which arable cropping for subsistence and livestock farming are important. People will be engaged in various on-farm manufacturing activities like woodcarving, basket weaving, carpentry, and charcoal production in order to earn additional income.

The landscape pattern is characterized by clumps of vegetation, and by groups of farms with a pattern of wind breaks, hedgerows, and reafforested areas the benefits of which are shared. The path system is focused on the centre of the area around the main dam and reservoir. This area is more important than the market places.

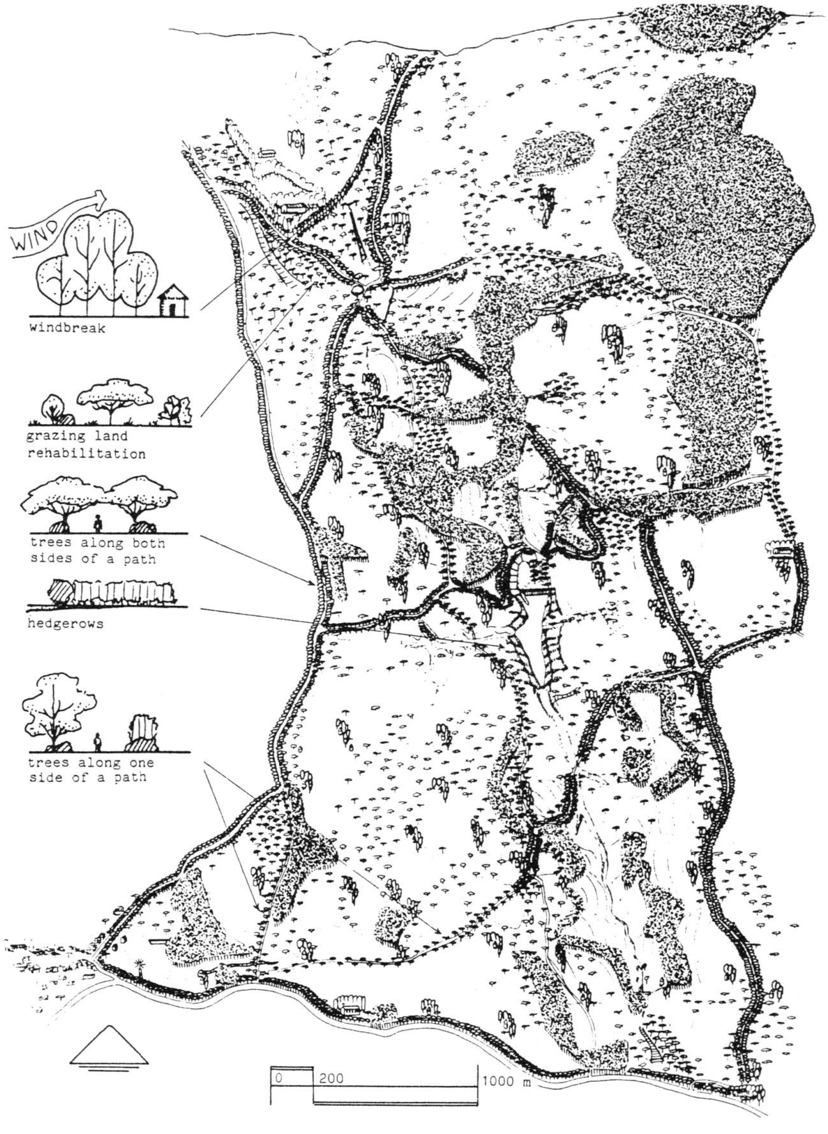

Fig. 5. "Independence" scenario.

(vii) <u>The implementation plan for scenario B</u>. This scenario provides some recommendations that may help guide the development of Kakuyuni towards a landscape in which farm families work together in self-help groups to establish communal wood resources, shelterbelts, and soil and water conservation measures in an integrated form. The plan is based on the assessment of priority areas for intervention, and on the location of the land of cooperating farming families. In this way, the concept is designed for the specific target group, second generation families, and is primarily focused on the rehabilitation and development of the valley landscape.

The measures proposed in this plan are fencing of paths and drinking places; protection of the dam area; the reservoir and its catchment area; reafforestation of severely eroded areas; and the creation of shelterbelts for wind swept places. A combination of private and public places and paths will be treated at the same time. Additionally, important places, such as the drinking points for animals, the dam area, schools, and markets are protected with shade trees, wind breaks and living fences. In many places, grazing land rehabilitation and gully control can be combined with path stabilization and the creation of shelter against wind and solar radiation.

The DAR project will concentrate its activities on field trials with integrated soil and water conservation methods and agroforestry techniques. Self-help groups and cooperating farmers can be trained during the implementation of the field trials. On the one hand, the trials will be part of the implementation of a short-term plan, and on the other hand they will be objects for research (Fig. 6).

(viii) <u>Short-term plan</u>. The short-term plan, which covers the period from 1985 to 1990, should be viewed as a first step in a strategy for the rehabilitation of the Kakuyuni catchment area. Strong regulations on land use and conservation from the government are not anticipated. Therefore, it is assumed that land degradation in Kakuyuni would continue. The major processes likely to continue include: more land fragmentation, conversion of grazing into cropland, decrease of primary and secondary production, increase of cash shortages, pressure on land as a result of overgrazing, aggravation of soil erosion, increasing wood shortages, and continuing pressure on women's labour capacity as more and more young men seek off-farm employment.

The most important issues that need attention in the short-term plan are the prevention of soil erosion and the siltation of the main reservoir; the protection of paths and grazing lands against degradation of soil and vegetation; and the encouragement of self-help, cooperation, training, and extension services. These goals fit best in the cooperation scenario. This

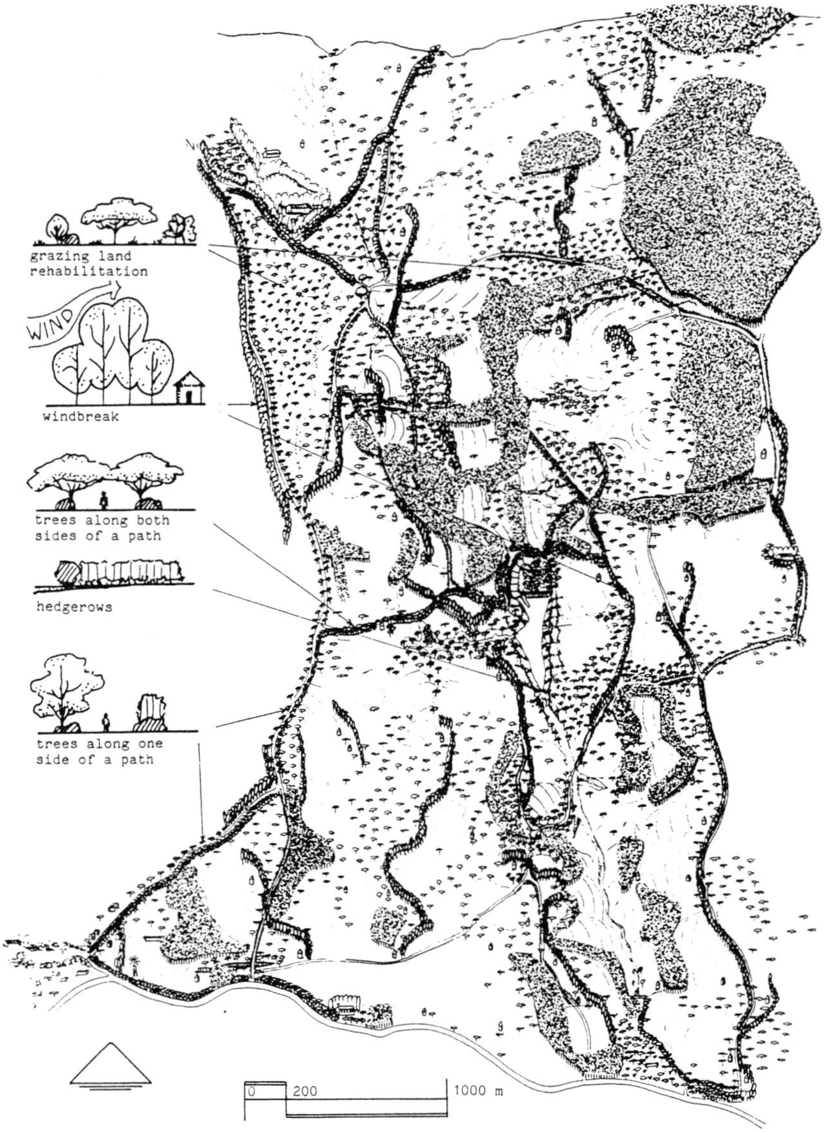

Fig. 6. "Cooperation" scenario.

scenario may eventually be followed by a period with a planning strategy based on the independence scenario. However, permanent evaluation and reassessment of goals, opportunities, constraints, needs, or strengths and weaknesses of occurring events will be indispensable for further strategic planning and design.

Planning according to the cooperation scenario means that special efforts will be made to accommodate the needs farm families whose land has been most seriously deteriorated. This is particularly the group of second generation farmers. An advantage of approaching this group is that they are in fact already actively participating in self-help activities and project guided fields trials. This cooperation, which includes training, can very well contribute to local organization of the farming community. In this way, better continuity will be provided for planned activities on erosion control and tree planting. Moreover, short-term benefits from the project will go especially to this group of people, which has the highest actual needs of the total community.

The short-term plan basically follows the priority areas with the highest emergency class. These areas are the slopes and steep paths around the reservoir and dam, and various steep, bare grazing lands, and paths elsewhere in Kakuyuni (Fig. 7).

Implementation should start with plots situated at the upper parts of the slopes and preferably those in the corners of path crossings. Fencing of the plots under rehabilitation is necessary to prevent game and livestock damage to the earth works and to the newly planted and regenerated vegetation. If the short-term plan is implemented on small plots and through a step-by-step approach, the farmers do not have to be deprived of all their land. The field trials have proved that after a period of one to two growing seasons the tree seedlings will be strong and tall enough to survive in spite of browsing animals. Grazing rotation in paddocks, repair of hedgerows, and soil-conservation measures will be indispensable for sustainable land use. Finally, the use of indigenous and locally adapted tree species must be encouraged, since the farmers are well aware of their functions and ecological requirements. In this way, adaptation of the species both to the environment and the land-use system will be facilitated.

Further planning activities depend on the resources of the project and the development of intensive production systems in the area. In this respect, the potentials of intensive arable cropping (maybe in the form of alley-cropping) and cattle breeding (based on zero-grazing) or their combination have to be evaluated (Mwendandu, 1989). Also the local

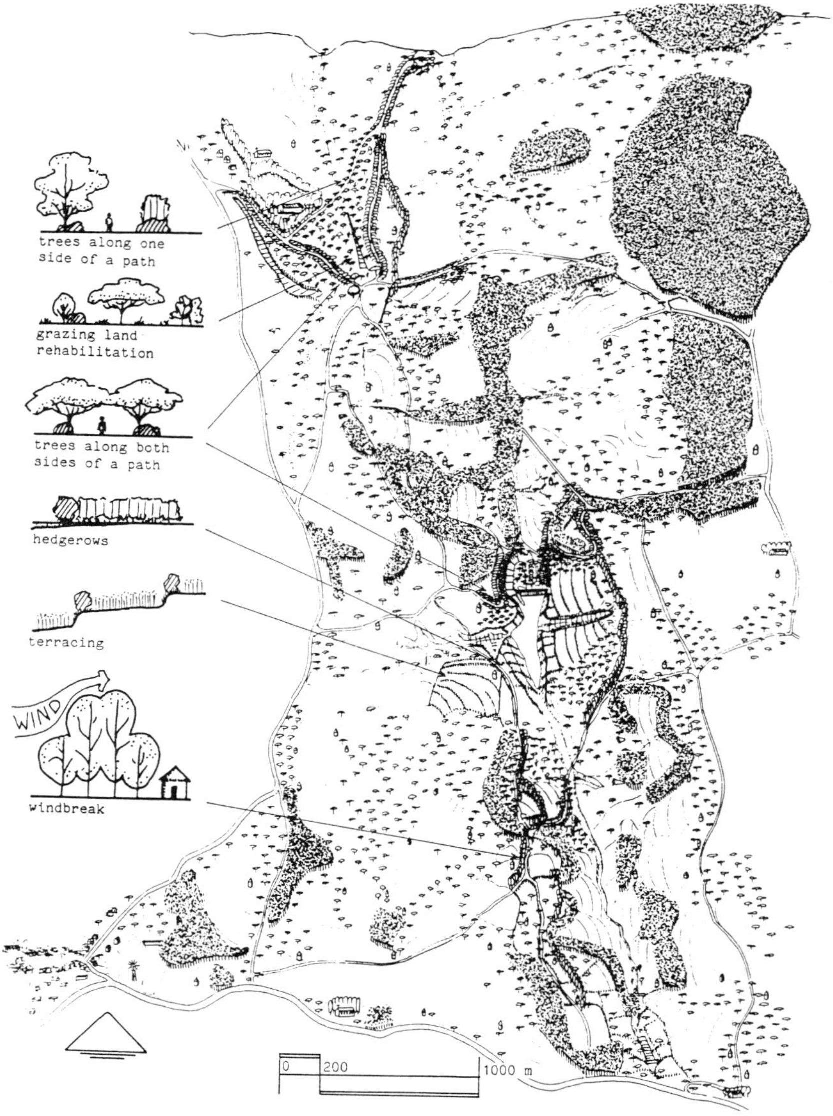

Fig. 7. Short-term plan.

production of wood products, vegetables, and fruits may offer interesting opportunities for the farming systems in Kakuyuni.

3 DISCUSSION

3.1 Summary

In the present landscape of Kakuyuni, annual resource production of the area does not meet present and future needs of humans and animals. Lack of vegetation and increased soil erosion are indicative of land use practices which are not based on sustainable productivity. The main reason for this situation is that too many people are trying to live in an area that cannot support them under their current land use practices. The only feasible way to improve this situation is to encourage the people to use more efficiently the opportunities provided by the landscape. Soil and water-conservation measures plus agroforestry techniques can possibly offer solutions for improved land management. Establishment of field trials on the land of farming families who are willing to cooperate with the project has the potential to be a successful approach. Techniques can be implemented and tested, people trained, and researchers will have opportunities for realistic experiments and evaluations. Continuous evaluation and adjustment is indispensable for the success of these field trials.

The general lack of information and need for continuous adjustment of interventions means that it will be impossible to make one definitive, final plan for the area in which all the problems can be solved. Such a plan will never create a single path to the future. A plan can only be a tool for the exploration of solutions and techniques, and for the generation of new goals for the community. This open-ended planning attitude is based on the opinion that planning must guide development processes rather than try to control development.

Landscape planning with the application of agroforestry techniques is not enough to guide the future of the community. A landscape development plan is only one segment in a total development strategy in which other elements, such as extension work, grazing, arable-cropping, seed research, soil management, water management, employment, development of banking and credit facilities, development of energy resources, family planning, health care, and education must be addressed.

3.2 Evaluation

This case-study presents a landscape design procedure for planning and implementation of agroforestry techniques for the rehabilitation of the upper part of a water catchment. The procedure attempts to formulate design options which can guide discussions on the improvement and development of the landscape with the application of agroforestry techniques.

The discussion is guided by distinguishing three different active systems at the level of the water catchment (the ecosystem, the farming system, and the major land-use system), and by the design of two scenarios which show the consequences of the selection of a specific target group among the farming families. Major opportunities, restrictions, and consequences of implementation of agroforestry techniques can be tested in this way. A land capability map, for instance, will provide information on techniques or tree species which are ecologically feasible for a specific area. The map of farming systems will show the type of farming system that will benefit from proposed techniques in a certain area. Accordingly, consequences on adoptability, maintenance, and fitness can be anticipated.

Landscape planning can help to raise questions on the consequences of the implementation of agroforestry techniques at the level of a water catchment at even higher levels. Such questions can address, for instance, the availability of water resources to serve the introduced tree species, as well as the effect of a large scale introduction of trees on the occurrence of harmful insects or on the general practice among women to sift maize in the wind. A landscape approach can also help to integrate agroforestry systems with rural planning and implementation at levels above farm level. At these higher levels, agroforestry techniques will not have a protective role on private farm land only, but also serve public grounds (e.g. shade, shelter, and soil conservation on paths, school yards, and market places), while the trees are planted mostly on private land and are private property. In this way planning of agroforestry techniques contributes to upgrade the multi-purpose character of agroforestry.

Early design and sketching of possible spatial solutions can be a great help in the search and selection of relevant information and in the collection creative of ideas for scenarios. More basic research is needed, however, to integrate the local plan with district and national development policies and plans. These linkages can serve as examples in research, planning, and training within the local project.

The study of landscape units based on the landscape patterns and farming systems requires more research on whether farming systems can be deduced from the landscape patterns. If landscape-pattern studies can contribute to the assessment of landscape units within particular farming systems, it will be a great aid to farming-systems classification and rapid rural appraisal studies.

The combination of an analysis of the dominant land-use system and the socio-economic analysis of the farming system with a tentative land evaluation has proven to be quite useful. Although it is still difficult to quantify the relevant factors in the system, the method used was sufficient to design scenarios for future development. In order to test the feasibility of the scenarios and the short-term plan and to assess the potentials and restrictions of agroforestry techniques, more investigation is needed on indigenous land-use and management techniques and tree use, on local decision-making and land-adjudication rules, on micro-climate variation within the area, on hydrology with a view to the water-harvesting potential in the area, and on opportunities and constraints regarding arable cropping.

Testing various techniques through integrated field trials has proven to be a helpful method, both for planning and implementation. It has to be made clear, however, which techniques need to be tested and further developed and which techniques can be implemented safely. This distinction is especially important for research purposes and in official demonstration plots. These trials can only be successful if the local people are willing to cooperate and labour input is secured. In the fourth phase, the method has to be expanded with several steps that were left out of this case study for practical reasons (e.g., the implementation plan).

The training and extension component is important too. The local inhabitants have to become familiar with the recommended changes in their landscape and with agroforestry techniques. Moreover, regular checks of the plans with the needs of the inhabitants can provide feedback to the planners and as well as to the inhabitants. An organizational structure related to this training component has to be created.

4 ACKNOWLEDGMENTS

Several people contributed greatly to this chapter. I am very grateful for the help I received from Janet Muller with editing of the manuscript. Supervision was provided by I. Duchhart of the Department of Landscape Planning and Design and L.A.A.J. Eppink of the Department of Land and Water Use, Agricultural University Wageningen, in The Netherlands, F.K. Arap Sand of the KARI and D.E. Rocheleau of ICRAF in Nairobi, Kenya.

I also would like to acknowledge I. Duchhart and F. Steiner for their valuable comments on both the scientific aspects and the formulation of this case study. R. Mwendandu, the principal field research officer of the DAR project and my counterpart during the field study in Kakuyuni, has ever since provided me with the most recent information about the progress of the project and has supported me greatly with his information and his comments on the manuscript. I like to express a special acknowledgment to the International Institute for Aerospace Survey and Earth Sciences for providing the facilities and funds that supported my contribution to this book. Finally, this study would not have been possible without the everlasting support I received from my parents.

5 REFERENCES

Duchhart, I. and Nauta, A., 1986. Landschapsplanning in district (sub) tropen: een verkenning naar District mogelijkheden, wegen, resultaten; verslag van een studiedag. Department of Landscape Architecture, Agricultural University Wageningen, NVTL/BNT, Amsterdam, The Netherlands.

Hoek, A. van den, 1983. Landscape planning and design of watersheds in the Kathama Agroforestry Project, Kenya. MSc thesis for the Department of Landscape Planning, Agricultural University Wageningen, Wageningen, The Netherlands, 60 pp.

Hoekstra, D.A., Torres, F., Raintree, J.B., Darnhofer, T. and Kariuki, E., 1984. Agroforestry systems for the semi-arid areas of Machakos district, Kenya. (Working Paper Number 19) ICRAF, Nairobi, 28 pp.

Jansens, J.W., 1986. A landscape development plan for Kakuyuni, Kenya: landscape planning and design with the application of agroforestry in the rehabilitation of eroded grazing lands. MSc thesis for the Department of Landscape Planning, Agricultural University Wageningen, Wageningen, 165 pp.

Mwendandu, R., 1989. MSc thesis (in preparation) on potential land use on Yatta plateau, Machakos district, Kenya, Toronto University, Toronto.

Rocheleau, D.E. and Hoek, A. van den, 1984. The application of ecosystems and landscape analysis in agroforestry diagnosis and design: a case-study form Kathama sub-location, Machakos District, Kenya, ICRAF, Nairobi, Kenya, 92 pp

Sang, F.K. Arap, 1984. Annual technical and financial report, 1983-1984, Dryland Agroforestry Research Project, Machakos (unpubl.), 50 pp.

Schork, W., 1980. Die Entwicklung marginaler Gebiete in Kenia; Das Fallbiespeil Machakos, Beitrage zum Ecodevelopment, Heidelberg, BRD.

Ssekabembe, C.K., 1984 Soil moisture retention around micro-catchments in the grazing lands of semi-arid areas of Machakos District, Kenya, 13 pp. Appendix in: Sang, F.K. Arap (Editor), 1984. Annual technical and financial report 1983-1984, Dryland Agroforestry Research Project, Machakos (unpubl.).

Tjia, J.G.J. and Tjiook., Y.H., 1985. A landscape plan for Cilanang watershed, a subcatchment of Citarum river, near the Saguling hydropower project, West Java, Indonesia. MSc thesis for the Department of Landscape Architecture, Agricultural University Wageningen, The Netherlands.

Chapter 13

SOCIAL FORESTRY IN KARNATAKA STATE: INDIA

J.H. BALDWIN
Department of Planning, Public Policy and Management, University of Oregon, Eugene, Oregon, (United States of America)

D. BANDHU
School of Planning and Architecture, Center for Environmental Studies, Indraprastha Estate, New Delhi 110002 (India)

ABSTRACT

Deforestation is a serious problem in India diminishing the health, safety, and welfare of millions of people. To provide fuel, fodder, and wood products to the rural poor, the Indian government, along with the World Bank and Overseas Development Administration, initiated the Karnataka Social Forestry Programme. However, the vast majority of programme benefits have been realized by large landholders converting agricultural lands to eucalyptus plantations. These plantations provide considerable timber and pulp wood to urban and industrial markets, but provide few benefits to redirect the programme to better achieve its worthy goal of assisting the rural poor.

INTRODUCTION: THE DECLINING FORESTS OF INDIA

With recent growth in population and industrialization in India, the demand for forest products for domestic and industrial uses has increased dramatically. According to Bandhu and Garg:

> The extent of forest lands diverted to non-forestry purposes during the last 25 years is reported to be over two million hectares. This does not take into account unauthorized and illegal encroachments in forest areas which are a serious problem to the forest administration in many areas of the country. In addition, shifting cultivation has laid bare vast stretches of forest lands, especially in north-east India. All over India, forests in the vicinity of villages are being subjected to excessive illicit felling by the local population, mainly to meet their fuel wood requirements (1986, p. 155).

In the early 1980s in India, state and national forestry departments were claiming an annual deforestation rate of 0.15 million hectares. Many scientists, however, criticized these figures as significantly low.

TABLE 1

Forest Cover in the Major States and Union Territories of India (Paryavaran, 1984) *448,305 = 4.48 "lac" square kilometers

State/Union Territory	Total Forest Area (in Sq.Km.) 1972-75	1980-82	Change in Forest Area in Seven Years (Sq.Km.)	
1. Andhra Pradesh	49,049	40,435	(-)	8,614
2. Assam	21,055	19,796	(-)	1,259
3. Bihar	22,687	20,139	(-)	2,548
4. Gujarat	9,459	5,057	(-)	4,402
5. Haryana	757	401	(-)	356
6. Himachal Pradesh	5,075	9,130	(-)	5,945
7. Jammu & Kashmir	22,335	14,361	(-)	7,974
8. Karnataka	29,480	25,655	(-)	3,825
9. Kerala	8,611	7,376	(-)	1,235
10. Madhya Pradesh	108,568	90,215	(-)	18,353
11. Maharashtra	40,682	30,350	(-)	10,332
12. Manipur	15,090	13,572	(-)	1,518
13. Meghalaya	14,390	12,458	(-)	1,932
14. Nagaland	8,154	8,095	(-)	59
15. Tripura	6,330	5,972	(-)	1,192
16. Orissa	48,383	39,425	(-)	8,958
17. Punjab & Chandigarh	1,120	499	(-)	621
18. Rajasthan	11,294	5,972	(-)	5,322
19. Tamil Nadu	16,676	13,187	(-)	3,489
20. Uttar Bengal	25,869	21,022	(-)	4,847
21. Paschim Bengal	8,347	6,483	(-)	1,864
22. Sikkim	1,761	2,883	(+)	1,122
23. Arunachal Pradesh	51,438	58,104	(+)	666
24. Delhi	18	10	(-)	8
25. Goa, Daman & Diu	1,221	1,139	(-)	82
26. Mizoram	13,860	11,971	(-)	1,889
Total	551,709	448,305*	(-)	103,404 or 18.8%

Confirmation of these concerns came in 1984 with the publication of data from a national land-use remote sensing survey conducted by the Indian National Remote Sensing Agency (NRSA). The survey (see Table 1 and Fig. 1) indicated that in the seven-year interval between 1972-1975 and 1980-1982, India lost 1.3 million hectares of forest -- over eight times the government forestry agency estimates (Centre for Science and Environment, 1985; Paryavaran, 1984).

According to Bandhu and Garg:

> While the forests are shrinking and getting decimated at an alarming rate, the demand for timber, fuel wood and various other forest products are increasing and the gap between demand and supply is

widening. Timber...is getting scarcer and the prices have been rising prohibitively (1986, p. 155).

Remote sensing data from the NRSA indicate that nearly 19% of the forests existing in the period 1972-1975 were lost in the next seven years! Thus, today in India, a mere 13.7% of the land is forested. The national forest policy of India suggests a minimum goal of 33% of the land under forest (Bandhu and Garg, 1986).

Further evidence of the significance of deforestation in India is provided by calculating a forest area/capita ratio. India has 16% of the world's population and 2% of the world's land area. Table 2 compares the per capita forest cover in India (0.066 hectares per capita) with figures from several countries and the world average (Shetty, 1982).

TABLE 2
Per Capita Forest Area (in Hectares) (Shetty, 1982).

Country	Per Capita Forest Cover
1. World (average)	1.04
2. USSR	3.60
3. USA	1.40
4. Canada	20.00
5. Japan	0.24
6. India	0.066

The NRSA study broke down forest cover into three categories: "closed forests," "open or degraded forests," and "mangrove forests." Table 3 summarizes forest loss by category. In total, over the seven-year period, India lost 10.4 million hectares of closed forest, defined as "canopy covers <u>30%</u> of ground cover" (as compared to a conventional 40% definition in many international studies), 63,000 hectares of mangrove forest, and 1.29 million acres of closed forest were converted to degraded forest (Centre for Science and Environment, 1985).

Except for Tripura, Manipur, Meghalaya, and Mizoram, all the states and union territories in the NRSA study had less forest area than reported by the Indian Ministry of Agriculture. Thus, caution is warranted in using official national or state forestry department figures to characterize or manage deforestation in India.

With the new NRSA figures on deforestation in India at his disposal, Rashimi Mayur, Vice President of the Global Futures Network, estimates optimistically that India will be fortunate to have a 2% forest cover by the

Fig. 1. Major administrative divisions of India.

TABLE 3
Vital Forest Statistics (Paryavaran, 1984).

	1972-75	1980-82
Forest cover (million hectares)	55.52	46.35
Closed forests (million hectares)	46.42	36.35
Open forests (million hectares)	8.77	10.06
Mangrove forests (million hectares)	0.33	0.26
Forest cover (% of total land area)	16.89	14.10

turn of the century. A.K. Ganguly, of the Bhabha Atomic Research Centre, in writings about the emergence of the "Great South Indian" and "Great Himalayan Mountain" deserts, believes India will be fortunate to have any forest cover by the mid 1990s. It is in this contest that a dynamic new dimension has to be given to afforestation (Roy, 1986).

In the five years of the sixth development plan (1980-85), India spent 40% more on forestry that it spent in the previous 30 years. However, the increased annual plantation rate of 0.4 million hectares targeted for the sixth plan still meant a million hectares of forests lost each year (Centre for Science and Environment, 1985). By 1990, India hopes to be planting nearly 2 million hectares per year at an annual cost of RS 600 crore (1 crore = 10 million RS, "RS" stands for "Rupees;" 14 RS = $1).

In summary, of all of the environmental problems facing India, deforestation has received both the most public attention and criticism. The controversies surround government programmes that promote commercial and industrial planting, often at the expense of critically needed food, fodder, and fuel for domestic needs. This chapter takes a critical look at the social forestry programme of India to detail and discuss programme successes and failures. The majority of information presented was obtained from independent scientific and policy papers.

2 THE SOCIAL FORESTRY PROGRAMME

Social forestry programmes were initiated in the mid 1970s by several states to promote afforestation of private farm or village commons non-forest land. The term "social forestry" was first used by the National Commission on Agriculture in 1976 to promote tree planting programmes to supply firewood, fodder, small timber, and minor forest produce to rural populations. After over a decade of development, social forestry has become one of the most controversial initiatives of the Indian government.

According to T.H. Sankar, former member-secretary of the Indian planning commission's working group on energy policy:

> I have seen many programmes launched in the name of the poor but soon distorted to benefit the upper class. But no programme has been diverted further away from its objective than social forestry... The forest departments never seem to ask the question: wood production for whom or for what? (Centre for Science and the Environment, 1985, p. 51).

The latter question is the important one, for if one looks at social forestry from the standpoint of tree production (regardless of benefactors), one could conclude that the programme is one of considerable achievement. The Union (Indian) government proclaims social forestry as a new "green revolution" in the offing. From 1950 to 1980, the government spent RS 483.22 crore for reforestation, while the sixth development plan (1980-85) spent RS 692.49 crore for forestry. The Indian government and World Bank are discussing a possible $600 million proposal for a nationwide social forestry programme to cover 13 states (Centre for Science and Environment, 1985).

Ambitious social forestry programmes have already been initiated by several state governments, most of which were financed by foreign aid agencies. The Indian agricultural ministry in its 1983-84 annual report claims the social forestry programmes have been a "spectacular success" and that all targets were exceeded. The report continues that, "The year 1983-84 witnessed an unparalleled enthusiasm among the people for tree planting." The report points out that since 1980, over 1000 crore number of seedlings have been planted on 1.2 million hectares with the number of seedlings and area planted growing by 50 to 70% per year (Ministry of Agriculture, 1983-84).

Despite these achievements, mounting criticism from a number of interests --ranging from peasant farmers to atomic scientists-- has peaked recently in domestic violence and acts of civil disobedience. The critics of the social forestry programme charge that programmes to reforest private, village and non-forest land, in conjunction with large tracts of degraded government lands, overemphasize the needs of commercial, industrial, and urban markets at the expense of food, fuel, and fodder for the rural poor. This concern is reflected in the preferred species planted--a hybrid *Eucalyptus tereticornis*. Over one-half million hectares of *eucalyptus* has been planted to date, but the species is not popular with environmentalists, independent forestry scientists, or the rural poor (the theoretical recipient of programme benefits.) In the state of Karnataka, protesters have uprooted eucalyptus plantations to keep eucalyptus roots away from their crops. This followed the publication of a report that croplands were being replanted with eucalyptus plantations that were destroying soils and depleting water tables (Centre for Science and Environment, 1985).

A closer look at programme implementation provides some useful insights. Table 4 provides data from the Gurjarat Forestry Department on beneficiary farmers and land planted under the social forestry programme for the years 1982-1983. The tables indicate the trend of benefits to larger farmers planting trees on prime agricultural lands because of the high profits of the resulting wood products. In at least one state, this has resulted in a call to ban tree farming on prime irrigated agricultural lands.

This trend can also be seen in data from the World Bank on programme progress in Uttar Pradesh. In Uttar Pradesh, the World Bank financed social forestry programme overshot its large farm forestry target by an astounding 3430 percent, while the rural community self-help programme achieved only 11 percent of its target (Centre for Science and Environment, 1985).

The primary reason for the distortions of social forestry programmes is the combination of very low incomes of Indian citizens (averaging less the $200/capita/year) and the acute shortage (and therefore high price) of building poles and pulpwood. As practiced in India, large agroforestry concerns have used the social forestry programme as a financial windfall. A conversion from short-term cash cropping to long-term tree planting (especially with relatively unpalatable and valueless [to rural residents] hybrids such as Eucalyptus) significantly reduces labour, materials, and monitoring costs. For example, cotton formers in the Punjab plant Eucalyptus, with free seedlings, low maintenance, and high profits. Villagers, who used to earn wages caring for the cotton, lose income and the cotton sticks that were once used regularly for fuel (Centre for Science and Environment, 1985).

Advocates of large agroforestry programmes in India assert that the benefits in poles and pulpwood offset the losses in food, fuel, and fodder. By increasing the amount of wood available for urban and industrial uses, demand will ultimately be met, prices will fall, farmers will diversify their plantations, pressures on natural forests will be reduced, and forest products will be available to the poor. Three fatal assumptions compromise this "trickle down scenario."

TABLE 4

Beneficiary Farmers and Types of Land Used for Tree Farms in Gujarat (1982-83).

Beneficiary Farmers Category	#Farms	Area Planted (Ha)
Large Farms (>4 Ha)	45,000	11,339
Intermediate Farms (2-4 Ha)	82,500	17,688
Small Farms (<2 Ha)	22,500	2,469
Totals	150,000	31,496

First, that ultimately demand for pole and pulpwood can be met. The Indian Development Council on Paper, Pulp, and Allied Industries, for example, estimates that the forest raw material requirements of the paper industry will increase from 2.85 million tonnes to 4.4 million tonnes between 1981 and 1991, going up to 6.70 million metric tonnes by the year 2000. Similar demand increases are predicted in the rayon and pole markets as population and industrialization continue to increase. Can this sort of growing demand ever be met? One must remember that tree plantations require a minimum of five years to mature for the market.

Second, there is an assumption that the rural poor (80% of the Indian population with annual incomes under $200/year) can pay anything for forest product food, fodder, and fuel. A recent study (Environmental Defense Fund, 1987), found that 100% of poor families in Karnataka depend substantially on commons lands for biomass and 84% for food. The calculated monetary value of the biomass alone to a rural family was between RS 530 and 830 depending on the district. The study concluded that the biomass value alone exceeded the average income generated by the rural poor.

The final assumption is that when surpluses occur, agroforestry officials will tolerate diminished prices and promote diversification. For example, expanded food production in the 1980s in India did not result in reduced prices and diminished hunger. Wealthy farmers simply began exporting food on the lucrative international market. Consequently despite having one of the highest rates of hunger and malnutrition in the world, India, today, is a net exporter of food. There is no reason to believe a similar phenomenon will occur, should India someday produce a wood product surplus. The same actors and market conditions are involved.

This, in essence, is stated in a model study of the market for wood products in Guyarat provided to the World Bank by the International Institute for Environment and Development (Dewees, n.d.). The study, which was used as a model for planning social forestry projects of the World Bank concluded:

> The existing markets for wood, particularly for construction poles, are probably much larger than had been originally anticipated in the Bank's mid-term review Mission of November-December 1982.
> In addition, enterprising and innovative individuals are actively considering potential markets that could be tapped, should any surpluses become apparent. The demand for wood products in India is exceptionally high, and the lack of a sufficient raw material base has greatly hindered market development. If the raw material become available through social forestry activities, it can be expected that new forest product industries will develop, particularly pulp and paper, wood-based panels, and saw milling industries. Because a relatively large amount of wood will become available over a short period of time, there is some concern that the current marketing structure may fail to protect vulnerable smaller farmers with little experience.

The study continues, advising the World Bank to alter management decisions to:

> increase the size of the farm forestry programme, to reduce expectations under the village self-help component and to increase the forest department supervised village woodlot component.

In other words, if self-help woodlot programmes have not met expectations, "forget the poor and switch to large scale farm forestry" (Centre for Service and Environment, 1985). Needless to say, in these authors' opinions, this is bad advice. We question the propriety of international financing of forestry programmes that benefit only the wealthy. If large farm social forestry is so profitable, why then should there be any subsidies?

Finally, a major failing of social forestry is the lack of involvement of the poor--especially rural women. Rural women in India spend most of their time collecting food and fodder (and clean water) and should be the main beneficiaries of a true social forestry programme. A programme reorientation to focus on the needs of rural women would help in providing for the basic needs of the poor.

3 THE KARNATAKA SOCIAL FORESTRY PROGRAMME (KSF)

The state of Karnataka is located on the southwest coast of India, adjacent to Goa (see Fig. 1). As indicated in Table 1, the state of Karnataka is experiencing rapid deforestation, losing 3825 hectares of forested land (or 13%) in the interval between the sampling periods of 1972-1975 and 1980-1982. The result has been severe shortages of forest products and community conflicts in local villages.

In response to this rapid deforestation, the five-year Karnataka social forestry (KSF) programme was initiated in 1983. The total project is estimated to cost RS 550 million (RS 55 crore) or approximately $40 million, $27 million of which would be financed by the World Bank and the Overseas Development Administration (ODA). The KSF programme targeted 1.45 lac (1,490,000) hectares of land for reforestation.

The overall objective of the social forestry programme in initial planning documents emphasized meeting the basic needs of the rural poor and soliciting their active involvement in programme development. For example, the official programme summary from the Karnataka Government Secretariat states:

> [Social forestry is] for the benefit of the community through active involvement and the participation of the community. In the process, the rural environment will improve, rural migration will reduce, rural unemployment will substantially cease. Thus by the end of 1985-86, some half-a-million hectares of Government land--forests, pasture-lands and other open cultivable areas have been brought under _eucalyptus_ mono-culture, in the name of afforestation, essentially (1980, p. 17).

These intensive plantings (Table 5) were more frequently than not being accomplished in prime agricultural lands. A study by the Directorate of Economics and Statistics of the State of Karnataka concluded:

> From 1975-76 up to the end of 1982-83, 611.31 lakhs (1 lakh = 1,000,000) of _eucalyptus_ seedlings in Kolar District and 247.63 lakhs of _eucalyptus_ seedlings in Bangalore District have been distributed by the Forest Department. The study has undoubtedly revealed that the cultivators who have taken to farm forestry have converted their agricultural lands to farm forestry. Out of the total extent of farm forestry raised by the agricultural households, 63.3 percent of farm forestry cultivated area has been raised on agricultural land in Kolar District, and 49.6 percent in Bangalore District (1984, p. 16).

TABLE 5

Plantations Raised in the State of Karnataka Up to 1985-86 (Ha) (Forest Department, 1985-86 Table 13).

	Up to 1984-85	During 1985-86	Total
1. Teak	120,769	1,762	122,531
2. Soft	51,201	1,381	58,582
3. Cashew	38,169	651	38,820
4. _Eucalyptus_	107,612	5,136	112,748
5. Fuel and Mixed	340,869	29,461	370,330
6. Bamboo	3,467	515	3,982
7. Lavender	1,097	----	1,097
8. Red Oil Palm	9	----	9
9. Cocoa	1,595	----	1,595
10. Rubber	5,527	----	5,527
11. Spices	74	----	74
12. Coffee	20	----	20
13. Tea	7	----	7
14. Fruit Orchards	568	101	699
15. Sandal (pure)	2,262	316	2,578
16. Avenue (kms)	2,945	1,226	4,171
17. Canal-side (kms)	846	284	1,130

A subsequent study by the Directorate of Economics and Statistics (1986) found that:

> Between April 1984 and July 1985, around 96.2 percent of _eucalyptus_ produced in Bangalore district and 97.5 percent of _eucalyptus_ produced in Kolar District has been marketed. The entire quantity of _eucalyptus_ sold in Bangalore District and 97 percent of the quantity sold in Kolar District was consumed as industrial raw material by M/s. Harihar Polyfibers; and hardly 3 percent sold in Kolar District has been used as fuel.

According to Chandrashekhar _et al_.:

> These findings of the Karnataka Government's official agency itself should leave no doubt that what has been done, and is being done even now with the World Bank and ODA support, by the Government in the name of "Social Forestry" and "Farm Forestry" is only diverting farmers' lands on an increasing scale from food crop cultivation to raising _eucalyptus_ trees for serving industrial interests (1987, p. 6).

On November 28, 1985, the Federation of Voluntary Organizations for Rural Development in Karnataka (FEVORD-K) sent a letter to the president of the World Bank, head of the ODA, and chief minister of Karnataka asking these officials to explain, justify, and alter their forest management practices in

Karnataka. In his reply to FEVORD-K of January 6, 1986, C.P.R. Nottidge, chief, General Agriculture II Division, South Asia Projects Department, World Bank, Washington, D.C. stated:

> As you noted in your memorandum, the primary objectives of the Social Forestry Project in Karnataka was to increase the supply of fuel wood to rural and semi-urban areas through the establishment of some 150,000 hectares of plantations located throughout the State. It is important to note that 80% or 120,000 hectares of this target comprises tree planting on private land. If the Farm Forestry target is achieved, which it has been so far, it means that the majority of the trees planted under the project will be planted by the farmers themselves. This leaves the choice of area, number of trees and species planted to the individual concerned. Present plantings reflect a response to market conditions which, no doubt, will change over time, resulting in, amongst other things, changes in species planted.
>
> The remaining 20% (30,000 ha) is to be planted on public lands of which some 23,000 hectares have moderate bearing on the daily lives of people. Here, however, it is certainly within the scope of the project to ensure that trees planted on public land meet the needs of local people for fodder, fuel, small timber, and raw materials for cottage industries. It is here that voluntary agencies like yours can play an important role (emphasis added).

One can openly wonder how the term "social forestry" can be applied to a programme with a goal to "increase the supply of fuel wood to rural and semi-urban areas," using densely planted (5000 seedlings/hectare) mono-cultured *eucalyptus* (see Table 5) to convert food crop lands to large-scale tree farming?

3.1 Negative synergy: land reform and Eucalyptus plantations

As previously discussed, social forestry in Karnataka has failed to help, and may even be hurting the poor. The plight of the poor in Karnataka has further been exacerbated by another social programme, designed to assist the poor--land reform.

The Karnataka land reform acts of 1974 and 1977 set ceilings of land ownership to 54 acres for individual family units and abolished tenancy, with the ownership of a tenant-cultivated land taken away from the original owner and vested with the tenant. The ceiling never caused much of a problem because large landowners were allowed to "club" large aggregate land holdings with several adult family members (Chandrashekhar *et al*., 1987).

However, the abolition of tenancy and the transfer of ownership to *bona fide* tenants was a real threat to large landowners. Large landowners were faced with the problem of getting rid of tenants while keeping their lands in cultivation (to prevent government confiscation of lands uncultivated for three years--also a part of the Karnataka Land Reform Acts). Tenants were easy to evict because there were few written tenancy agreements, but now landowners would have to hire agricultural labourers to maintain existing crops (Chandrashekhar *et al*., 1987).

At this precise moment, the Karnataka Forestry Department stepped forward with a "spectacular offer of *eucalyptus* plantations." By planting *eucalyptus*, large (absentee) landowners could evict their tenants, would not need farm labour to annually tend food crops, would keep their land in cultivation, would reap windfall incomes without much maintenance effort and would receive free seedlings to boot. Little maintenance is required of *eucalyptus* plantations because the trees are relatively useless for rural residents and relatively unpalatable to roaming livestock (Chandrashekhar *et al*., 1987).

In a sample of 613 households in Karnataka, Reddy (1985) found that landowners of five acres and above accounted for 87% of the *eucalyptus* areas planted. Landowners with holdings of less than one acre planted no *eucalyptus* -- their tiny area of land being much too valuable for growing staple grains such as *ragi* and *jowar* for family needs.

The former chief justice of the state of Karnataka concludes:

> These lands that have been covered by single-stand dense *eucalyptus* plantations are not the marginal or degraded lands, as the Forest Department wants us to believe. These are the best lands under the given ecological milieu of the region, belonging to big and medium landowners, on which rain fed food crops were being cultivated. The fact must be explicitly recognized that lands that had been traditionally under food crop cultivation are no longer available for growing food, as a direct consequence of "farm forestry" (Chandrashekhar *et al*., 1987, p. 17).

3.2 Allocation of uncultivated lands in Karnataka

The Karnataka government's allocation and distribution of uncultivated lands has also met with a storm of protests, street, demonstrations, and petitions to the government (one signed by 6000 residents). These "uncultivated" lands (often called *panchayat* lands) are often permanent pastures and mixed forest lands that are vital to the region's natural heritage and are very valuable sources of fuel, food, and fodder for local villagers.

With the principal goal of social forestry being to provide food, fodder, and fuel for local villagers, one would assume that the Forestry Department of Karnataka would protect and promote sustainable utilization of these lands. To the contrary, as Chandrashekhar et al. state:

> It is this total neglect by the Government that has been responsible for the progressive deterioration of these grazing lands that are so precious for the indigenous systems of livestock and agriculture. And now these are labelled as degraded revenue lands "unfit for agriculture," classified as "C" and "D" lands, and being handed over to the Forest Department ostensibly to increase land under vegetation but in reality to raise eucalyptus plantations single stand, for meeting the needs of pulp-based rayon and paper industries (1987, p. 19).

The primary beneficiaries of these land transfers and development are a number of pulp and rayon mills in the district.

At last count the government had plans to provide uncultivated lands to 13 other forest-based industries in the state. According to Chandrashekhar et al., the net result of these industrial give-a-ways of uncultivated lands is:

> gradual extinction of grazing resources and fodder base (so essential for the indigenous stock of draft animals, milch cows, buffaloes, sheep and goats) and permanent destruction of these lands through raising eucalyptus plantations (1987, p. 20).

These numbers, however, pale by comparison to some new plans of the Karnataka forestry department to develop eucalyptus plantations. Under the auspices of the new government of India "Wastelands Development Programme," a five-year, 25 million-hectare programme, the state of Karnataka has targeted 2.55 million hectares of "wasteland" in the state for "development." Table 6 provides a summary of lands targeted for "plantation opportunities in Karnataka," the title of the article from which the table was obtained.

Finally, in January of 1986, a joint IDA/ODA mission visited Karnataka and acknowledged the legitimacy of village concerns about adverse environmental and community impacts. In their "Mid-Term Review of the Karnataka Social Forestry Project" (1986), the Overseas Development Administration states:

> Too much emphasis was placed on physical planting targets. These should have been balanced by recognizing two equally important objectives: a) that the planting programme had to be directed at meeting the forestry

needs of the rural and semi-urban poor; b) that the capability of the Karnataka State Forest Department, in collaboration with other agencies, had to be built up to enable it to meet the above needs.

TABLE 6

Wasteland Development Projections for the State of Karnataka (Kushalappa, 1986).

Degraded Forest Lands	Area Available (in acres)
Between 10 @ 20% canopy density	270,600
Non-tree clad area with less than 10% canopy density	981,890
Revenue lands transferred to Forest Department	
"C" class	182,577
"D" class	61,740
Fallow lands:	
Permanent fallows	510,501
Current fallows	252,270
Avenue roadside	77,919
Railway side	
Other lands-industries, schools, institutions, etc.	38,000
Grazing lands	72,622
Tank foreshore lands	100,000
Canal banks	2,500
TOTAL	2,552,859

3.3 Campaigns for the protection of village commons

The first village protest of the mono-cultural plantings of *eucalyptus* occurred in March of 1983. Farmers in the Tumkur district of Karnataka attributed the drying up of life-sustaining streams to the planting of *eucalyptus* plantations in water catchment areas. The farmers of the district, led by the Karnataka Rajya Sangha (state farmer's association) uprooted *lakhs* (hundreds of thousands) of *eucalyptus* seedlings and planted seeds of traditional trees (Ramaswany, 1986).

Similar protests, demonstrations, petition drives, and acts of "ecotage" have occurred in other districts as new forestry department initiatives were announced. For example, in the Chikmagalur District, eucalyptus planting in 1985 was totally stopped by village farmers. According to Chandrashekhar *et al.*:

> Ever since the formation of Karnataka Pulpwood Limited was finalized there have been continuous protests from all over Karnataka, and especially in Malnad which is directly affected. In answer to the

> Government's claim that only barren lands would be made over to the company, Arogya Vikasa Prakalpa, a voluntary agency of the region, sponsored a physical survey of "C" and "D" class lands in Chikmagalur and Shimoga Districts. Contrary to Government's claims, lands classified as "C" and "D" in this region have a tree density of 50-200 per acre on the average. There are also cultivated lands, lands allotted to Linganamakki dam evacuees, and even wells, roads, and houses. There are villages (e.g. Balapura in Koppa taluk, Chikmagalur District) in which lands have been granted to tribal families (1987, pp. 24-25).

Community organizations in Karnataka filed a law suit to halt the Karnataka Pulpwood Project. On March 25, 1987, the Indian Supreme Court issued a stay order preventing the Karnataka government from the transferring of land. The stay prevented further government action for three weeks until a social and environmental assessment could be completed for the Shimoga, Shikmaglar, Dharwad, and Balgaum districts. To date, the assessment has not been made (Environmental Defense Fund, 1987).

In many cases, government attempts to take over "uncultivated" or "waste" lands have been met with stiff resistance -- ranging from Ghandian type programmes of passive resistance to outright violence. For example, in Hariharapura, in mid-1985, the government tried to use police force to evict hundreds of families who had cultivated village lands for a decade or more. Similar confrontations took place in the Shimoga and Chikmagalur districts (Ramaswany, 1986). "Ecotage" activities ranged from the destruction of eucalyptus seedlings, to plowing fields, to uprooting forestry department survey stones.

Members of one group called the Ryota Sangha (Farmers' Association) have uprooted millions of seedlings in the Tumkur district of Karnataka. The powerful farmers group has warned the state government that it must stop planting mono-cultural eucalyptus plantations. One of the leaders of the organization is quoted in the press (Centre for Science and Environment, 1986):

> If the government does not accede to our demand, we need only two days to ensure that eucalyptus is wiped out in Karnataka.

On February 25, 1985, a group called the Mannu Rakashana Koota (Save Soil) was formed to develop a sustained and ongoing campaign against the Karnataka social forestry programmes. Mass campaigns were developed, involving meetings, street plays, symbolic burning, media, and legislative events, and even a three-day (February 16-18, 1986) "alternative development strategy" workshop with participation from several adjoining states.

In addition to the Mannu Rarashana Koota, a large number of village, taluk, district, and state level organizations were formed to develop environmental education programmes. These included:

> the Karnataka unit of Citizens for Democracy, the Dharwad-based Samaja Parivartana Samudaya (Transnational Centre for Nonviolent Social Change-TNC), Rajya Raita Sangha, the State unit of Bharsteejras Kisan Sangh, etc. Several village-level organizations have initiated voluntary afforestation and nursery-raising programmes in recent years -- e.g. Aregya Vikasa Prakalpa in Shimoga District, Gramevikas in Mulbagal, Kolar District, India Development Service in Medleri, Dharwad District, Institute for Youth and Development in Bangalore District. The Seva Sagar Trust and other organizations in Sagar Taluk in Shimoga District have undergone a massive tree-planting programme in that taluk called "Vriksha-Laksha Andolana" (Chandrashekhar et al., 1987, p. 26).

On November 25, 1986, the Karnataka High Court directed the government of Karnataka "not to prevent the villagers of Sagar Taluk from the use and enjoyment of 'C' and 'D' class lands owned by the State earlier reserved for the use of the public." (Order pronounced by Justice P.O. Bopanna.)

Finally, the state level FEVORD-K has been carrying on the dialogue with various levels of state and national governments as well as with the World Bank and the Overseas Development Administration.

4 THE GREAT EUCALYPTUS DEBATE

According to the Centre for Science and Environment (1976), the hottest environmental issue in India is the controversy over the massive mono-cultural plantings of *Eucalyptus tereticornis*, a *eucalyptus* hybrid tree. Private tree farmers and government foresters (called "pro-eucalyptans") favour the use of the tree because it grows well in dry conditions, does not get browsed (is relatively unpalatable to livestock), is fast growing and straight, coppices well, and fetches a high price (RS 17-100/tree) on the pulp and pole wood markets (for predominantly urban uses). Thus, over 80% of the plantings of India's social forestry programme are of *eucalyptus*. To date, nearly half a million hectares of Indian lands have been planted with the exotic *eucalyptus* hybrid. According to the Centre for Science and the Environment:

> Many farmers have become hardened *eucalyptus* cultivators. D.N. Mishra, a senior forester from Uttar Pradesh, claims that "in this state we

even insist that 10 percent of plantations should comprise of species other than eucalyptus. But farmers in western Uttar Pradesh want only eucalyptus." Shyamsundar, "Last year some 5 million seedlings (of other species) were wasted in our nurseries due to lack of demand (1986, p. 64).

The "Great Eucalyptus Debate" has strong ecological, political, and social components. On the ecological side, "anti-eucalyptans", such as J. Bandyopadhy (1985), contend that the eucalyptus hybrid consumes excessive soil, water, and nutrients, increases the acidity of soils, reduces the development of soil humus, and discourages understory growth useful for food, fuel, and fodder. Further, if one were to try to revert a eucalyptus plantation to a plowed agricultural use, the extensive surface root systems remaining after harvesting the eucalyptus would make clearing and plowing economically unfeasible. This ecological debate must be taken in the context of alternative land uses. Compared to natural forests, eucalyptus monocultures are more damaging to soil moisture, fertility systems, and understory growth, but compared to intensive agricultural practices, the mono-culture is, perhaps, less damaging. One must remember that eucalyptus plantations, according to "pro-eucalyptan" V.J. Patel, "will last 30

years... and get some six or seven coppices." What then?

In its midterm review report (ODA, 1986), the ODA and the World Bank acknowledge many of these concerns in Paragraph 3.3:

> [Eucalyptus trees] do not lend themselves well to intercropping, are competitive, consume large volumes of water, have no value as fodder, and, because of its value as an industrial raw material, part of the produce may not reach the firewood market.

The general government consensus today is that eucalyptus should continue to be planted in suitable, marginal, and degraded lands to meet India's growing wood needs. T.N. Khoshoo, Secretary to the Department of Environment (1984) states:

> Eucalyptus can be a blessing and a captive plantation... eucalyptus plantations can be grown successfully on marginal or degraded land without any major disadvantage.

However, there is growing government concern about the diversion of prime irrigated agricultural lands to eucalyptus plantations. Some experts have even called for national legislation to prevent this diversion. S.K. Roy (1986), a former member of the Indian National Planning Committee, states:

> The time has come for the enactment of an act to prevent the transfer of prime agricultural lands for other purposes...the encroachment of

> eucalyptus on agricultural land is a significant, not to be ignored danger. We are going to need every bit of cultivable land, especially prime irrigated land, in order to produce the additional 100 million tons of food grains to meet the population of 100 crores in 17 years. Linked with this is another danger arising out of encouraging industries to raise their own raw materials. In many areas, the poor marginal farmers are being persuaded to sell their lands for plantations and are thereby themselves rendered an addition to the numbers of the indigent.

Scientists and some forestry officials also openly worry about the long-term effects of fast rotating tree plantations on soils--a subject for which there is little scientific data available. M.P. Gupta of the Himachal Pradesh Forest Development Corporation states:

> Knowledge of eucalyptus is ad hoc. As a forester, I am skeptical about exotics. It is indeed a fact that less is known in India about indigenous species than about eucalyptus. Probably no programme in the world has been taken up for ecological restoration and for meeting peoples' needs in greater ignorance of both ecology and local needs. (Centre for Science and Environment, 1986, p. 69)

As mentioned previously, the "Great Eucalyptus Debate" also has important political and social dimensions. The political dimension is best summarized by the Centre for Science and Environment:

> The decision to plant eucalyptus in the existing socio-economic situation marked by extremely high prices for wood in urban areas and in industries and an acute shortage of fuel and fodder faced by poor rural populations who cannot afford to purchase anything, displays a clear choice in favour of earning a quick buck by meeting the needs of the rich in society and not the rural poor. It is not surprising that whenever rural women have been organized to plant trees, in almost all those few cases their choice has been in favour of mixed plantations of fruit, fuel, and fodder species. Tree farmers and foresters repeatedly point out that they favour eucalyptus over other species because it cannot be browsed. But it is ironic that a species which cannot be browsed is being planted on such a large scale when the fodder shortage may be even more acute than the fuel shortage, and production of fodder was, in fact, one of the primary reasons for the National Commission on Agriculture to recommend social forestry programmes in 1976. Yet social forestry programmes riding on the eucalyptus wave have today totally turned their back on fodder production (1986, p. 70).

The sociological aspect of the "Great Eucalyptus Debate" involves the impacts of eucalyptus on food and wood product availability to the rural poor and program impacts on tribals. The conversion of croplands to tree farms has already been discussed, but the conversion of uncultivated lands to plantations also adversely affects the poor. The Centre for Science and the Environment concludes:

> There is strong reason to believe that eucalyptus plantations could worsen the energy situtation for the really energy-starved rural areas. Energy scarcity afflicts mostly those who do not have adequate land, to obtain crop wastes or trees, or any cattle, to obtain their own cow dung. These people are also too poor to purchase kerosene to cook. In other words, it is the landless peasants and small and marginal farmers who really suffer from energy scarcity. Though the quality of available data is poor, the latest report by the National Council of Applied Economic Research shows that the proportion of those families which have to collect their energy from others' lands is growing in northern India for which past data are available.
>
> Even if degraded public lands with nothing but seasonal weeds growing on them are converted to eucalyptus plantations, the energy crisis of the poor will be immediately intensified. Similarly, if a big farmer turns over extensive crop lands to eucalyptus, the landless farmer's surplus crop wastes for fuel would be deprived of it (1986, p. 70).

The impacts of eucalyptus on tribal population is particularly bad. In the state of Kerala, natural forests were cleared to make way for eucalyptus plantations. Roy concludes:

> The position of the tribal communities is increasingly unfortunate. Not only are the forests on which they depend shrinking rapidly according to Forest Department Officials, more the 1 million hectares of forest land is lost annually, but their struggling efforts to undertake social forestry receive little or no cooperation from state forest officials in remote areas (1986, p. 5).

Many believe the greatest casualty of the "Great Eucalyptus Debate" is the forestry departments themselves who are now widely viewed as anti-people and anti-environment and as essentially pole and pulpwood producing departments which have adversely affected the poor. The continued insistence of the foresters that they alone have the knowledge and right to manage forests has further alienated many citizens.

5 CONCLUSIONS AND RECOMMENDATIONS

Karnataka currently has a firewood demand of 4 million tonnes of which only 0.6 million is being provided by the forest department. The social forestry programme of Karnataka is simply not meeting its intended goal of providing fodder, fuel, and food for the rural poor. Rather, the programme has become a mono-cultural eucalyptus plantation programme for poles and pulpwood. Roy (1986) states:

> From the data that are available, it is clear that the most significant beneficiaries are the paper factories and the Forest Department. The former get the raw material they need, having significantly depleted all available sources in the vicinity, including some of the richest bamboo forests in the world, and the latter obtain the revenue demanded by their political masters with substantial budget increases for implementation of the schemes. Subsidiary beneficiaries are the rural elites and the owners of medium-sized plots. There is almost nothing in the present policy for the most needy, including the tribal and hill dwelling peoples.

Either the stated goals should be changed (with the distinct possibility of loss of funding from international development agencies) or the programme needs to be redirected to achieve its worthy goals and objectives. We believe the latter policy is in order.

India prides itself at being the largest democracy in the world. It is quite literally a melting pot of politics, religion, language, history, and culture. Thus, it is not surprising that a massive bureaucratically controlled mono-cultural eucalyptus tree plantation programme of this nature is very controversial. Some simple recommendations apparent from the case study include:

(a) One key to mobilizing villages is to promote multipurpose trees to meet village fuel, fodder, and food needs. Particularly promising is a true agroforestry scheme of planting a combination of crops and a multiple of tree species. Nitrogen-fixing trees planted as shelter belts or interspersed with rows of crops can enhance soil moisture and fertility and reduce wind and water erosion. Through proper pruning (coppicing and pollarding), a single tree can provide five to ten times the wood volume of a felled plantation tree (Anderson and Fishwick, 1984; Spears, 1983).

(b) India, as well as other developing countries, should severely restrict the practice of a subsidized conversion of prime/irrigated agricultural lands to urban and industrial pulp and pole wood plantations.

(c) Reforestation must become a development priority in developing countries. Yet, this is one of the critical needs of the poor to be addressed. For example, from 1980-1984, the major development banks allocated less than one percent of their annual financing to reforestation programmes and the United Nations Development Programme (UNDP) only approximately two percent (Harrison, 1987). Two new initiatives, however, hold promise for the future. First, the $8 billion tropical forest action plan was initiated in 1985 jointly by the FAO, UNDP, the World Resources Institute, and the World Bank. The programme is expected to increase forestry funding from a collective contribution of $600 million in 1984 to over $1 billion in 1988 (World Resources Institute, 1987). Second, Prime Minister Rajiv Gandhi in India's seventh five-year plan (1985-1990) tripled funding (RS 6925 million to RS 18,593 million) for forestry and created a National Wastelands Development Board to distribute funds directly to schools, NGOs, women's groups for tree planting, and the development of "people's nurseries" (Postel and Heise, 1988). Still needed, however, is a considerable augmentation of funding for research on forest protection, ecosystem stability, and forestry genetics research.

Our two major recommendations are: 1) to decentralize and democratize the Indian social forestry programme (even to the point of removing them from the forestry departments -- discussed later); and 2) to dramatically improve trust, communications, and the manner in which the programme is evaluated.

5.1 Decentralization and democratization

Publicly sponsored reforestation programmes in India should always have a social context to them to focus benefits on the rural poor (80% of India's population), and to generate local support and motivation to successfully engage a project. Postel and Heise state:

> Tree planting programmes are most effective when local people are involved in their planning and implementation and perceive their own interest in success. If fodder is a critical need, for example, a project that promotes an unuseable species like eucalyptus will receive little popular support. Knowledge of villagers' access to cash, seasonal pattern of labour, and preferences for tree species is also crucial. Designing a reforestation project without local input is like letting a doctor prescribe treatment without asking the patient what hurts...Much could also be gained by reinforcing the efforts of small, local NGOs. Around the world, women's associations, peasant collectives, and church groups have taken up tree planting; in Kerala,

India, alone some 7,300 organizations are involved. Because local groups reflect the true needs, abilities, and limitations of their community, their projects are more likely to succeed, given adequate resources and technical assistance (1976, p. 98).

The suggested use of religious leaders to promote programme development and support is an excellent tool, for the religious leaders in a rural village are often the most trusted. In many cases, rural people are reluctant to participate in social forestry programmes because of uncertainties of how the benefits are to be allocated. Postel and Heiss state:

Often their [rural poor] needs and ideas were not solicited, nor were they involved in the selection of tree species to be planted. Perhaps the greatest lesson of the first generation of fuel wood projects was that villagers rarely were motivated to plant trees exclusively for fuel wood (1976, p. 89).

To greater involve village people in reforestation programmes, there must be clearly stipulated rights to the reforested lands or the planted trees (a programme termed by OXFAM "a tree tenure system"). An example of a tree tenure system is found in West Bengal where the forestry department supplied free seedlings, fertilizer, insecticides, technical assistance, and small financial incentives (based on the number of saplings surviving for three years). The landless villagers were not granted title to the reforested lands but rather title to the reforested trees. After five years, over 5000 hectares of previously denuded forestlands were harvested, providing landless families sufficient income to purchase small parcels of land for farming. The programme produced fodder, fuel wood, and ultimately added income to poor villagers (Foley and Barnard, 1984; Chowdry, 1987).

According to Roy (1976), financial incentives in village forestry programmes could take two forms: 1) direct annual payments to panchayats on the basis of each growing sapling; and 2) a programme for food for work in planting, weeding, and watering the growing saplings.

Another fundamental reason for the failure of social forestry to help the most needy is the unrealistic degree of collaboration and cooperation between villagers and the rural poor who use the forests for fuel, food, and fodder and forestry department personnel who have traditionally governed and policed the forests. According to the Centre for Science and Environment:

Foresters too, do not relish the new role thrust upon them of wooing the rural public as partners in afforestation. A young divisional forest officer in Madhya Pradesh states: "We used to be kings (of the forest), now we have to be beggars (to persuade people to plant

> trees)." It is clear there has to be a fundamental change in objectives and attitudes if the social forestry programme is to make a lasting impact in areas where such an impact is of prime importance (1986, p. 59).

When queried in 1986 about the possibility of taking the Karnataka social forestry programme away from the forestry department by the group FEVORD-K, the response from World Bank, however, was interesting (Nottidge, 1986):

> You also recommended the establishment of a separate agency under the aegis of the Department of Ecology and Environment to implement the Karnataka Social Forestry Project. We have strong reservations about this proposal. One of the basic objectives of IDA-assisted Social Forestry Projects in India is not only to plant target numbers of hectares of trees, but to gradually orient the Forest Departments in Karnataka and elsewhere to be able to effectively provide long-term support to forestry development, be it for social or territorial forestry as presently delineated. In the long run, both social forestry and the development of reserved forests should address social and ecological needs of the states. Eventually the delineation within the Forest Departments between the Social Forestry Wing and that for other (territorial) forestry development should be eliminated. It is also important to remember that forestry is a technical subject. We, therefore, view the Karnataka Forest Department as the most appropriate agency to handle this programme, although we entirely concur with your concern that it should plan and implement its projects in full consultation with environmentalists, agriculturalists, and sociologists. A better mechanism for more regular interaction between all these parties should be explored.

Experience in several Asian countries suggests that village "motivators" or agricultural extension agents may be more trusted by the locals than foresters to develop and promote reforestation programmes. World Bank projects in Kenya, India, and Malawi now use agricultural agents to motivate workers to plant multipurpose tree species. An approach targeted toward rural women has proven very successful (Agriculture and Rural Development Department, 1986).

Finally, one last issue under the heading of decentralization and democratization concerns the control of forest nurseries. In Karnataka, local villagers were allowed to establish small scale nurseries and sell a small number of saplings to the KSF project. This small programme, although difficult to administer, provides a small but significant income to villagers from the KSF project. It is interesting to note that in 1986, the Karnataka social forestry programme required 40% of the species ground in these decentralized "Kisan Nurseries" to be eucalyptus. This requirement should be eliminated and the nursery programme expanded (Clark, 1987; Environmental Defense Fund, 1987).

5.2 Communications and programme evaluation

Desai (1987) cogently argues in his paper "Role of Communication in Wasteland Development," for improved communications of development projects. He lists a number of improvements and suggestions, as follows:

-wasteland development information is badly needed by the masses for a mass movement of restoration,

-communication should be improved on all levels from the World Bank to the rural villager in appropriate forms and languages for proper understanding,

-there should be a consistent and continuous flow of information and feedback from the field to the laboratory to the financing agencies,

-tree-growing cooperatives in villages are a very useful source of information flow,

-materials and humanpower must be made available for nursery training,

-extensive documentation and communication of field experiences is important for programme success, and

-successful demonstration projects should be widely published.

We further suggest that more appropriate measures of programme failure and success be developed more directly tied to the overall primary goal of social forestry -- that of providing fuel, food, and fodder to the rural poor. Postel and Heise observe: "...in rural areas where fuel wood is not part of the cash economy, the cost of increasing scarcity is measured in women's time, something that may have little value to male decision makers" (1988, p. 89). Since the rural poor do not participate in the cash economy for food, fuel, or fodder, a non-monetary measure may be most appropriate. For example, a more appropriate measure of programme success would be the change in the amount of time rural women spend gathering food, fodder, and fuel products. Such an unconventional approach would seem appropriate in an economic system so different from Europe and North America.

In conclusion, a massive social forestry programme will be very difficult to develop in India with such a huge, poor, and diverse population. But such a programme could generate enormous indirect benefits including land for the landless, homes, food, fuel, fodder, soil retention, and ecological stabilization. For example, the Centre for Science and Environment estimates that a true social forestry programme could provide land to between 5 and 10 million families, making social forestry the country's largest land reform movement. Postel and Heise (1988) conclude:

> Successfully reforesting large areas of degraded lands, however, will require much more than financial commitments from governments and international lending agencies. It will take a shift in emphasis from government foresters establishing and maintaining commercial plantations to the much more complex tasks of starting nurseries in thousands of villages and encouraging the planting of multipurpose trees along roads, on farms, and around houses. Only by gathering the knowledge, support, and human energy of rural people themselves -- and planting to meet their basic needs -- is there any hope of success.

Finally, the Environmental Defense Fund (1987) states:

> Gravely flawed planning and inadequate supervision plagued the Bank-financed Karnataka Social Forestry Project. The Bank's loan of more than $25 million will bear adverse economic, social, and ecological impacts unless the Bank implements substantial reforms. The problems in this project also raise disturbing questions about others ongoing and elsewhere, which in many instances have similarities in species choice and design. If the Bank plans to double its forestry lending, it will be a tragedy if a substantial number of projects promote ecological and social deterioration, rather than environmental enhancement and sustainable growth.

6 REFERENCES

Agriculture and Rural Development Department, 1986. World Bank Financial Forestry Activity in the Decade 1977-86: A Review of Key Policy Issues and Implications of Past Experience to Future Project Design. World Bank, Washington, D.C.

Anderson, D. and Fishwick, R., 1984. Fuel wood Consumption and Deforestation in African Countries. (Staff Working Paper No. 704) World Bank, Washington, D.C.

Bandhu, D. and Garg, R.K. (Editors), 1986. Social Forestry and Tribal Development, Indian Environmental Society, New Delhi.

Bandhu, D. and Berberet, W.G., 1986. Social Forestry and Tribal Development. Indian Environmental Society, New Delhi.

Bandhu, D. and Berberet, W.G., 1987. Environmental Education for Conservation and Development. Indian Environmental Society, New Delhi.

Centre for Science and Environment, 1985. The State of India's Environment 1984-85. New Delhi.

Chandrashekhar, D.M., Krishna Murti, B.V. and Ramaswamy, S.R., 1987. Social Forestry In Karnataka: An Impact Analysis. New Delhi.

Chowdry, K., 1987. Wasteland and the rural poor: essentials of a policy framework. Forest News 14:2.
Clark, J., 1987. Letter to C.L. Robless, Chief, India Division, South Asia Programme Department, World Bank, Washington, D.C. on Behalf of OXFAM, Oxford, England.
Department of Forestry, 1986. Mid-term review report: Karnataka social forestry project. Government of Karnataka, Bangalore, India.
Desai, K., 1987. Role of communication in wasteland development. In: Bandhu, D. and Berberet, W.G. (Editors), Environmental Education for Conservation and Development. Indian Environmental Society, New Delhi, pp. 270-273.
Dewees, P. A., n.d. Gujarat community forestry project-wood balance and market studies: A preliminary assessment. International Institute for Environment and Development, Washington, D.C.
Directorate of Economics and Statistics, 1984. Impact of social forestry programme on land use in Kolar and Bangalore Districts. Government of Karnataka, Bangalore, India.
Directorate of Economics and Statistics, 1986. Marketing of farm forestry products of Eucalyptus and Casuarina in Kolar and Bangalore Districts in 1986. Government of Karnataka, Bangalore, India.
Dutt, A. K. and Geib, M.M., 1987. Atlas of South Asia. Westview Press, Boulder, Colorado.
Environmental Defense Fund, 1987. The failure of social forestry in Karnataka. Ecologist 17(15).
Foley, G. and Barnard, G., 1984. Farm and community forestry, International Institute for Environment and Development, Washington, D.C.
Forest Department, 1985-1986. Annual report. Government of Karnataka, Bangalore, India.
Harrison, P., 1987. The Greening of Africa. Viking/Penguin Inc., New York, Part 1.
Karnataka Government Secretariat, 1980. Project for social forestry with World Bank assistance, Karnataka. Forest Department Government of Karnataka, Bangalore, India.
Khoshoo, T.N., 1984. Speech to the national seminar on Eucalyptus in Peechi, Kerala, Secretary of the Department of Environment, New Delhi, India.
Kushalappa, K.A., 1986. Plantation opportunities in Karnataka. Conservator of Forests (Research). Government of Karnataka, Bangalore, India.
Ministry of Agriculture, 1983-84. Annual report. Department of Agriculture and Cooperation, New Delhi, India.
Nottidge, C.P.R., 1986. Letter to FEVORD-K, Chief, General Agriculture II Division South Asia Projects Department, World Bank, Washington, D.C.
Overseas Development Administration, 1986. Mid-term review of the Karnataka social forestry project. United Kingdom.
Paryavaran, S., 1984. Forest Cover in India: A Survey by National Remote Sensing Agency. Volume 1, no. 2.
Postel, S. and Heise, L., 1988. Reforesting the earth. In Brown, L.B. (Editor), State of the World 1988. Worldwatch Institute, W.W. Norton and Company, New York, pp. 83-100.
Ramaswany, R.S., 1986. People's Response to Decertification of Karnataka. Presented at the National Seminar on Control of Drought, Decertification and Famine, Sponsored by the Indian National Trust for Art and Cultural Heritage, New Delhi and the Research Foundation for Science Technology and Natural Resource Policy, Dehra Dun, India International Centre, New Delhi, India.
Reddy, S.T.S., 1985. Who prefers Eucalyptus and why. Social Action, 35.
Roy, S. K., 1986. Social forestry--for whom? In: Bandhu, D. and Garg, R.K. (Editors). Social Forestry and Tribal Development. Indian Environmental Society, New Delhi, pp. 1-6.

Shetty, B.K.A., 1982. Social forestry. In: Some Aspects of Forestry In India. ESG Publishers, pp. 44- 53.
Spears, J., 1983. Replenishing the world's forests: Tropical reforestation, an achievable goal? Commonwealth Forestry Review 2:3.
World Bank, 1983. India Karnataka Social Forestry Project. Staff Appraisal Report, General Agriculture Division, South Asia Projects Department.
World Resources Institute, 1987. The tropical forest action plan: Background information and update, Washington, D.C.

Chapter 14

A CONSPECTUS FOR AGROFORESTRY

W.W. BUDD
Program in Environmental Science and Regional Planning, Washington State University, Pullman, Washington, 99164-4430 (United States of America)

L.H. HARDESTY
Department of Natural Resource Sciences, Washington State University, Pullman, WA, 99164-6410 (United States of America)

and W.T. HINDS
Technology, Planning, and Policy Analysis Center, Battelle-Northwest, Washington, D.C., 20024 (United States of America)

ABSTRACT

This chapter provides a conspectus of several issues facing planning in agroforestry as discussed during the "Planning for Agroforestry" symposium, held at Washington State University, Pullman, WA, USA April 24-27, 1989. Four major areas of planning are considered: evaluation strategies and methods, institutional concerns, ecological considerations, and the integration of research and development.

INTRODUCTION

> A healthy farm will have trees on it--woodlands, where trees are native, but also fruit and nut trees, trees for shade and for windbreaks. Trees will be there for usefulness: for food, lumber, fence posts, firewood, shade, and shelter. But they will also be there for comfort and pleasure, for the wildlife that they will harbor, and for their beauty. The woodlands bespeak the willingness to live that keeps wildness flourishing in the settled place (Berry, 1977).

Developing strategies for enhancing the sustainability of agricultural systems is an important task, both for developing and developed nations. Agroforestry is a land use generally regarded as having much to offer in these efforts. The contributors to this volume have presented illustrations of theoretical perspectives on agroforestry, methods which have been used to affect changes in land use, and illustrations of the impacts of those changes. This chapter focuses on questions for future planning of agroforestry and implications for research.

Planning is a process through which human needs are expressed, the means for achieving these needs evaluated, and decisions made and implemented to reach those expressed objectives. Research provides the understanding needed to move reliably (one hopes) from a present state to an improved future; hence planning to some extent defines research needs and the availability of knowledge constrains the effectiveness of planning. Paramount to any discussion of planning and the relationship between planning and research is the issue of the planner-client relationship. Who is the client? The ultimate audience of planners and scientists may be the same, but the proximal audience, whose influence is most easily exerted and most strongly felt, can have a disproportionate influence on immediate activities. Thus planners can be responding to local or governmental imperatives while researchers are responding to the scientific community, making the integration of research and planning a complex undertaking. Within the domain of agroforestry, social and economic needs must be met within finite, but often unknown ecological capabilities. The time scale of these activities is also problematic; science's reductionist approach to the exploration of complex multi-dimensional problems rarely proceeds fast enough to meet urgent needs for both community development and ecological stabilization.

Agroforestry has been promoted as both a land use and a development strategy that borders on a philosophy. Obviously, it can be useful as either, but the value content of planning varies with the degree to which one view or the other predominates, hence this too should be clarified early in the planning process. One objective of the "Planning for Agroforestry" symposium held in Pullman, Washington (USA) on April 24-27, 1989 was the sharing of experiences in an evaluative atmosphere where generalities, if they exist, could be derived from a variety of case studies. A broad range of viewpoints, from village-level development to commercial applications, were presented. For those involved in agroforestry planning at all levels, planning was generally felt to be successful to the extent to which all involved parties were truly part of the planning process. The obstacles to "successful" planning are many, and well documented, and the case studies related in this volume offer some meaningful models for resolution in a variety of settings. The actual litany of obstructions to planning "success" is a familiar one. There is a need that agroforestry schemes address the interests of all involved, but not at the expense of long-range economic and environmental objectives. Achieving this balance between short- and long-range motivation is a challenge in all natural resource management activities, but might, erroneously, appear to be less critical in cases of extreme urgency. The need for longer range activities was frequently stated, as was the consensus that solution to this obstacle lies in the political realm among those who sponsor projects designed to produce benefits that can be attributed to a particular administration. Equity issues among local land users and distant beneficiaries of their land stewardship and economic activities are complicated by the range of economic systems that exists and the fact that ecological and geopolitical boundaries are often incongruent.

Communication, in myriad contexts, was often the subject of discussion. The need to record, utilize, and better understand indigenous knowledge and traditional agroforestry systems was frequently voiced as was improvement of the research-extension linkage. Discussions of the latter seemed to be underlain by the assumption that all research results are ready for extension, a questionable assumption in some cases. The perception that not all reliable knowledge is in use is undoubtedly correct and again emphasizes the urgent nature of the demand for improved land management practices.

There also was concern among some participants in the symposium that existing planning methods were not being optimally used. For others there was concern that method *per se* was not as important as the details of the technology being developed or implemented. Yet others saw the emphasis on specific methods and not process to be the major issue facing planning. A pressing need expressed by many participants was better access to existing information. Clearly there are both planning and information needs yet to be met, many of which are discussed below.

2 ISSUES FACING AGROFORESTRY PLANNING

The need to incorporate both analytical judgements and techniques with individual participation is well documented in planning. Agroforestry planning is no different in this regard. Indeed, because of the complexities of the natural systems with which agroforestry deals and the conditions of cultural richness within which agroforestry operates, the planning of agroforestry is faced with serious challenges in both domains.

Several planning methodologies and hybrids of methodologies are being used in agroforestry. It is not apparent, nor do any of the authors of chapters in this book suggest, that any single approach is superior to another. Land evaluation, farming systems, and agro-ecosystems approaches have contributed to resolving many the challenges of implementing agroforestry. However, as was noted in the introduction, each approach has limitations (Duchhart, *et al*.). There would appear to be at least four major needs in planning which have to be addressed, and which are briefly dicussed below. The first of these deals with understanding the effects of planning decisions. This will involve improving evaluation criteria and improving the integration of those criteria with "on-the-ground" implementation of agroforestry. The second deals with the political aspects of agroforestry planning. Here the concerns are institutional and focus on the role which agroforestry as an activity can play within the broader contexts of social change. Third, there are technical issues dealing with the ecological aspects of agroforestry planning and the notion of sustainability which need to be addressed. These issues focus both on current needs as expressed by the contributors to this book, but also on future concerns involving perhaps global effects of agroforestry. Finally, but not of least importance, there needs to be stronger integration of research and development activities.

3 EVALUATION OF AGROFORESTRY PLANNING

Throughout this volume, sustainability has been noted as a justification for agroforestry adoption and as a criteria for evaluating agroforestry plans. It is an idea whose day has indeed come as evidenced by its presence as a cornerstone of the World Conservation Strategy. There is, however, some division among planners and decision-makers as to (a) what the elements of sustainability are and (b) what constitutes an operable definition of sustainability that can be the basis for the planning, implementation, and evaluation of land use. Indeed, a major planning concern is with developing appropriate mechanisms for evaluation of sustainability. For agroforestry these include questions concerning the limitations of agroforestry applications, the compatibility of various land uses, the appropriateness of agroforestry within certain cultural frames, the elements of sustainability (e.g. social and ecological), population relationships vis-a-vis productivity, and timing and scale effects of implementation.

With calls for, and proclamations of, sustainability, planners need to examine the adequacy of the techniques used for measuring sustainability in its broadest sense. In addition, the means for assessing the development of agroforestry projects within these criteria will also be needed. As noted by the symposium participants, the availability of information, the quality of information, and the type of information available are major impediments to the success of planning. Consequently, where agroforestry is defined and justified as a sustainable land-use system, it becomes critical that (a) a functional set of criteria be established to evaluate sustainability within these systems and (b) the adoption of such systems include consideration for assessing these criteria. But to achieve this end, comparative data are needed.

Moreover there needs to be a realistic evaluation of what agroforestry can and cannot achieve ecologically. The prevailing hypothesis concerned with the sustainability of agroforestry systems is that appropriate agroforestry systems can improve soil physical properties, maintain soil organic matter, and promote nutrient cycling (Sanchez, 1987). Recent data, however, suggest that some of these claims, in particular maintenance of soil organic matter, are not substantiated over time (Lal, 1989a, 1989b, 1989c). Further evaluation of the potential capabilities and limitations of agroforestry systems is necessary. In addition, planners need to be able to assess these elements along a broader range of settings than is currently available. Planners have now begun to recognize these gaps, and several of the authors in this volume have identified ongoing efforts to address these needs. That agroforestry systems were not developed for scientific evaluation is an important consideration here (see section 5 below; Franz, 1989). However, there remains a critical need for scientific evaluation of the potential and limitations for these techniques to form bases for realistic evaluation of specific policies and projects.

The cases presented in this book demonstrate that implementation of agroforestry programs is as multi-faceted an activity as the agroforestry systems that are being considered for adoption. Moreover, it appears that a vital planning process, one which includes widespread local participation and which begins to address the issues raised above, contributes to designing the ideal agroforestry system. The work at CATIE (Beer et al., Chapter 3) best illustrates this relationship. At the heart of this program is an emphasis on learning, as opposed to problem-solving. The distinction between a learning-systems approach and a problem-solving approach is an important one. In virtually all of the cases presented, the use of indigenous knowledge and cultural resources has been stressed. However, to reach these objectives a process must be established through which evaluations, designed for use in these systems, can be integrated with local action. There are a variety of institutional frames which can meet this need for feedback. In the CATIE case, it was the close interaction with an established local organization that was the key, while in other instances the use of de facto community D&D or agro-ecological approaches, sometimes combined with landscape planning, appeared to be effective. The critical point is that where these approaches were effective, consideration of the close interrelationship between evaluation and implementation took place.

4 INSTITUTIONAL ISSUES

As the illustrations presented in this volume show, there is strong political will and institutional support for agroforestry in many nations and agencies. The challenges which successful programs have faced, and which successors must also face, are considerable. Agroforestry adoption often requires transcending cultural and political boundaries. There are no formulae for dealing with these concerns, and consequently planners must be constantly involved in eliciting information concerning these boundaries and attempting to negotiate solutions that address these concerns.

The interdisciplinary nature of agroforestry is a challenge to the rigidly discreet disciplinary domains of many governmental organizations. Within this context, the issue of institutional control of agroforestry planning activities needs to examined. This is, to be sure, an introspective process within individual nations and institutions, but the location of planning activities within an organizational structure does have an impact on the nature of the planning activity. This issue also impacts the opportunities for representation of participants ranging from local residents, land managers, extension agents, researchers, to government representatives, and support agencies in the planning process.

Many problems being addressed by planners involved in agroforestry deal with the representation of local interests within a complex mix of national, institutional, and agency interests: local participation in the implementation of the agroforestry project; incorporation of indigenous knowledge; and the design of workable agroforestry systems. Clearly, these concerns are not mutually exclusive. For example, the representation of local interests and local participation in adoption of an agroforestry project are clearly linked. Within this context, agroforestry may be viewed philosophically as only one element of social change and rural development, representing not only a change in land use, but also a change in attitude toward the land and toward the use of other factors of production. Consequently, the quest for participation, as Raintree (Chapter 2) so eloquently points out, must be a dynamic interaction and exchange of ideas between those who must embrace the change as good and those who want to promote it. This challenge will continue to play a prominent role in future agroforestry planning.

5 ECOLOGICAL ISSUES OF AGROFORESTRY PLANNING

5.1 Complexity

Perhaps the key characteristic of agroforestry thought and practice is its application in the midst of complicated environments. The upshot of this characteristic is that agroforestry as a discipline ends up faced with site-specific requirements (Gregersen and McGaughey 1987). This "bewildering" (Goodman 1987) degree of site-specificity of course reflects non-ecological issues, such as demographic or socio-political causes, which may in fact be viewed as human ecological issues. But ecological and topographical causes of site-specificity abound. In this context, Goodman's admonition to avoid simplistic explanations and sweeping generalizations is well taken.

One of the major ecological determinants of complexity, at least in tropical and sub-tropical environments, is the number of species potentially involved. Agroforestry is blessed with an abundance, even super-abundance, of potentially valuable tree species, numbering in the hundreds (Huxley 1986, quoting J. Burley from an ICRAF workshop), which may be combined with an almost infinite variety of herbaceous species. Further complexity derives from the nature of the habitats characteristic of low-latitude environments. The issue for both rain-forest and seasonal tropical environments is the ecological observation that, on the average, the ranges of organisms diminishes with decreasing latitude. Stevens (1989) summarizes the extensive ecological literature that reflects this relationship, so it is sufficient to note the salient point for agroforesters: tropical (and to a lesser extent sub-tropical) species may be moved out of their range by slope or aspect contrasts, or by elevation changes measured in dozens of meters. Thus, if useful species number in the hundreds, then one may be sure that optimizing their deployment in agroforestry designs is a task of immense complexity.

Ecologically, this is an interesting focus for research, but in an agroforestry context, complexity of this magnitude is as much an adversary as a benefit. Partly, this is because some of the concerns that face agroforestry require a solution right away. Thus, in the face of complexity, one must search for simplifying principles or approaches. Experience in the field suggests that agricultural research is necessary but not sufficient; ecological principles must be added, if priorities are to be properly established.

5.2 Priorities among the components of agroforestry systems

Priorities among the "agro-" portions of agroforestry may differ from the "-forestry" portions, as suggested by the need for wood fuels. Although analyses vary with the analyst, one analyst has argued strongly that "the basic [global] forestry issue" is facing a radical change, from industrial wood supply to fuel wood supply (Goodman, 1987). This aspect of forestry is central to agroforestry, considering the potential utility of trees for either fuel wood or food supply, but usually not both.

The existence of "wild wood" influences the economics of forestry, and agroforestry, in a fundamental way, so long as "free" wood exists:

> [T]here is no way a farmer could be induced to plant for wood fuel. Hence, current deforestation cannot be reversed. ... Wood growing costs to farmers are nevertheless very modest and would surely represent an attractive alternative if "free" trees came to an end. If a properly designed agroforestry system were to be implemented, [trees]...would not compete...against maize (Goodman, 1987).

Goodman's conclusion is sobering:

> Perhaps the main message here is that planting for wood fuels may only succeed in crisis situations of acute shortage. It may turn out to be impossible to anticipate and head-off the total loss of indigenous woody cover...apart from a few rigorously guarded woodland areas, the only way to restore trees would be by replanting after all the wild-tree cover had been destroyed.

Consequently, it would seem that local ecosystems face hard times, and possible extermination, regardless of global attention to the consequences of deforestation. The potential fate of forests is similiar to wildlife in other habitats, such as the great whales. The economics of the resources' existence makes it most profitable to exploit them to extinction, then go on to the next resource. This is not a purely "capitalistic" enterprise, by any means, because the same actions can be seen in all economies. This observation has implications for investigation and experimentation.

Wood fuels are not the only cause for concern regarding deforestation. The impact on downhill and downstream communities can be great, although the connections can be controversial. For example, the correlation between deforestation and flooding has been described as strongly positive (cf Clark, 1987), non-existent (cf Hewlitt, 1982, quoted in Hamilton and Pearce, 1987), and only partial (cf Hamilton, 1988). Technically, this disparity of conclusions regarding flooding and deforestation may be correct, because headwater streams can be sensitive, whereas the lower reaches of large drainages may not be. However, the Amazon River may be showing signs of such a correlation between deforestation and increased peak flow, and it is hardly a headwater stream.

But water yield from cut-over watersheds is not the only issue, because as discussed earlier, soil loss is often a major concern. Probably the keenest insight into the connection between vegetative cover and erosion comes from Wiersum's table of soil loss under various cover conditions (Wiersum, 1984; and quoted in Hamilton and Pearce, 1987). Considering the disparate sources and methods summarized, close or quantitative interpretation of these data is unwise. However, a simple conclusion is warranted: it is not so much tree cover as ground cover that is important in reducing erosion. This is an important generalization, especially if Goodman's pessimistic appraisal that forests are doomed is even approximately correct. Techniques that will maintain ground cover (including agronomic enterprises) can suffice to reduce erosion substantially. Again, this is a generalization that points toward specific research issues for agroforestry.

5.3 Scale in agroforestry research

Research issues that merit attention abound, as noted above, because of the ecological complexity of agroforestry. These complexities come on all scales, from the micro-scale of specific plot management, to the potential for hemispheric-scale changes in weather patterns due to climate change, or perhaps deforestation.

On the smaller scale, the fruits of successful research and development are evident in the chapters presented earlier in this volume. However, research needs a proper context to appraise its utility and priority, so here the scientific foundations for agroforestry are briefly examined, in relation to the actual practice of agroforestry research.

Scientifically, agroforestry represents a hard case. It is an endeavor for understanding that is fraught with difficulty, with abundant uncertainty as to its long-term ability to ameliorate the environmental conditions it seeks to address. This is most clearly seen when one considers how the classic scientific method cannot apply to agroforestry. For one thing, most agroforestry investigations are not scientific, so much as instructive. It is positively wrong to cast instructive, descriptive, or observational efforts in a hypothesis-testing framework. Where processes involving long time-scales are involved, especially in complex systems, devising even one testable hypothesis is difficult, much less alternative working hypotheses. Furthermore, the issues that are virtually untestable in ecology are exactly the issues that agroforestry needs to address. Obvious examples, which clearly do not exhaust the list, include community organization, interspecific competition, predation, insect population dynamics, and human-nature interactions.

Many of the chapters presented in this book make it clear that hypothesis testing is not the point for most agroforestry research efforts. The hypothesis-testing mode of agroforestry is still at least partly in the future, because agroforestry is currently a developmental, rather than merely investigatory, activity.

On the other hand, the authors of these chapters illustrate the great value associated with experimental and observational data. For this reason, agroforestry depends upon insight and experience, rather than solely upon deductive thought. No successful substitute for experience is yet at hand for agroforestry, nor will one likely emerge, but a lot can be done to improve the wisdom and beneficial application of that experience. The data needed range widely in both time and space scales, and over a wide spectrum of complexity as discussed below. The global scale issues are arguably impossible to resolve in the next few years, but their importance requires that they too be considered. More tractable experiments will then be discussed.

On a global scale, two major issues need investigation: climate change and rainfall patterns. The climate change concern is not global climate warming so much as the more difficult issue of regional climates and seasonal changes. For example, rainfall patterns may be intimately involved with deforestation patterns, as well as climatic change. As a matter for research, the correlation between deforestation and rainfall needs more thought. Hamilton and Pearce (1987) recommended that this topic be dropped from the research agenda, except perhaps for the Amazonian Basin. This is a significant caveat, given the evidence for recycled evapotranspiration there (Salati and Vose, 1984). An important issue is that, if, over the next several decades the entire Amazon Basin ends up more or less treeless, one may surmise that the Amazonian climate would change, irrespective of any global change. If large tracts are deforested, more rainfall will run off rather than being transpired in situ. The ecologic and climatic effects of such changes are unknown at this time. The potential impacts of such deforestation may be quantifiable in the next several years, depending upon the competence of large-scale meteorological analysis and hydrological modeling to handle such analyses. The best analyses to date are not encouraging--see, for example, Lean and Warrilow (1989).

On smaller scales, it is not hard to find issues where both insight and information are needed. These concerns include relatively simple notions, such as examining the many species to identify the "very useful" from among the "potentially useful" (Huxley, 1986). However, the transition from identifying species to actually growing them in place, with other species in the community, is both necessary (Mergen, 1986) and arduous (Buck, Chapter 5). Amplifying the utility of multi-species research is the ecological observation that species together may be mutualists rather than competitors. The existence of competitive influence is not doubted, as when alley-cropped plantings show reduced yield under trees (Logan and Cooperband, 1987). But, the existence of beneficial groupings of species is well-known, as well. Shade-tolerant leguminous ground covers are an example.

Searching for multiple-species synergisms, and relying on sophisticated understanding to yield increased productivity are part of what agroforesters can anticipate with pleasure. The converse issue is that in the midst of complexity, specific topics must be selected. It is an open question whether these should be agronomic, optimizing growth and yield of a few species; or ecological, putting many species together to maximize useful outputs at a farm or village level. In either case, precedents exist, with the simpler agronomic approach currently popular (cf Oldeman, 1983; Huxley, 1986; Mergen, 1986; Logan and Cooperband, 1987). However, the existence of relatively sophisticated, species-rich rural gardens which have not yet been successfully emulated using architectural or biological principles (Michon, 1981) gives hope that success in multi-species research may yet be attainable, and useful.

6 PLANNING AS A MEANS OF INTEGRATION

No greater challenge faces planners than meeting the various needs of multiple parties. Whether the issue is the siting of a hazardous waste treatment facility; the promotion and construction of low-income housing; or the promotion of land-use practices that will hypothetically improve soil quality, reduce erosion, and thereby enhance the quality of life of those adopting these techniques; the process of eliciting values and integrating those expressed needs into the decision-making process remains the fundamental issue.

Linking scientific research with development needs is a theme which runs concurrent with all of the issues discussed above. The political, ecological, and institutional challenges facing agroforestry planning are imbedded in a debate which forms the very context of planning.

The confrontation of the dominant philosophical foundations of science and development is perhaps captured best in the different uses of the term agroforestry. To some agroforestry refers to a technology, and therefore planning becomes a matter of selecting among a range of possible tools for the one that best fits site-specific needs. To others, agroforestry holds a more ideological significance. The task in this context is one of social choice and not technical choice. The two views offer differing perspectives on planning as an activity and on what is meant by "successful planning". In the former, the argument is often for a technocratic approach to planning. While decision-makers listen, the responsibility of finding a technique, method, or solution rests primarily with scientists or analysts. In the technocratic model, it is the responsibility of analysts to tell society what is "right" and how it should be done. In the latter, the quest is for incorporating the ideas of a wider public into the process of change and adoption. Here the challenge is one not necessarily of providing the choices, but one of establishing a framework to allow individual needs to be expressed and integrated into decision-making.

What are the implications of this ongoing debate? Some of the implications in terms of institutional and political concerns have already been raised. The actual extent of the impacts that these often divergent perspectives have on planning is too broad to be covered adequately in this forum. Two general comments do warrant some consideration here. First, while there are apparent needs in determining appropriate criteria for ideas such as sustainability (as noted above), these efforts will be of little value if there is no consideration given to the implementation of agroforestry within the context of these new criteria. Examining the integration of evaluation schemes and implementation needs to be a major focus of future planning activities in agroforestry. The challenge, to paraphrase Pressman and Wildavsky (1984), will be to blur the distinction between evaluation and implementation. The reason for this is that where the criteria for evaluation are lacking or poorly defined, often there will be poor implementation. Many of the obstructions to planning, identified both in various chapters of this book and in the ideas presented by participants of the agroforestry symposium, can be represented as problems of either "good" science but "bad" development, or "good" development strategies and "little or no" science. In the eyes of the beholder, both represent "successful" planning, yet clearly both can be improved.

The second comment stems from the first. While there is a need to build analytic capability, planners should recognize that many of the problems which are addressed by agroforestry do not have clear and explicit objectives associated with them. That is, the task is not necessarily one of solving a problem, but rather one of establishing a context in which social learning can take place. The implications of this difference are already being felt. The emphasis on centralized approaches in resource allocation and decision-making, as discussed by several authors but most notably Raintree (Chapter 2), is a direct result of adhering too tightly to a strict problem-solving paradigm for planning. Indeed, many of the revisions and modifications to existing planning methodologies discussed in this volume are targeted to addressing these needs. The losses, however, from failing to take such actions can be significant. Ideas and information which cannot be amalgamated into a measurable format can be dismissed, and at worst lost. Under the problem-solving paradigm, the world becomes a much simpler place, and far less real valuable.

In light of this, planning and planners need to reassess what it is they do. To some degree the methodological shifts that are occurring in D&D, in design analysis, and agro-ecosystem analysis represent a subtle readjustment in perspective. It is difficult to assess whether any interest in addressing the role of planning and the planner exists. With the institutionalization of many policies and planning approaches it also remains open as to what, if any, changes in planning can occur. But if nothing else, planning and planners need to recognize the implications of adhering to approaches that are unable to deal with values and concepts that cannot be expressed in a measurable form. This should not be interpreted as a condemnation of what has already taken place. As documented by many of the chapters in this book, the accomplishments of planning activities in agroforestry have been significant and noteworthy. Planners, extension specialists, and researchers have made significant inroads to affecting change.

Despite these achievements, however, there needs to be a recognition that, as with technology, changes in the way information is handled and used can also occur. These can and do affect what is considered of value, and what can or cannot be integrated into decision-making. The result is often a loss of traditional information, of perspective, or, in the most tragic cases, of culture. So long as the process of making changes recognizes and addresses these impacts, planning will have done its job. Being able to make that claim, however, is indeed the "rub". It is a task which planners must constantly be reminded of and engaged in, and one which the authors of this book have shown to be most crucial. There are no technological panaceas, nor will there be. This fact is perhaps the best testimony to the challenges that planning in agroforestry will yet have to face.

7 ACKNOWLEDGMENT

The contribution of Hinds was supported by the U.S. Environmental Protection Agency under an Inter-Agency Agreement with the U.S. Department of Energy's Pacific Northwest Laboratory, Contract DE-AC06-76RLO 1830. The Pacific Northwest Laboratory is operated for the Department of Energy by Battelle-Northwest. He wishes to acknowledge Allison Cook, Coordinator for Global Change Program for American Scientific International Inc., for providing both insight and critical thought during the development of this paper.

8 REFERENCES

Berry, W., 1977. The Unsettling of America: Culture & Agriculture. Sierra Club Books, 228 pp.

Clark, C., 1987. Deforestation and floods. Environmental Conservation 14(1):67-69

Dickinson, R. E., 1986. How Will Climate Change? In: The Greenhouse Effect, Climatic Change, and Ecosystems, B. Bolin, B.R. Doos, J. Jager, and R. A. Warrick, (Editors) John Wiley and Sons, New York, pp. 207-270 .

du Toit, R. F., 1985. Soil loss, hydrological changes, and conservation attitudes, in the Sabi catchment of Zimbabwe. Environmental Conservation, 12(2):157-166.

Franz, E.H., 1989. Tropical Agroforestry. Background paper for the International Workshop on the Contribution of Greenhouse Gas Emissions from Agricultural Systems to Climate Change, International Panel on Climate Change, Response Strategies Working Group, December 12-14, Washington, D.C.

Goodman, G.T., 1987. Biomass energy in developing countries: problems and challenges. Ambio 16(2-3):111-119.

Gregersen, H. and McGaughey, S.E., 1987. Social Forestry and Sustainable Development. In: Southgate, D. D. and J. F. Disinger, (Editors) Sustainable Resource Development in the Third World. Westview Press, Boulder, Colorado, pp. 7-20.

Hamilton, L.S., 1988. The recent Bangladesh flood disaster was not caused by deforestation alone. Environmental Conservation, 15(4):369-370.

Hamilton, L.S. and Pearce, A.J., 1987. What are the Soil and Water Benefits of Planting Trees in Developing Country Watersheds? In: Southgate, D. D. and J. F. Disinger, (Editors) Sustainable Resource Development in the Third World. Westview Press, Boulder, Colorado, pp. 39-58

Hinds, W.T., 1980. Dangerous Devegetation and Monocultures, A Panel Discussion. In: N. Polunin, (Editor), Growth Without Ecodisasters? Proceedings of the Second International Conference on Environmental Future, John Wiley and Son, New York, pp. 229-230.

Huxley, P.A., 1986. The prediction of biological productivity and sustainability of tree-crop mixtures. Tropical Agriculture, 63(1):68-70.

Kemph, G. S. and Hernandez, A. 1987. Evolutionary Conservation Project Planning and Implementation: NARMA in the Dominican Republic. In Southgate, D. D. and J. F. Disinger,(Editors) Sustainable Resource Development in the Third World. Westview Press, Boulder, Colorado. pp. 113-128.

Lal, R., 1989a. Potential of agroforestry as a sustainable alternative to shifting cultivation: concluding remarks. Agroforestry Systems, (8):239-242.

Lal, R., 1989b. Agroforestry systems and soil surface management of a tropical alfisol: III. Changes in soil chemical properties.

Agroforestry Systems, (8):113-132.
Lal, R., 1989c. Agroforestry systems and soil surface management of a tropical alfisol: IV. Effects on soil physical and mechanical properties. Agroforestry Systems, (8):197-215.
Lean, J. and Warrilow, D.A., 1989. Simulation of the regional climatic impact of Amazon deforestation. Nature, (342):411-413.
Lee, J. C. and Kramer, P.J., 1987. Forestry Research Needs and Strategies. pp. 295-302 in The Greenhouse Effect, Climate Change, and U. S. Forests, W. E. Shands and J. S. Hoffman, (Editors) The Conservation Foundation, Washington, D.C. 304 pp.
Loehle, C., 1987. Hypothesis testing in ecology: psychological aspects and the importance of theory maturation. Quarterly Review of Biology 62(4):397-409.
Logan, T.J. and Cooperband, L.R., 1987. Soil Erosion on Cultivated Steeplands of the Humid Tropics and Subtropics. In Southgate, D. D. and J. F. Disinger, (Editors) Sustainable Resource Development in the Third World. Westview Press, Boulder, Colorado. pp. 21-38.
Ma, S., 1985. Ecological engineering: application of ecosystem principles. Environmental Conservation, 12(4):331-335.
Mergen, F., 1986. Agroforestry - an overview and recommendations for possible improvements. Tropical Agriculture, 63(1):76-80
Michon, G., 1983. Village-Forest-Gardens in West Java. In: P. A. Huxley (Editor), Plant Research in Agroforestry, ICRAF, Nairobi, pp 13-24.
Oldeman, R.A.A., 1983. The Design of Ecologically Sound Agroforests. In: P.A. Huxley (Editor), Plant Research in Agroforestry, ICRAF, Nairobi, pp. 173-207.
Pressman, J.L. and Wildavsky, A., 1984. Implementation. University of California Press, Berkeley.
Salati, E. and Vose, P.B., 1984. Amazon Basin: a system in equilibrium. Science, 225:129-138.
Sanchez, P.A., 1987. Soil productivity and sustainability in agroforestry systems. In: H.A. Steppler and P.K.R. Nair (Editors), Agroforestry: A Decade of Development, ICRAF, Narobi, pp. 205-233.
Southgate, D.D. and Disinger, J.F., (Editors), 1987. Sustainable Resource Development in the Third World. Westview Press, Boulder, Colorado, 177 pp.
Stevens, G.C., 1989. The latitudinal gradient in geographical range: how so many species coexist in the tropics. The American Naturalist, 133(2):240-256.
Wildavsky, A., 1979. Speaking Truth To Power: The Art and Craft of Policy Analysis. Little, Brown and Company, Boston.
Winterbottom, R. and Hazelwood, P.T., 1987. Agroforestry and sustainable development: making the connection. Ambio, 16(2-3):100-110.

INDEX